Agricultural Precedents Handbook

Second edition

Agricultural Precedents Handbook

Nigel Davis

Graham Smith

Angela Sydenham

JORDANS

Published by
Jordan Publishing Limited
21 St Thomas Street
Bristol BS1 6JS

British Library Cataloguing-in-Publication Data

A catalogue record for this book is available from the British Library.

ISBN 978 1 84661 061 5

Typeset by Letterpart Ltd, Reigate, Surrey

Printed in Great Britain by CPI Antony Rowe, Chippenham, Wiltshire

PREFACE

It is fair to say the three of us have been quite amazed at the success of the first edition of the Agricultural Precedents Handbook, not to mention the enthusiasm there has been for a new edition incorporating additional precedents and, or more specifically, updated versions of some of the 'old favourites' as a result of the introduction of the CAP Single Payment Scheme in 2005, now recently amended and updated by Council Regulation (EC) No 73/2009, as well as the changes brought about by the provisions of the Regulatory Reform (Agricultural Tenancies) (England and Wales) Order 2006, and the Agricultural Land Tribunal (Rules) Order 2007.

Hopefully what we have incorporated here will assist both the specialist and the non-specialist to deal with everyday problems, as well as one or two of the more erudite ones.

As with the original edition, this edition has taken longer to come to fruition than any of us would have hoped for a whole host of different reasons which I will not bore you with, most recently of course the changes to the Single Payment Scheme, with the new Council Regulation only published at the end of January 2009. Suffice to say the second edition is now here. The law is correct as at 1 March 2009.

My sincere thanks once again to Graham Smith and Angela Sydenham for their invaluable contributions. Thanks also to Graham's team members at Roythornes, Vember Mortlock, Tom Foottit, Rachel McKillop, Helen Peck and Roxane Lamch, and my own colleagues Wanda Goodhead, Vivienne Hollick, Sophie Baker and Rachel Capstick for all their skill and assistance in dealing with the many and various legal and technical issues.

I would also like to thank the long-suffering Tony Hawitt at Jordans for his support and patience, and last but by no means least our respective spouses and families for their forbearance during the gestation period of this second edition.

Nigel Davis
Sedbergh, Cumbria
March 2009

CONTENTS

Preface v
Table of Cases ix
Table of Statutes xi
Table of Statutory Instruments xv
Table of European Legislation xvii
List of Abbreviations xix
How to use this book xxi

Chapter 1
Single Payment Scheme 1

Chapter 2
Sale and Acquisition of Agricultural Land 15

Chapter 3
Tenancies under the Agricultural Holdings Act 1986 99

Chapter 4
Tenancies Under the Agricultural Tenancies Act 1995 201

Chapter 5
Occupation of Farm Buildings 311

Chapter 6
Residential Tenancies 341

Chapter 7
Grazing Agreements 375

Chapter 8
Alternative Land Uses 399

Chapter 9
Sporting Rights 417

Chapter 10
Milk Quota 431

Chapter 11
Horses **443**

Chapter 12
Partnerships, Contracting Agreements and Share Farming **469**

Index **539**

TABLE OF CASES

References are to page numbers.

CIR v Forsyth-Grant 1943 SC 528, 25 TC 369 376

Davis v H & R Ecroyd [1996]2 EGLR 5, [1996] 30 EG 97, [1996] EGCS 77 470

Farmer v IRC (1999) STC (SCD) 321 312
Faulks v Faulks [1992] 1 EGLR 9, [1992] 15 EG 82 470

Harrison-Broadley v Smith [1964] 1 All ER 867, [1964] 1 WLR 456, 108 Sol Jo 136 469
Holdcroft v Staffordshire County Council [1994] 2 EGLR 1, [1994] 28 EG 131 432

Mannai Investment Co Limited v Eagle Star Life Assurance Co Limited [1996] 1 All ER 55, [1995] 1 WLR 1508, 71 P & CR 129, [1996] 1 EGLR 69, [1995] 31 LS Gaz R 34, [1996] 06 EG 140, [1995] EGCS 124 104
McCall and Keenan v HMRC [2009] NICA 12 376

Notting Hill Housing Trust v Roomus [2006] EWCA Civ 407, [2006] 1 WLR 1375, [2006] All ER (D) 432 (Mar) 343

Oxford County Council v Oxford City Council and Robinson [2006] UKHL 25, [2006] 2 AC 674, [2006] 4 All ER 817, [2006] 2 WLR 1235, [2006] LGR 713, [2006] 22 EG 177 (CS), (2006) *Times*, 31 May, 150 Sol Jo LB 706, [2006] All ER (D) 340 (May) 17

Pennell v Payne [1995] QB 192, [1995] 2 All ER 592, [1995] 2 WLR 261, [1995] NPC 151, [1995] 1 EGLR 6, [1995] 06 EG 152, [1994] EGCS 196 104
Posthumous v Oostervoud [1992] 2 CMIR 336 432
Puncknowle Farms v Kane [1985] 3 All ER 790, [1986] 1 CMLR 27, [1985] 2 EGLR 8, 275 Estates Gazette 1283 432

R (on the application of Godmanchester Town Council (Appellants) v Secretary of State for the Environment, Food and Rural Affairs (Respondent); R (on the application of Drain) (Appellant) v Secretary of State for the Environment, Food and Rural Affairs (Respondent) [2007] UKHL 28, [2007] 4 All ER 273, [2007] 3 WLR 85, [2008] 1 P & CR 210, [2007] 26 EG 163 (CS), (2007) *Times*, 22 June, 151 Sol Jo LB 859, [2007] All ER (D) 201 (Jun) 400
R v Oxfordshire County Council ex parte Sunningwell Parish Council [2000] 1 AC 335, [1999] 3 All ER 385, [1999] 3 WLR 160, [1999] LGR 651, 79 P & CR 199, [1999] 29 LS Gaz R 29, [1999] NLJR 1038, [1999] 31 EG 85, [1999] EGCS 91 17

Sabella v Montgomery (1997) 77 P & CR 431, [1998] 1 EGLR 65, [1998] 09 EG 153, [1997] EGCS 169, 75 P & CR D41 104
Street v Mountford [1985] AC 809, [1985] 2 All ER 289, [1985] 2 WLR 877, 50 P & CR 258, 17 HLR 402, 129 Sol Jo 348, [1985] 1 EGLR 128, [1985] LS Gaz R 2087, [1985] NLJ Rep 460, 274 Estates Gazette 821 376

Yeandle v Reigate BC [1996] 1 EGLR 20, [1996] 14 EG 90 104

TABLE OF STATUTES

References are to page numbers.

Agricultural Act 1986
Sch 1 432
Agricultural Credits Act 1928 114, 129, 213
Agricultural Holdings Act 1986 8, 36, 52, 67, 84, 85, 100, 107, 141, 142, 143, 160, 171, 173, 174, 191, 202, 469, 471
s 2 184, 197
s 3 185, 197
s 4 101
s 5 185, 197
s 6 103
s 6(1) 163
s 10 164
s 12 104, 165
s 13 166
s 14 121
s 26 37, 54, 68, 169
s 27 37, 54, 68
s 28 37, 54, 68
s 34 100
s 35 100
s 36 101
s 36(3) 184
s 36(6) 101
s 37 102
s 39 102, 176, 181, 184, 186, 191
s 45 176
s 49 100
s 50 101
s 53 176, 198
s 53(1) 191
s 64 164
s 66 164
s 67 37, 54, 68, 167
s 69 100, 104
s 72 167
s 73 100, 104
s 83 168
s 84 105
Sch III 37, 54, 68
Sch 1 103
Sch 2 101
Sch 3 169, 170, 172, 175
Sch 6 179, 180, 184, 185, 194, 197
Sch 9 100, 105
Sch 11 105, 171, 174
Agricultural Tenancies Act 1995 8, 100, 150, 202, 204, 245, 258, 267, 469, 471
s 1(4) 237
Agricultural Tenancies Act 1995—*continued*
s 4 100, 101, 201
s 4(1)(d) 102
s 5 203
s 6 203
s 7 203
s 8 232
s 9 203
s 10 203, 308
s 10(6) 259
s 11 203
s 12 203
s 13 203
s 13(2) 204
s 14 203
s 17 219, 220
s 18 219
s 19 264, 280, 288, 295, 301
s 19(1) 273
s 20(1) 204
s 36 236
Agriculture Act 1947
s 11 120, 219
Sch 2 121
Agriculture Act 1986 434, 435, 436
s 13 438
Sch 1 438
Animal Welfare Act 2006 379, 426
Arbitration Act 1996 7, 10, 83, 85, 96, 98, 105, 134, 152, 161, 380, 384, 389, 425, 437, 439, 440, 448, 480, 483, 529, 536

Commons Act 2006 16
s 15 16
Commons Registration Act 1965 16
Companies Act 2006
s 994 497
Contracts (Rights of Third Parties) Act 1999 90, 92, 425
Control of Pollution Act 1974 34, 49
Countryside and Rights of Way Act 2000 400
s 51 16
s 57 400
Sch 5, para 2 16
Sch 6 400, 401

Environment Act 1995 315

Environmental Protection Act 1990 34, 49, 65, 123, 224, 315, 426
s 78 48

Ground Game Act 1880 109, 239, 396, 419
Ground Game (Amendment) Act 1906 109, 239, 419

Highways Act 1980
s 31(6) 400, 412
s 31(7) 400
s 31A 401
Housing Act 1980 341
s 89 343
Housing Act 1988 341, 342, 345, 364
s 6(2) 357, 360, 368, 373
s 6(3) 359
s 8 363
s 13 342, 356
s 13(2) 366, 369, 370, 373
s 19A 345
s 20 342
s 21 343, 345, 363
s 22 342
s 24 355
Sch 2 346, 354, 363, 364
Sch 2A 345, 353, 354
Sch 3 354
Housing Act 1996 253, 342, 343, 364
s 97 342
s 151 363
Sch 7 353, 354
Housing Act 2004 347
s 213(5) 345

Insolvency Act 1986 448, 473, 487, 498, 506
s 1 231, 266
s 214A 501
s 253 231, 266, 275, 282, 290, 297, 303, 394
s 267 394
s 268 231, 266, 275, 282, 290, 297, 303, 394
s 273 394
s 286 394

Land Registration Act 2002 70, 205
s 3 417
Sch 1 22, 39, 56, 78
Sch 2 22, 418
Sch 3 22, 39, 56
Sch 12 39, 56
Landlord and Tenant Act 1927
s 19 322
Landlord and Tenant Act 1954 311, 317, 321, 333, 425
s 24 317, 331
s 25 317, 331
s 26 317, 331
s 27 317, 331
s 28 317, 331
s 38 331
Landlord and Tenant Act 1985 352
s 18 369, 373
Landlord and Tenant Act 1987
s 48 163
s 48(1) 106, 207
Landlord and Tenant (Covenants) Act 1995 317, 318
s 5 322
s 11 322
s 17 320
Law of Property Act 1925 425
s 62 79, 331
s 146 37, 54, 228, 265, 274, 281, 288, 296, 302, 320, 334, 352
s 146? 68
s 147 228, 265, 274, 281, 289, 296, 302, 320
s 196 330
Law of Property (Miscellaneous Provisions) Act 1994
s 3 206
s 6(2)(a) 79
s 6(3) 79
Limited Liability Partnership Act 2000 470, 491
s 2 471
s 10 470
s 11 470
s 12 470
Local Government and Rating Act 1997
s 2 417

Mental Health Act 1983 473, 513

Partnership Act 1890 478
s 1 469
Pests Act 1954 427
Planning (Consequential Provisions) Act 1990 209, 316
Planning (Hazardous Substances) Act 1990 209, 316
Planning (Listed Buildings and Conservation Areas) Act 1990 209, 316
Planning and Compensation Act 1991 209, 316
Public Health Act 1936 34

Recorded Delivery Act 1962 425
Rent Act 1977 341, 370, 374
Rent (Agriculture) Act 1976 341
s 27 19

Statutory Declarations Act 1835 413

Town and Country Planning Act 1990 209, 316
s 55(2) 443
s 366(1) 443
Treasure Act 1996 403

Value Added Tax Act 1994 71, 317

Water Industry Act 1991 315
Water Resources Act 1991 223, 315, 392
Wildlife and Countryside Act 1981 120, 422, 426
s 12 109, 239, 396

Wildlife and Countryside Act 1981—*continued*
s 16(1) 423
s 16(5) 423
s 53B 16
Sch 2 109, 239, 396

TABLE OF STATUTORY INSTRUMENTS

References are to page numbers.

Agricultural Holdings (Forms of Notice to Pay Rent or to Remedy) Regulations 1987, SI 1987/711 136
Agricultural Land Tribunals (Rules) Order 2007, SI 2007/3105 176
Agriculture Holdings (Arbitration on Notices) Order 1987, SI 1987/710 171
Agriculture (Maintenance Repair and Insurance of Fixed Equipment) Regulations 1973, SI 1973/1473 116

Civil Procedure Rules 1998, SI 1998/3132
rr 55.11–55.18 343
Common Agricultural Policy Single Payment and Support Schemes (Appeals) (England) Regulations 2004, SI 2004/2689 1
Common Agricultural Policy Single Payment and Support Schemes (Cross Compliance) (England) Regulations 2004, SI 2004/3196 1
Common Agricultural Policy Single Payment and Support Schemes Regulations 2005, SI 2005/219 1
Common Agricultural Policy Single Payment Scheme (Set-aside) (England) Regulations 2004, SI 2004/3385 1
Commons Registration (General) (Amendment) (England) (No 2) Regulations 2007, SI 2007/2404 16
Countryside and Rights of Way Act 2000 (Commencement No 8 and Transitional Provisions) (Wales) Order 2006, SI 2006/1279 16
Countryside and Rights of Way Act 2000 (Commencement No 9) Order 2005, SI 2005/2459 16

Dairy Produce Quotas Regulations 2005, SI 2005/465 157, 434, 436

Groundwater Regulations 1998, SI 1998/2746 123, 224

Housing (Tenancy Deposits) (Prescribed Information) Order 2007, SI 2007/797 345

Limited Liability Partnership Regulations 2001, SI 2001/1090 470, 493

Nitrate Pollution Prevention Regulations 2008, SI 2008/2549 124, 225

Public Rights of Way (Register of Applications under section 53(5) of the Wildlife and Countryside Act 1981) (England) Regulations 2005, SI 2005/2461 16
Public Rights of Way (Registers) (Wales) Regulations 2006, SI 2006/42 16

Regulatory Reform (Agricultural Tenancies) (England and Wales) Order 2006, SI 2006/2805 100, 102, 104, 105, 201
art 4 100
art 5 101
art 6 100
art 8 101
art 12 100
art 13 203
art 14 204
art 15 204
art 16 204
Sch 1 331
Regulatory Reform (Business Tenancies) (England and Wales) Order 2003, SI 2003/3096 366, 369, 370, 373
art 22 311
Sch 1 311
Sch 2 311

Spring Traps Approval Order 1995, SI 1995/2427 427

Town and Country Planning (General Permitted Development) Order 1995, SI 1995/418 443
Town and Country Planning (Use Classes) Order 1987, SI 1987/764 318

Treasure (Designation) Order 2002, SI 2002/2666 403

Welfare of Farmed Animals (England) Regulations 2007, SI 2007/2078 379

TABLE OF EUROPEAN LEGISLATION

References are to page numbers.

Commission Regulation (EC) No 795/2004 laying down detailed rules for the implementation of the single payment scheme provided for in Council Regulation (EC) No 1782/2003 establishing common rules for direct support schemes under the common agricultural policy and establishing certain support schemes for farmers 1

Commission Regulation (EC) No 796/2004 laying down detailed rules for the implementation of cross-compliance, modulation and the integrated administration and control system provided for in of Council Regulation (EC) No 1782/2003 establishing common rules for direct support schemes under the common agricultural policy and establishing certain support schemes for farmers 1

Council Regulation (EC) No 73/2009 establishing common rules for direct support schemes for farmers under the common agricultural policy and establishing certain support schemes for farmers, amending Regulations (EC) No 1290/2005, (EC) No 247/2006, (EC) No 378/2007 and repealing Regulation (EC) No 1782/2003 1, 2, 5, 10, 80, 93, 96, 207, 377, 381, 387, 445, 473, 508

Art 34(2) 11

Council Regulation (EC) No 1782/2003 establishing common rules for direct support schemes under the common agricultural policy and establishing certain support schemes for farmers 1

LIST OF ABBREVIATIONS

AHA 1986	Agricultural Holdings Act 1986
ALA	Agricultural Law Association
ATA 1995/1995 Act	Agricultural Tenancies Act 1995
CLA	Country Land & Business Association
CROW Act 2000	Countryside and Rights of Way Act 2000
DEFRA	Department for the Environment Food and Rural Affairs
ERDP	England Rural Development Programme
FBT	Farm business tenancy
GAEC	Good agricultural and environmental condition
IACS	Integrated Administrative and Control System
LLP	Limited liability partnership
NFU	National Farmers' Union
RPA	Rural Payments Agency
RRO 2006	Regulatory Reform (Agricultural Tenancies) (England and Wales) Order 2006, SI 2006/2805
SDA	Severely Disadvantaged Area
SMR	Statutory Management Requirements
SPS	Single Payment Scheme
TDS	Tenancy deposit scheme
UELS	Uplands Entry Level Stewardship
VAT	Value added tax
VATA	Value Added Tax Act 1994

HOW TO USE THIS BOOK

Purpose

The purpose of this book is to provide in one volume a comprehensive collection of precedents which will be useful to the specialist and non-specialist adviser alike. It is hoped that the accompanying practical commentaries and notes will assist practitioners when considering the precedents and will highlight points of concern.

Contents

The book has been structured as follows:

Chapter 1: Single Payment Scheme

Chapter 2: Sale and Acquisition of Agricultural Land

Chapter 3: Tenancies under the Agricultural Holdings Act 1986

Chapter 4: Tenancies Under the Agricultural Tenancies Act 1995

Chapter 5: Occupation of Farm Buildings

Chapter 6: Residential Tenancies

Chapter 7: Grazing Agreements

Chapter 8: Alternative Land Uses

Chapter 9: Sporting Rights

Chapter 10: Milk Quota

Chapter 11: Horses

Chapter 12: Partnerships, Contracting Agreements and Share Farming

Structure

At the front of each chapter is a listing of all the precedents in that chapter.

There then follows an introductory commentary giving practical advice and guidance specific to the area covered.

The precedents are drafted to cover both the usual as well as some optional clauses. They can all be adapted to suit individual requirements by amending the text supplied on the accompanying free disk.

In many cases there are also notes to the precedents which provide additional guidance.

Chapter 1

SINGLE PAYMENT SCHEME

INTRODUCTION

PRECEDENTS

1 Sale of entitlements (without land)

2 Deed of variation

INTRODUCTION

The Single Payment Scheme (SPS) was introduced in 2005, following a period of proposals and consultations. Council Regulation (EC) No 1782/2003 set out the framework of the reform proposals, and Commission Regulations (EC) No 795/2004 and (EC) No 796/2004 'fleshed out' the Council Regulation. In England statutory instruments SIs 2004/2689, 2004/3196, 2004/3385 and 2005/219/2005 established the relevant framework. Council Regulation (EC) No 1782/2003 has been repealed by Regulation (EC) No 73/2009. The new Commission Regulations and statutory instruments here in England are still awaited.

The Government's method of implementation of the reforms was announced in Parliament by Margaret Beckett on 12 February 2004, and the changes came in on 1 January 2005.

With effect from 2005 there has been a single payment (SP) made each year.

Arable area payments, sheep annual premium, suckler cow premium and the like have been abolished.

In England, a hybrid scheme was introduced comprising a flat rate regional payment and a payment based on subsidy payments received by a farmer in 2000, 2001 and 2002 ('the Reference Period'). Eventually, all agricultural land in England will be paid at a single flat rate. Wales decided to pay the SP based on historic claims during the period from 2000 to 2002. Scotland also chose the historic payment method, while Northern Ireland has a hybrid between the historic and the regional flat rate systems.

A National Reserve was created in order to allocate entitlements to new farmers, established farmers whose business had changed since the reference period for calculating historical element of entitlements, or those who intended to invest or expand their businesses.

Farmers who were eligible to receive the SP were allocated 'entitlements' in 2005. These constituted a right to receive the SP each year. There had to be an eligible hectare of land for each entitlement and the number of entitlements was based on the qualifying land contained in a farmer's IACS (Integrated Administration and Control System) claim in 2005. If not claimed in that year, any entitlements were lost.

In 2005 there was no need to establish entitlements on all land farmed. In some cases farmers registered their owner occupied land but not land they tenanted. Land without entitlements has become known as 'naked acres'.

Up-to-date information on the SPS can be obtained from the Rural Payments Agency's (RPA's) guidance notes and/or its website (www.rpa.gov.uk), from the Department for the Environment Food and Rural Affairs (DEFRA) at www.defra.gov.uk, or from the websites of the National Farmers' Union (NFU) (www.nfu.org.uk), Country Land & Business Association (CLA) (www.cla.org.uk) or Agricultural Law Association (ALA) (www.ala.org.uk). Readers are strongly advised to keep a close eye on these websites so that they are able to obtain copies of the Commission Regulations and statutory instruments at the earliest opportunity, as well as ensuring that all of the changes which occur from time to time are monitored.

The SP has been 'decoupled' from production and is now being paid through the new system. The link between production and payment has gone. Payment will be received regardless of the farming of the land. SP is not tied to land, it is held by a farmer, as defined. Farmers can farm free of the restraints of the various previous quota regimes (save for milk quota which presently is scheduled to remain until 2014), or simply keep the land 'nice and tidy'.

Entitlements do not attach to land. This means that they may be sold with land, or without land or they may be leased out. Transfers need to be to a farmer.

Only a farmer could apply for the allocation of the entitlements in 2005, and only a farmer can claim SP based on his entitlements. Non-farmers could not claim, or have entitlements allocated to them or transfer them to themselves. That said, to comply with the definition of farmer was and is not difficult – perhaps as little as having land pasture topped once a year? Certainly there is no requirement to undertake any agricultural production.

Who or what is a farmer? Article 2(a) of Council Regulation (EC) No 73/2009 defines 'farmer' as:

'... a natural or legal person, or a group of natural or legal persons, whatever legal status is granted to the group and its members by national law, whose *holding* is situated within the Community territory, as referred to in Article 299 of the Treaty, and who exercises an *agricultural activity*.'

What is a Holding? Article 2(b) of Council Regulation (EC) No 73/2009 defines it as:

'... all the production units managed by a farmer situated within the territory of the same Member State.'

What is agricultural activity? Article 2(c) of Council Regulation (EC) No 73/2009 defines it as:

'... the production, rearing or growing of agricultural products including harvesting, milking, breeding animals and keeping animals for farming purposes, or *maintaining the land in good agricultural and environmental condition* as established in Article 6'.

Land used to claim must presently be at a farmer's disposal on just one day during the claim year, which runs from 1 January to 31 December. In the UK this is 15 May. *This requirement has applied from 1 April 2008 and replaces the previous requirement for farmers to have land at their disposal for a period of 10 months*. It is important to remember that the signing of a property transfer document by a seller and buyer does not effect the transfer of entitlements. A form RLE 1 must also be signed and submitted to the RPA, *and* the RPA requires a minimum of 6 weeks' prior notification to transfer the entitlements. This means that, in order to ensure that a buyer can claim the payment in any scheme year, the last effective date for notifying the transfer of entitlements is 2 April.

Calculation

As already mentioned, the entitlement to the single payment comprised two elements, namely the historic payment and the regional flat rate payment, and the percentages constituting these two elements change with the passage of time. In the first year of the SPS the payment comprised 90% historic and 10% regional flat rate, with those percentages changing over the years to 2012, so that in that year there will be 0% historic and 100% regional flat rate. The intention of government was to have a gentle transition from historic payment, based on the Reference Period, to the regional payment system. Set-Aside Entitlements only attracted a flat rate and this continues after 1 January 2009.

The element based on the historic payment was calculated using the farm's average arable area payments and livestock headage claims submitted during the Reference Period, together with any dairy premium. Ultimately, there will be a single rate on all eligible agricultural land in England – the regional flat rate. By 2012 all land in England that is 'farmed' is to receive a flat rate payment regardless of cropping. Once the entitlements had been allocated following the 2005 claim the historic element and regional payment could not, and cannot, be separated. They are indivisible. It is also important to note that entitlements are an asset in the hands of the claimant in 2005, whether landowner or tenant.

Sugar beet compensation was added in to the value in 2008 for relevant growers.

Cross compliance

To receive the SP farmers have to observe a plethora of cross compliance rules governing how land can be farmed. These are the EU's standards relating to the broader welfare of the land and include:

- The management requirements, which comprise a number of existing items of EU legislation dealing with issues such as the environment, public and animal health, disease notification, animal welfare and the like.

- The good agricultural and environmental conditions which are intended to protect soils from erosion, protect soil organic matter, and soil structure. There are also minimum levels of maintenance and minimum livestock stocking rates, retention of landscape features, prevention of encroachment of unwanted vegetation and the like.

There is a regular update of information in connection with cross compliance, which can be located on the RPA website.

Farmers have to abide by these requirements on all their farmed area for the SP to be paid. The RPA is responsible for policing the cross compliance obligations.

Modulation

The reforms have introduced this concept whereby a percentage of direct aid is removed each year and the monies that have been removed then go into rural development spending.

PRECEDENTS

PRECEDENT 1 SALE OF ENTITLEMENTS (WITHOUT LAND)

[*In this example the farmer retains the land but sells the entitlements.*]

THIS AGREEMENT is made the [ ] day of [ ] 20[ ]

BETWEEN:-

(1) [*name*] of [*address*] ('the Seller')

and

(2) [*name*] of [*address*] ('the Buyer')

WITNESSES as follows:-

1. Definitions and interpretation

1.1 'the Regulations' means Council Regulation (EC) No 73/2009 ('the Council Regulation') and all relevant subsequent Commission Regulations administrative provisions statutory guidance or codes of practice implemented by the EC and/or the UK Government pursuant to the Regulations

1.2 'Entitlements' means the following entitlements to the single income payment held by the Seller in his name pursuant to the Regulations namely:-

Type(s)	Number(s)	Value(s)

and reference to Type(s) of entitlements has the same meaning(s) as that contained in the Regulations

1.3 'DEFRA' means the Department for the Environment Food and Rural Affairs or any successor department

1.4 'RPA' means the Rural Payments Agency or any successor body carrying out similar functions

1.5 'farmer' shall have the same meaning as that contained in Article 2(a) of the Council Regulation

2. Sale and Purchase

In consideration of the sum of [ ] pounds [£] plus VAT @ 15% if such shall be applicable ('the Purchase Price') the Seller shall sell and the Buyer shall buy the Entitlements on the terms of this Agreement

3. Payment

[*Note: leave out if clause 5 applies.*]

The Buyer shall pay the Purchase Price to the Seller within five working days of the Buyer receiving notification from the RPA that it has registered the transfer of the Entitlements from the Seller into the name of the Buyer in its records

AND in default of his making the payment of the Purchase Price in accordance with the aforementioned time scale the Buyer shall pay interest to the Seller on the Purchase Price at the rate of 4% per annum above the Nat West Bank plc base rate from time to time from the date the payment was due until the date of actual payment

4. Warranties

4.1 The Seller warrants to the Buyer that:-

4.1.1 he will not seek to claim any of the Entitlements that are intended to be transferred to the Buyer pursuant to this Agreement nor will he by any act or omission on his part permit any third party so to do

4.1.2 he knows of no facts, and has not done or omitted to do anything, that would in any way invalidate the Entitlements or prevent them being transferable to the Buyer pursuant to the Regulations

4.1.3 he has not received any communication from the RPA to indicate there are any problems with the Entitlements or that would in any way prevent them being transferred to the Buyer in accordance with the terms of this Agreement AND in the event of his receiving any communications from the RPA in connection with this or any other matters he will deal promptly with them

4.1.4 he will indemnify the Buyer in respect of any shortfall in the number of Entitlements that may actually be transferred to the Buyer as opposed to those warranted to be transferred in accordance with the terms of this Agreement and also against any loss of single payment arising from such shortfall whether or not such shortfall is occasioned by any act of negligence omission or default on his part

4.1.5 if he is notified that the transfer has not been accepted by the RPA he will promptly inform the Buyer

4.1.6 he will provide the Buyer with a valid VAT invoice in respect of the Purchase Price

5. Payment of Purchase Price to the Seller

[*Note: leave out if clause 3 applies.*]

5.1 On the signing of this Agreement the Buyer will deposit the Purchase Price with his solicitors to be held by the Buyer's solicitors as stakeholders and placed in a designated deposit account in the joint names of the Buyer and the Seller [or their respective solicitors as the parties shall agree]

5.2 The Purchase Price shall be paid out of the designated deposit account by the Buyer's solicitors in the following circumstances:-

5.2.1 on receipt of official confirmation of the completion of the transfer of the Entitlements into the name of the Buyer by the RPA the Purchase Price shall be paid to the Seller together with all accrued interest

5.2.2 on receipt of official confirmation of the completion of the transfer of some but not all of the Entitlements into the name of the Buyer by the RPA a proportion of the Purchase Price (including a proportion of any accrued interest) shall be paid to the Seller such proportion to be calculated by dividing the number of Entitlements transferred by the Entitlements and the percentage the one bears to the other being applied to the Purchase Price, the remainder of the Purchase Price (and the remainder of the accrued interest) being paid to the Buyer

5.2.3 in the event that none of the Entitlements are transferred to the Buyer by the (*date*) the Purchase Price (together with all accrued interest) shall be paid to the Buyer and this Agreement shall be at an end

6. Further agreements between Seller and Buyer

It is hereby further agreed between the Seller and the Buyer that:-

6.1 they will act in fairness and co-operate in good faith with each other and take all (if any) reasonable steps to ensure the transfer of the Entitlements from the Seller to the Buyer in accordance with the terms of this Agreement from the earliest possible date

6.2 if after the date of this Agreement there are changes in the Regulations or in the interpretation of one or other or all of them by DEFRA or the RPA or otherwise before the Seller has fully complied with the terms of this Agreement in such a way that the intention of the parties as set out in this Agreement would be frustrated without amendment to it either party may give notice to the other, in the absence of agreement between them on the content of any amendments or subsequent agreement to achieve their objectives as herein evidenced, to appoint an arbitrator in accordance with the provisions of clause 6.3 below to redraft the terms of this Agreement in such a way as may fairly achieve the objectives of the parties and the costs of such appointment shall be borne equally by the Seller and the Buyer

6.3 if any dispute arises under this Agreement either party may give notice to the other in writing specifying the dispute and requesting that agreement be reached on the identity of a single arbitrator to be appointed to determine the dispute in accordance with the provisions of the Arbitration Act 1996 and if no arbitrator has been appointed by agreement within one month of the notice either party may make a written request to the President for the time being of the Royal Institution of Chartered Surveyors for him to appoint an arbitrator to determine the matter in dispute

SIGNED by
in the presence of:-

SIGNED by
in the presence of:-

PRECEDENT 2 DEED OF VARIATION

[*The likelihood is that there is either nothing at all in a tenancy agreement whose provisions are governed by the Agricultural Holdings Act 1986 which will embrace the SPS, or, that the wording refers to matters that could not under any interpretation fall within the SPS.*

For example:-

> 'IT IS FURTHER AGREED between the Landlord and the Tenant:-
>
> that the Tenant shall before quitting the holding nominate as his successor the Landlord or incoming tenant in respect of any basic quota of which he has the right of nomination and that he shall not allow any quotas to the holding under any agricultural marketing scheme or otherwise to lapse by his default and that he shall use his best endeavours to obtain for the holding suitable quotas under any new scheme and to maintain those already attached to the holding …'

The above is the wording taken from the Estates Gazette *form of tenancy agreement and cannot in the authors' opinion be taken to embrace the SPS and thereby require the tenant to transfer any entitlements he may hold to the landlord at the end of the term.*

Assuming the tenancy agreement contains either the above or something similar, or, as is more likely to be the case, nothing at all, what then can be done/needs to be done?

In short, the parties will need to talk to see if agreement can be reached between them. If it can then what follows is what the parties should be encouraged to enter into to confirm their agreement.

This can be used to vary the terms of either an Agricultural Holdings Act 1986 tenancy or an Agricultural Tenancies Act 1995 tenancy.]

THIS DEED made on the [ ] day of [ ] 20[ ]

BETWEEN:-

[*name*] of [*address*] ('the Landlord')

and

[*name*] of [*address*] ('the Tenant')

WITNESSES as follows:-

In this Deed

1. Definitions and interpretation

1.1 'the Tenancy Agreement' means the Tenancy Agreement dated the (*date*) and made between the Landlord (1) and the Tenant (2)

1.2 'the Holding' means ALL THAT property known as [ ] situated at [ ] full particulars of which are contained in the Tenancy Agreement

1.3 Where any party to this Deed comprises two or more persons obligations expressed or implied to be made by or with that party are deemed to be made by or with the persons comprising that party jointly and severally

1.4 Words importing one gender include all other genders; words importing the singular include the plural and vice versa and words importing persons shall be construed as importing a corporate body or a partnership and vice versa

1.5 Reference to any clause, sub-clause or schedule without further designation shall be construed as a reference to the clause sub-clause or schedule in this Deed so numbered

1.6 The clause paragraph and schedule headings do not form part of this Deed and shall not be taken into account in its construction or interpretation

2. Recitals

2.1 This Deed is supplemental to the Tenancy Agreement

2.2 By the Tenancy Agreement the Landlord let to the Tenant the Holding from the (*date*) for the term as therein defined ('the Term')

2.3 In consideration of the terms hereinafter contained the Landlord and the Tenant have agreed that the terms of the Tenancy Agreement shall be varied as hereinafter appearing

3. Consideration [if any]

[*NB: as this is a deed consideration is not required although the parties may well prefer to show in this Deed what has been agreed for the avoidance of doubt at a later date.*]

[In consideration of [ ] it is mutually agreed that the provisions set out in the Schedule hereto shall be incorporated into the Tenancy Agreement and take effect from the date hereof as if they were incorporated in the Tenancy Agreement]

4. Tenant's covenant

The Tenant covenants with the Landlord to observe and perform the obligations set out in the Schedule hereto as from the date of this Deed as if they were incorporated in the Tenancy Agreement

5. **Landlord's covenant**

The Landlord covenants with the Tenant to observe and perform the obligations set out in the Schedule hereto as from the date of this Deed as if they were incorporated in the Tenancy Agreement

6. **Confirmation of the Tenancy Agreement**

The Tenancy Agreement shall continue in full force save as modified by this Deed and shall operate as if the provisions contained in the Schedule were incorporated in it. The covenants conditions and provisions of the Tenancy Agreement shall have effect as though the provisions contained in this Deed had been contained in it from the commencement of the Term

7. **Subsequent alteration**

If after the date of this Deed there are changes in the Regulations or in the interpretation of them by DEFRA or the RPA or in any other way the intention of the parties as set out in this Deed would be frustrated without amendment to this Deed either party may give notice to the other, in the absence of agreement between them on the content of any subsequent Deed to achieve their objectives as herein evidenced, to appoint an arbitrator to redraft the terms of this Deed in such a way as may fairly achieve the objectives of the parties and the costs of such appointment shall be borne equally by the parties

8. **Arbitration**

If any dispute arises under this Agreement either party may give notice to the other in writing specifying the dispute and requesting that agreement be reached on the identity of a single arbitrator to be appointed to determine the dispute in accordance with the provisions of the Arbitration Act 1996 and if no arbitrator is appointed by agreement within one month of the notice either party may make a written request to the President for the time being of the Royal Institution of Chartered Surveyors for him to appoint an arbitrator to determine the matter in dispute

IN WITNESS whereof the parties hereto have set their hands to this Deed the day and year first before written

SCHEDULE

1. In this Schedule:-

1.1 the Regulations' means Council Regulation (EC) No 73/2009 and all relevant subsequent Commission Regulations administrative provisions statutory guidance or codes of practice implemented by the EC and/or the UK Government pursuant to the Regulations

1.2 'Entitlements' means such entitlements to the single income payment and any other entitlement to financial support of whatever kind deriving in any way from the Regulations as are held by the Tenant in his name

1.3 'DEFRA' means the Department for the Environment Food and Rural Affairs or any successor department

1.4 'RPA' means the Rural Payments Agency or any successor body carrying out similar functions

1.5 'Eligible Hectare' shall have the same meaning as in Article 34(2) of Council Regulation (EC) No 73/2009

2. The Tenant further covenants with the Landlord:-

2.1 to maintain the Entitlements upon the maximum eligible area of the Holding in accordance with the Regulations

2.2 at all times during the Term to do all acts and things which may be required to maintain the maximum number of Entitlements under the Regulations

2.3 during the Term to claim payment upon the Entitlements in accordance with the Regulations in such a way that neither the Entitlements nor any part of them are lost or charged or whereby they are removed from the Tenant whether permanently or temporarily

2.4 to use his best endeavours to obtain for the Holding any new or further entitlements or additional payment value for the Entitlements which become available under the Regulations or otherwise whether existing at or introduced after the commencement of the Tenancy and whether they become available by reference to the Entitlements, the Tenant's occupation of the Holding or otherwise

2.5 to keep and maintain records of all entitlements allocated to or transferred or leased in or out by the Tenant and their use together with any changes in their number type or payment value as may arise as a result of any official action and to permit the Landlord and any person nominated by him to inspect and take copies of that record in so far as it relates to the Entitlements

2.6 at all times during the Term to carry out all obligations imposed on the Tenant as occupier or producer with regard to all of the requirements of the Regulations

2.7 within fourteen days of a written request being received to furnish the Landlord (in writing if so required) with all information which the Landlord may reasonably request concerning the Entitlements or concerning other farming activities of the Tenant which may affect or tend to affect the Entitlements whether such farming activities are carried out by the Tenant directly or indirectly and whether they relate to the Holding or to other land

2.8 if so requested in writing prior to submitting any forms to DEFRA and/or the RPA in any year to agree with the Landlord the information required by the forms and the contents thereof and to obtain the Landlord's consent to that information and to the contents of the forms

2.9 not to dispose of transfer lend lease charge or otherwise deal with the Entitlements or any of them so as to result in the Entitlements lapsing or reducing in number or value during the Term or being transferred off the Holding or otherwise becoming unavailable for use in full on the Holding on the termination of the Tenancy

2.10 not by any direct or indirect act or omission in respect of any land or property not included in the Holding to allow the Entitlements to pass to any other person or become claimable in respect of land not comprised within the Holding

2.11 not to make or permit anyone claiming on his behalf or through him to make any claim either during or at or following termination of the Tenancy which may result in the Entitlements or any of them ceasing to belong to or be available to the Landlord and if as a result of any such claim the Entitlements or any of them are lost to the Landlord to indemnify the Landlord against all losses which the Landlord suffers thereby including (but not necessarily limited to) all costs and expenses incurred in the acquisition of replacement entitlements which are comparable in all respects to those lost

2.12 on or after quitting the Holding to take all such steps and do all such acts as may be required to procure the transfer of the Entitlements to the Landlord or such person or persons as the Landlord nominates in writing

delete as appropriate

*[at a consideration to be agreed between the parties being the open market value and in default of agreement to be determined by arbitration]

*[without payment]

*[the Landlord paying the Tenant compensation in the sum of one pound (£1)]

*[being a sum equivalent to the open market value of that part of the Entitlements that would represent the historic element of each of them at that time in the same way as if such were capable of being separately valued notwithstanding the fact that the Entitlements are not divisible AND in default of agreement to be determined by arbitration]

*[being [ ]% of the open market value of the Entitlements at the date of termination of the Tenancy]

2.13 as the provisions of clause 2.12 also apply where the Tenancy is terminated in respect of any part of the Holding in that case to procure the transfer pro rata to the Landlord of the number of entitlements that the Entitlements bear to the number of Eligible Hectares in respect of which the Tenancy is terminated compared with the number of Eligible Hectares comprised in the Holding and the compensation payable shall also be calculated pro rata on the same basis

2.14 during the final year of the Tenancy to obtain the Landlord's consent in respect of all matters concerning the management of the Entitlements to include cross compliance claiming the single payment and effecting the transfer of the Entitlements as the Landlord may direct

2.15 in so far as they remain to be observed and performed and are capable of taking effect after the termination of the Term howsoever determined the provisions of this Schedule shall remain in full force and effect notwithstanding the termination of the Tenancy

2.16 the Landlord covenants with the Tenant to comply with the terms of the Single Payment Scheme after the termination of the tenancy throughout the then subsisting calendar year

SIGNED as a Deed by
in the presence of:-

SIGNED as a Deed by
in the presence of:-

Chapter 2

SALE AND ACQUISITION OF AGRICULTURAL LAND

INTRODUCTION

PRECEDENTS

1 Enquiries before contract
   1.1 Agricultural – general enquiries (long form)
   1.2 Agricultural – general enquiries (purchase of bare land with vacant possession)
   1.3 Agricultural – supplemental to commercial property enquiries (purchase of bare land)
2 Agreement for sale and purchase (long form)
3 Single Payment Scheme contract clauses
   3.1 Sale of land with entitlements
   3.2 Sale of land without entitlements

INTRODUCTION

Since this does not purport to be the definitive work on conveyancing in general, it is not really appropriate to provide full precedents of every form of contract, enquiry or the like. There are set out below samples of specific enquiries and specific contractual clauses of relevance in agricultural matters. It is assumed that the contract will also contain all necessary clauses relevant to the property itself, including the grant and reservation of rights and a general clause ensuring that the terms will remain binding beyond completion. Those clauses, so far as relevant, will be carried forward into the transfer.

The most important change since the last edition of this book has been the introduction of the Single Payment Scheme which came into force on 1 January 2005 as a result of the European Commission's review of agricultural subsidies. The Enquiries before Contract include questions relating to this subsidy and also deal with the new requirements on cross compliance and environmental regulations. Although not fully covered in regulations at the time of writing, it is assumed that the health check proposals (eg the abolition of set-aside) will come into force.

There are precedents for contractual clauses where land is sold with and without single farm payment entitlements. The precedents include mutual warranties and indemnities relating to cross compliance. Transfers of entitlements have to be registered with the Rural Payment Agency and the parties to a transaction must ensure that transfer of land is coordinated in time with transfer of entitlements, or that there are relevant retention provisions in a contract. Any consideration for entitlements is not liable to stamp duty land tax but may be chargeable to VAT, depending on the circumstances of the sale.[1]

Aid schemes cause many practitioners particular problems. There is a wide variety of such schemes affecting agriculture; many of these involve undertakings and penalties on breach of obligations or on early termination including the recovery of aid paid. It is therefore essential for the advisers to a Seller of land to ascertain from their client details of any such schemes which are relevant. Where liability under a scheme remains with the Seller (or until a formal transfer of the scheme to the buyer is completed) the contract should include provisions that the buyer will observe any obligations imposed by the scheme(s) for the remainder of the duration of the term, and will indemnify the Seller for any breaches or non compliance. A buyer, on the other hand, will be well advised to request contractual warranties from a Seller that the terms of any scheme(s) have been complied with, as well as indemnities in respect of any breaches that may have occurred during the Seller's ownership and which may give rise to a demand for a repayment of subsidy.

Optional enquiries relating to public rights of way, common land and town and village greens should always be made on Con 29 Enquiries of Local Authorities. Further enquiries should reveal if there are any applications pending. The Countryside and Rights of Way Act 2000 imposed a new duty on surveying authorities to keep a register with prescribed information on application orders for modifications to the definitive map.[2] It is no longer possible to make a Commons Registration search.[3]

In addition, the Seller should be asked for any information relating to the use of the land by members of the public or local inhabitants. The Commons Act 2006, which is being brought into force in stages, repeals the Commons Registration Act 1965. Under the provisions of the 2006 Act, where local inhabitants have used land for lawful sports and pastimes for 20 years they may

---

1 HMRC, *Tax Bulletin Special Edition*, June 2005 explains the Revenue's approach to VAT liability on sale of entitlements.

2 Section 51, Sch 5, para 2, introducing s 53B into the Wildlife and Countryside Act 1981. Brought into force in England on 27 September 2005 by the Countryside and Rights of Way Act 2000 (Commencement No 9) Order 2005, SI 2005/2459 and in Wales on 11 May 2006 by the Countryside and Rights of Way Act 2000 (Commencement No 8 and Transitional Provisions) (Wales) Order 2006, SI 2006/1279 (W 124). The regulations are the Public Rights of Way (Register of Applications under section 53(5) of the Wildlife and Countryside Act 1981) (England) Regulations 2005, SI 2005/2461 and the Public Rights of Way (Registers) (Wales) Regulations 2006, SI 2006/42 (W 8).

3 Commons Registration (General) (Amendment) (England) (No 2) Regulations 2007, SI 2007/2404.

make a claim to a village green.[4] The effect of registration of a green means that development, even where planning permission has been obtained, is prevented. The phrase 'lawful sports and pastimes' has been held to include informal activities such as picking blackberries and flying kites.[5] Moreover in Lord Hoffmann's opinion low-level agricultural activities are not inconsistent with use for sports and pastimes.[6]

4 Section 15. One of three circumstances must apply in addition to 20 years' use: 1 – the land continues to be used for this purpose at the time of the application to register; 2 – if the use for that purpose ceased before the application to register is made, but after 6 April 2007, the application must be made within 2 years of the cessation of use; 3 – if the use for that purpose ceased before 6 April 2007, the application must be made within 5 years of the cessation of use.

5 *R v Oxfordshire County Council ex parte Sunningwell Parish Council* [2000] 1 AC 235.

6 *Oxford County Council v Oxford City Council and Robinson* [2006] UKHL 25, at [57].

PRECEDENTS

1 ENQUIRIES BEFORE CONTRACT

[*The enquiries in this chapter relate predominantly to the agricultural aspect of the transaction and are meant to be supplemental to general commercial property enquiries which cover in more detail issues such as planning, taxation, occupiers, employees, rates and outgoings (although there may be some overlap). A commonly used version of commercial enquiries is the CPSE suite of documents, prepared by the London Property Support Lawyers Group (Practical Law Company), and endorsed by the British Property Federation. The extent and nature of both commercial and agricultural enquiries may depend on the complexity of a particular transaction and the precedents are not. Precedent 1.1 is suitable for transactions where the property being sold includes land and buildings. Precedents 1.2 and 1.3 are more appropriate for bare land.*]

PRECEDENT 1.1 AGRICULTURAL – GENERAL ENQUIRIES (LONG FORM)

ENQUIRIES BEFORE CONTRACT
(AGRICULTURAL)

PROPERTY:

SELLER:

BUYER:

Reference	:	Reference	:
Buyer's Solicitors		Seller's Solicitors	
Dated	:		:

Dwellings

1. Are there any dwellings on the Property?

   If so, please specify which are sold with vacant possession

2. In respect of any part not to be sold with vacant possession, please:-

2.1 supply copies of any tenancy agreements and other relevant documents

2.2 supply a plan of the area occupied, unless this is self-evident on the ground

3. Insofar as they are not shown by any copy document supplied please:-

3.1 supply full details of any rights enjoyed or suffered by the dwelling

3.2 specify the commencement date of the current occupation

3.3 specify the period of the letting

3.4 specify the respective repairing and other obligations undertaken by the owner and the occupier

3.5 state who is responsible for emptying any septic tank

3.6 give full details of any specific insurance

4. In relation to the rent or any other payment by the occupier:-

4.1 how much is it?

4.2 on what dates is it payable?

4.3 is it payable in advance or in arrear?

4.4 please say when and how it was last fixed and supply a copy of any relevant document

5. If the tenancy was created before 15 January 1989 is the rent registered? If so, please supply a copy of the registration and any notice of increase

6. What was the rateable value of the dwelling on 31 March 1990?

7. Has there been a succession to the tenancy? If so, please supply full details

8. Has the tenant been supplied with the statutory particulars of the local rent allowance scheme? If so, how?

9. Has the landlord received payment of the tenants rent allowance direct from the local authority?

10. Has the Seller, or any predecessor in title, applied to the housing authority to re-house the tenant, under the Rent (Agriculture) Act 1976 section 27? If so:-

10.1 what was the result?

10.2 please supply copies of the DEFRA assessment, the advice of the agricultural dwellinghouse advisory committee, and the decision of the authority

10.3 please give details of any tenancy deposit scheme operated

11. Is any dwelling on the Property affected by an agricultural user restriction under any planning permission? If so, please give details

12. Please provide details of the Council Tax band of each dwelling on the Property and the current amounts payable

13. State the date on which the local authority certified the dwelling as being fit for occupation after completion of the relevant works

14. Confirm that all the conditions of the grant were complied with throughout the period of five years beginning with the certified date and that no part of the improvement grant became or has become repayable

15. Is there a septic tank on or serving any part of the Property? If so, please mark its position on a plan, also showing the position of all associated pipes and the overflow

Occupancy

16. Please confirm that:-

16.1 vacant possession of the whole will be given on completion

16.2 no right of holdover will be required. If so, please state when and in what condition the Seller proposes to leave the Property after any crop now growing has been harvested

17. If vacant possession of the whole will not be given:-

17.1 please supply details of all persons other than the Seller now occupying or farming any part of the Property (other than any dwelling)

17.2 on what basis do those persons occupy or farm the part concerned?

18. Please specify any sporting rights which are not in hand and supply full details of, and a copy of any relevant document relating to, the exercise of any such rights

Rights of way

19. Is there any public footpath or other public right of way crossing the Property?

20. Has any footpath or other public right of way on the Property been ploughed up? If so, please supply details and a copy of the relevant notice

In-filling

21. Has there been any in-filling on the Property? If so, please:-

21.1 supply full details, including the dates and the nature of the in-filling

21.2 state the use to which the land had previously been put

Flooding/subsidence

22. Please give details of:-

22.1 any flooding that has affected the Property

22.2 any subsidence or heave which has affected any structure on the Property

Disease

23 Have any cattle which have been on the Property within the last six years suffered from bovine spongiform encephalopathy? If so, please give details

24. Have any animals on the Property been affected by foot and mouth disease, blue tongue or any other notifiable disease? If so, please give details

25. Is any part of the Property affected by disease, pests or persistent weeds, eg wireworm, rhizomania, blackgrass. If so, please supply details

Employees

26. Are any employees of the Seller employed wholly or mainly on the Property? If so, in relation to each, please:-

26.1 give his date of birth

26.2 provide a copy of his contract of employment

26.3 give full details of any terms of his employment which are not set out in the contract of employment

Adverse matters

27. Is the Seller aware of any past existing or proposed public or private rights of way claimed or exercised over the Property? If so, please supply full details, including copies of all orders and notices

28. Please supply full details of any of the following of which the Seller is aware:-

28.1 any other rights or informal arrangements affecting the Property

28.2 any restrictions affecting the Property, other than any referred to in the draft contract

28.3 any interests set out in Schedules 1, 3 and 12 to the Land Registration Act 2002

29. Does the Seller own or occupy any adjoining or nearby property? If so, please:-

29.1 specify the property concerned

29.2 supply details of all quasi-easements or analogous rights or privileges enjoyed by such property over or affecting the Property, with details of the part of that property which has the benefit of them

30. Is the Seller aware that any part of the Property is subject to any liability for chancel repair, corn rent or other unusual liability or payment? If so, please give full details

31. Has the land been used by the public or the local inhabitants for open air recreation? If so, give details

Copyhold

32. Was any part of the Property formerly copyhold? If so, please supply details

Planning

33. Please supply a copy of each planning permission which relates to the Property or any part of it

34. In relation to each agricultural building please state whether it was erected or altered:-

34.1 under a specific planning permission

34.2 in reliance on the terms of the General Development Order

35. If any part of the Property is used for a non-agricultural use (e g farm shop, cottage holiday lettings, commercial buildings etc), please supply full details

36. Is there an established use certificate relating to any part of the Property or has there ever been an unsuccessful application for such a certificate? If so, please supply a copy with its plan (if any)

37. Is any building on the Property a listed building? If so, please:-

37.1 specify the building

37.2 supply a copy of the listing

37.3 confirm that no listed items have been removed or altered

37.4 supply a copy of any listed building consent obtained

37.5 confirm that no works for which listed building consent was required by have been carried out without such consent

38. Is there any ancient monument on any part of the Property? If so, please supply particulars

39. Is the Seller aware that any part of the Property is, or is likely to be, designated as an area of archaeological importance? If so, please give details

IACS – Single Payment Scheme ('SPS')

40. For ease of reference 'the Seller' refers to the transferor of the Property and the SPS Claimant (if different). 'Property' refers to the land being transferred, while 'Holding' is the Seller's IACS holding. (It may assist if whoever completed the Seller's SPS application form also completes the reply to these enquiries)

Entities

41. Please confirm the identity of the SPS claimant in respect of the Property

42. If the SPS claimant is a separate entity, please confirm that that entity will be party to and sign the contract to transfer entitlements ('Entitlements') to the Buyer or his nominee

Entitlements

43. How many Entitlements (to 2 decimal places) are to be transferred?

44. Please confirm that the hectarage of the Property comprising SPS eligible land is not lower than the number of Entitlements to be transferred under the proposed transaction

45. Of the Entitlements to be transferred please provide details of:-

45.1 Normal; formerly Set-aside Entitlements[7]

45.2 Normal Entitlements

45.3 Special Entitlements

---

7 During 2009 Scheme year, the Set-aside Entitlements registered with RPA will be converted to Normal Entitlements.

46. Please confirm that all Entitlements to be transferred were validly applied for

47. Please provide details of the Entitlements for which payments are being claimed this year

48. What part of the purchase price is being apportioned to the Entitlements?

49. If the Property being transferred is the Seller's whole Holding what is the intention regarding this year's payment?

50. Please confirm that a transfer application will be submitted as soon as possible

51. Does any of the Property lie within English Upland SDA or English Moorland within the Upland SDA? If so please provide details

52. Please provide details of the SPS eligible agricultural areas of the Property

53. Please identify which areas of the Property are not eligible for payments under SPS (such as non grazable woodland, forests (but not commercial forests tree nurseries in woodland or land in eligible forestry scheme), peat production, non agricultural use)[8]

54. Please identify any areas of the Property which are permanent pasture

55. Please identify any areas of orchards or grazed woodlands

56. Have any of the Entitlements to be transferred been applied for on the basis of commons rights of the Seller (Annex 5c of the SP5a)? If so, please provide details

8 Following coming into force of the new Council Regulation (EC) No 73/2009, land under permanent crops, including permanent fruit and vegetables, nursery crops and vines, is eligible for SPS from 2009 SPS year.

57. Do any of the Entitlements have an historic element which derives in full or in part from dairy payments, and for which an uplift in value was awarded in 2006?

58. Please provide details of any non-agricultural use on eligible SPS land including, where relevant, specific dates and areas

59. Please provide copies of the Rural Land Register maps for the property. If these have not yet been finalised please provide copies of correspondence, forms and plans submitted to the RPA

60. Please provide a copy of the latest SP5a Form and any supplemental forms and documentation submitted to the RPA together with confirmation as to when this was received

61. Please provide full details of any disputes or anomalies relating to the Seller's claim to establish Entitlements and claim payments

62. Please provide full details of any licence agreements affecting the Property

63. Please provide details of any contracting arrangements affecting the Property

64. Please supply a copy of the latest statement of payments and allocation station from the RPA

Other IACS and ERDP payments

65. Are any of the following relevant to the Property:-

    (a) area payment for nuts

    (b) aid for energy crops

    (c) protein crop premium

(d) Hill Farm Allowance[9]

66. (a) Please provide details of any agri-environment, Environmental Stewardship or similar national schemes relevant to the Property including copies of the relevant agreements

(b) If such schemes relate to both the Property and other land not being transferred, what is the Seller's intention regarding such schemes?

Non-food contracts

67. Please provide details of any non-food crop contracts (whether in formerly set-aside land or otherwise[10])? What is the Seller's intention in relation to any such contracts (transfer to Buyer/agency arrangement)?

68. Is any part of the Property used to grow hemp for fibre? If so, what is the Seller's intention in relation to the supply contract and licence?

Cross-compliance and eligibility

69. Please provide full details of any breach of Statutory Management Requirements ('SMR'), standards relating to good agricultural and environmental condition ('GAEC'), or requirements relating to permanent pasture

70. Please provide details of any resulting reductions and/or exclusions from the SPS

71. Please provide details of any inspections that have been carried out under the scheme by the RPA or one of its designated specialist agencies

---

9 From 2010 SPS year, the Hill Farm Allowance will be replaced by Uplands Entry Level Stewardship (UELS). It is understood at the time of writing that UELS agreements can start from July 2010.

10 For enquiries in the Scheme year 2009.

72. Please provide full details of any appeals, representations or complaints made in respect of proposed reductions and/or exclusions

73. Please confirm the following and where the answer is affirmative provide full details:-

74. is any part of the Property a Special Protection Area (SMR1)?

75. will any substances listed in Schedule of List 1 or List 2 substances in the Environment Agency's Groundwater Regulation fact sheet have been disposed of on the Property during the current year (SMR2)?

76. will you have used sewage sludge on the Property during the current year (SMR3)?

77. is any of the Property within a Nitrate Vulnerable Zone (SMR4)?

78. is any of the Property within a Special Area of Conservation and/or are there to the Seller's knowledge any European protected species the Property (SMR5)?

79. are there any uncultivated or semi-natural areas on the Property (GAEC5)?

80. is any of the Property a designated Site of Special Scientific Interest[11]?

81. are there any Scheduled Monuments on the Property (GAEC7)?

82. are there any visible public rights of way on the Property (GAEC8)?

83. will any eligible agricultural land on the Property not have been used in agricultural production in the current year (GAEC12)? If so, please give details

84. are there any stone walls on the Property (GAEC13)?

---

[11] GAEC6; SMR1 and SMR5 (under the updated CAP policy these requirements have been removed from cross compliance where they are not relevant).

85. are there any hedgerows on the Property (GAEC14 & 15)?

86. are there any watercourses on the Property (GAEC14)?

87. are there any trees subject to a Tree Preservation Order on the property?

88. have any plant protection products been used on the Property? If so, please provide full details

89. have any substances having hormonal or thyrostatic action and beta-agonists in farm animals been used on the Property? If so, please provide details

90. has the Seller complied with SMR11 (Food and Feed Law)? If not, please give full details of all breaches

91. if there has been livestock on the Property please confirm that Statutory Management Requirements 12–15 have been complied with

92. please provide a copy of the Soil Protection Review relevant to the Property and any updates and details of all actions taken to comply therewith

93. please confirm that all qualifying fields on the Property have/will have the required 2 metre buffer zone around the hedges and watercourses

94. if the Buyer is to receive the payment arising from the Seller's SPS application:-

94.1 please provide details of any final or provisional notification of the value of Entitlements and any payments received to date

94.2 please confirm whether or not the Seller has yet received payment

94.3 has the Seller opted in writing to receive payment in Euros?

95. has any footpath or other public right of way on the Property been ploughed up? If so, please supply details and a copy of the relevant notice

96. is the Seller aware that any part of the Property is, or is likely to be, designated as an area of archaeological importance? If so, please give details

Milk quota

97. Does the Property have the benefit of any milk quota? If so, please:-

97.1 supply full details

97.2 confirm that the Seller will co-operate for the Buyer's accession to the quota concerned

Sugar beet contract tonnage

98. Does the Property have the benefit of any sugar beet contract tonnage? If so, please:-

98.1 supply full details

98.2 confirm that the Seller will co-operate to have the contract transferred to the Buyer

99. Is any sugar beet compensation payment outstanding, whether expected to be received under the Single Farm Payment Scheme or otherwise? If so, please supply details and confirm that it will be available to the Buyer

Water and drainage

100. Is any water used on or for the Property obtained otherwise than from a main supply? If so, please:-

100.1 supply a copy of the requisite water abstraction or other licence and any supplemental documents

100.2 confirm that the conditions of the licence(s) have been fully complied with, or supply details of any breach

100.3 confirm that the Seller will co-operate for the transfer of the Licence to the Buyer

101. If any non-mains water for which no licence is held by the Seller is used on the Property, please supply full details of:-

101.1 the source of such water

101.2 the amount consumed or used

101.3 the use to which the water is put

102. Is there any dam or weir on the Property which requires an impounding licence? If so, please:-

102.1 supply details and a copy of any licence held

102.2 confirm that the conditions of the licence have been fully complied with, or supply full details of any breach

103. If payments are made pursuant to any licence relating to water, are they made at the full rate? If not, please supply details of the payments made and a copy of any relevant document

104. Are there any water meters or water abstraction meters serving the Property? If so, please:-

104.1 supply a plan showing their positions

104.2 confirm that each of them is in good working order and condition

105. Is the Seller aware of any unsuccessful application for any licence relating to water? If so, please supply details

106. If the Property abuts upon a river, or if the Property is in any way affected by any watercourses, please:-

106.1 confirm that the ownership of the entire watercourse passes with the Property so far as co-extensive

106.2 confirm that there are no public or private rights along the co-extensive banks of the river or watercourse

106.3 confirm that the fishing is in hand

106.4 give the names and addresses of the appropriate Environment Agency office and Drainage Board and say which of the river and watercourses they maintain

106.5 give details of the rights of drainage into the river or watercourse

107. Please confirm that on completion plans of the land drains on the Property will be handed over

108. As to the drainage system serving the Property:-

108.1 does the Seller have any reason to doubt the right to use it? If so, please give details

108.2 who maintains the drains?

108.3 has the system been the subject of or caused any dispute? If so, please give details

Environment

109. Please provide a copy of any environmental audit, assessment or investigation of the Property that has been carried out by the Seller or anyone else

110. Is any part of the Property designated as Environmentally Sensitive Area, a Nitrate Vulnerable Zone, a Site of Special Scientific Interest or any other special area or zone? If so, please provide copies of all relevant documents

111. Have any proceedings been instituted or has the Seller any knowledge of contemplated proceedings in respect of any environmental or environment-related issues arising from the use or activity on the Property?

112. Has the Property been the subject of any formal or informal investigation by any regulatory body in respect of any environmental or environment-related issues arising from any use or activity on the Property?

113. Please confirm that the Seller warrants that so far as he is aware the Property has not been contaminated or in any way affected by the deposit or presence, however temporary, of any hazardous or toxic material or other pollutant whether such arose on the Property or from any adjoining or neighbouring property

114. Please supply copies of any documentary or other information produced by or on behalf of the Seller or in the possession or control of the Seller relating to any contamination or pollution which is or may be present in the Property or to the actual or intended use of the Property including any environmental or similar assessment

115. In relation to the Property, or any process or activity carried on there by the Seller or anyone else now or in the past, is the Seller aware of any potential or actual civil or criminal liability? If so, please provide full details

116. Is the Seller aware of any circumstances which might lead to the NRA, any Waste Regulation Authority, the Health and Safety Executive, or any other authority exercising any of its powers in respect of the Property? If so, please provide full details

117. Has any notice been served under section 51 of the Public Health Act 1936 in respect of either farm buildings or ancillary fixed plant (eg defective slurry tanks)? If so, please supply a copy

118. Is the Property affected by any matter under the Control of Pollution Act 1974? If so, please give details

119. Is the land subject to any environmental land management scheme? If so, please supply a copy, with any other relevant documents

120. Has there been grown on the Property to the Seller's knowledge any plant which has been genetically modified within the meaning of Part VI of the Environmental Protection Act 1990? If so, please give details

121. Please supply a copy of any authorisation in relation to the Property under the Ground Water Regulations

122. Please confirm that the Seller will serve notice within twenty one days of completion transferring any authorisation under the Ground Water Regulations to the Buyer

123. Please supply details of any disposal and tipping within the last five years

124. If the Property is in a Nitrate Vulnerable Zone, please supply details of Fertiliser and manure applied to the Property in the last six years on a field by field basis

125. Is the Seller aware of any disused pipes or cables, footings or other structures under the Property?

Woodlands

126. Please supply full details of:-

126.1 any Woodland Grant Scheme

126.2 any other forestry grant

126.3 any tree planting scheme carried out during the last five years

127. Please supply full details of all felling licences and felling directions affecting the Property and confirm that all their terms have been complied with

Grants

128. Except as already disclosed, has the Property or any part of it been included in any scheme under which moneys have been paid by way of grant or otherwise? If so, please reply to the enquiries under the heading 'Grants' in the attached supplemental enquiries

Taxation

129. Is any part of the purchase price subject to VAT? If so, please:-

129.1 confirm that the purchase price stated in the draft contract is inclusive of VAT

129.2 supply evidence of option to tax

130. If no option to tax has been made, please confirm that the Seller will not opt to tax in relation to the Property on or before completion

131. Are there any unused capital or agricultural buildings allowances? If so, please specify:-

131.1 the nature and amount of the relevant expenditure

131.2 the date the expenditure was incurred

131.3 the allowances used to date

General

132. Guarantees etc

Please confirm that the Seller will assign (if appropriate) the benefit of any guarantees etc of which he has the benefit

133. Services

Please supply a copy of any wayleave agreements relating to any electricity supply which crosses the Property

Investment property

134. Is any part of the Property (other than a dwelling) occupied by anyone who may by virtue of that occupancy ('the letting') be entitled to claim a tenancy under the Agricultural Holdings Act 1986 ('AHA 1986')? If so please reply to the following enquiries

134.1 Please supply:-

134.2 full details (including age) of anyone occupying the whole or any part of the Property under the letting.

134.3 a copy of any document evidencing all or any of the terms of the letting

135. If the letting is of only part of the Property, then insofar as they are not shown by any copy document supplied, please:-

135.1 supply a plan of the part of the Property comprised in the letting

135.2 supply full details of any rights enjoyed or suffered by that part

135.3 specify the commencement date of the letting

135.4 specify the period of the letting

135.5 specify the respective repairing and other obligations undertaken by the owner and the occupier

136. In relation to the rent or any other payment by the occupier:-

136.1 how much is it?

136.2 on what dates is it payable?

136.3 is it payable in advance or in arrear?

136.4 please say when and how it was last fixed, and supply a copy of any relevant document

136.5 have there been any arrears during the last two years? If so, please give details

137. Have the owner and the occupier concluded any agreement as to the amount of the rent or any other sum which is to be payable for any future period? If so, please give particulars

138. If the occupier insures the Property or part concerned, please give particulars of the existing cover, stating the name and address of the insurers and any brokers, the policy number, the insured name(s), the risks covered, for what amount, the premium and the date to which insurance runs

139. Has any notice to quit relating to the Property or any part been served? If so, please supply a copy of the notice and any relevant details

140. Are there any outstanding matters arising from any notice served by the landlord:-

140.1 under the Law of Property Act 1925 section 146?

140.2 under AHA 1986 sections 26–28 and Schedule III Cases D, E and F

141. Please supply details of all agricultural tenancy successions

142. Please supply a list of, and all matters agreed as to:-

142.1 tenant's fixtures on the Property

142.2 tenant's improvements on the Property

142.3 redundant buildings

143. Please supply copies of:-

143.1 all licences or consents granted for any tenant's improvements that have been carried out or are to be carried out

143.2 all notices to the Agricultural Land Tribunal or approval by the Tribunal (AHA 1986 section 67)

144. Have there been any other tenant's improvements for which consent was NOT granted? If so, please give full details

145. Has the owner refused any application for consent from the occupier? If so, please give details

PRECEDENT 1.2 AGRICULTURAL – GENERAL ENQUIRIES (PURCHASE OF BARE LAND WITH VACANT POSSESSION)

ENQUIRIES BEFORE CONTRACT
AGRICULTURAL – BARE LAND
VACANT POSSESSION

PROPERTY:

PARTIES:

Reference	:	Reference	:
Buyer's Solicitors		Seller's Solicitors	

Dated :

1. Bare land

Please confirm that there are no buildings (including dwellings) on the Property

2. Occupancy

2.1 Please confirm that:-

2.1.1 vacant possession of the whole will be given on completion

2.1.2 no right of holdover will be required. If holdover is required, please state when and in what condition the Seller proposes to leave the Property after any crop now growing has been harvested

2.2 Please specify any sporting rights which are not in hand and supply full details of, and a copy of any relevant document relating to the exercise of any such rights

Is there any public footpath or other public right of way crossing the Property?

Has any footpath or other public right of way on the Property been ploughed up?

Has there been any in-filling on the Property?

Please give details of any flooding that has affected the Property

Have any cattle which have been on the land within the last ten years suffered from bovine spongiform encephalopathy? If so, please give details

Are any animal carcasses buried on the Property?

3. Employees

Are there any employees of the Seller employed wholly or mainly on the Property? If so, please give details

4. Adverse matters

4.1 Is the Seller aware of any past existing or proposed public or private rights of way claimed or exercised over the Property? If so, please supply full details, including copies of all orders and notices

4.2 Please supply full details of any of the following of which the Seller is aware:-

4.2.1 any other easements rights or informal arrangements affecting the Property

4.2.2 any restrictions affecting the Property, other than any referred to in the draft contract

4.2.3 any other interests to which the Property is subject under Schedules 1, 3 or 12 to the Land Registration Act 2002

4.3 Does the Seller own or occupy any adjoining or nearby land? If so, please:-

4.4 specify the land concerned

4.5 supply details of all quasi-easements or analogous rights or privileges enjoyed by such land over or affecting the Property, with details of the part of that land which has the benefit of them

4.6 Does the Seller know of any matters which would be revealed by a Local Search and of which the Buyer should be aware, in particular any matters registered or capable of registration as a local land charge?

4.7 Is the Seller aware that any part of the Property is subject to any liability for chancel repair, corn rent or other unusual liability or payment?

4.8 Has the land been used by the public or the local inhabitants for open air recreation? If so, give details

5. Planning

5.1 Please supply a copy of each planning permission which relates to the Property or any part of it

5.2 If any part of the Property is used for a non-agricultural use please supply full details

5.3 Is there an established use certificate relating to any part of the Property or has there ever been an unsuccessful application for such a certificate? If so, please supply a copy with its plan (if any)

5.4 Is there any field/ancient monument on any part of the Property?

5.5 Is the Seller aware that any part of the Property is, or is likely to be, designated as an area of archaeological importance?

6. Copyhold

Was any part of the Property formerly copyhold?

7. IACS – Single Payment Scheme ('SPS')

For ease of reference 'the Seller' refers to the transferor of the Property and the SPS Claimant (if different). 'Property' refers to the land being transferred, while 'Holding' is the Seller's IACS holding. (It may assist if whoever completed the Seller's SPS application form also completes the reply to these enquiries)

8. Entities

8.1 Please confirm the identity of the SPS claimant in respect of the Property

8.2 If the SPS claimant is a separate entity, please confirm that that entity will be party to and sign the contract to transfer entitlements ('Entitlements') to the Buyer or his nominee

9. Entitlements

9.1 How many Entitlements (to 2 decimal places) are to be transferred?

9.2 Please confirm that the hectarage of the Property comprising SPS eligible land is not lower than the number of Entitlements to be transferred under the proposed transaction

9.3 Of the Entitlements to be transferred please provide details of:-

9.3.1 Normal; formerly Set-aside Entitlements[12]

9.3.2 Normal Entitlements

9.3.3 Special Entitlements

9.4 Please confirm that all Entitlements to be transferred were validly applied for

9.5 Please provide details of the Entitlements for which payments are being claimed this year

---

[12] During 2009 Scheme year, the Set-aside Entitlements registered with RPA will be converted to Normal Entitlements.

9.6 What part of the purchase price is being apportioned to the Entitlements?

9.7 If the Property being transferred is the Seller's whole Holding what is the intention regarding this year's payment?

9.8 Please confirm that a transfer application will be submitted as soon as possible

9.9 Does any of the Property lie within English Upland SDA or English Moorland within the Upland SDA? If so please provide details

9.10 Please provide details of the SPS eligible agricultural areas of the Property

9.11 Please identify which areas of the Property are not eligible for payments under SPS (such as non grazable woodland, forests (but not commercial forests tree nurseries in woodland or land in eligible forestry scheme), peat production, non agricultural use)

9.12 Please identify any areas of the Property which are permanent pasture

9.13 Please identify any areas of orchards or grazed woodlands

9.14 Have any of the Entitlements to be transferred been applied for on the basis of commons rights of the Seller (Annex 5c of the SP5a)? If so, please provide details

9.15 Do any of the Entitlements have an historic element which derives in full or in part from dairy payments, and for which an uplift in value was awarded in 2006?

9.16 Please provide details of any non-agricultural use on eligible SPS land including, where relevant, specific dates and areas

9.17 Please provide copies of the Rural Land Register maps for the Property. If these have not yet been finalised, please provide copies of correspondence, forms and plans submitted to the RPA

9.18 Please provide a copy of the latest SP5a Form and any supplemental forms and documentation submitted to the ROA together with confirmation as to when this was received

9.19 Please provide full details of any disputes or anomalies relating to the Seller's claim to establish Entitlements and claim payments

9.20 Please provide full details of any licence agreements affecting the Property

9.21 Please provide details of any contracting arrangements affecting the Property

9.22 Please supply a copy of the latest statement of payments and allocation statement from the RPA

10. Other IACS and ERDP payments

Are any of the following relevant to the Property:-

10.1 area payment for nuts

10.2 aid for energy crops

10.3 protein crop premium

10.4 Hill Farm Allowance[13]

10.5 (a) Please provide details of any agri-environment, Environmental Stewardship or similar national schemes relevant to the Property

13 From 2010 SPS year, the Hill Farm Allowance will be replaced by Uplands Entry Level Stewardship (UELS). It is understood at the time of writing that UELS agreements can start from July 2010.

(b) If such schemes relate to both the Property and other land not being transferred, what is the Seller's intention regarding such schemes?

11. Non-food contracts

11.1 Please provide details of any non-food crop contracts (whether in formerly set-aside land or otherwise[14])? What is the Seller's intention in relation to any such contracts (transfer to Buyer/agency arrangement)?

11.2 Is any part of the Property used to grow hemp for fibre? If so, what is the Seller's intention in relation to the supply contract and licence?

12. Cross-compliance and eligibility

12.1 Please provide full details of any breach of Statutory Management Requirements ('SMR'), standards relating to good agricultural and environmental condition ('GAEC'), or requirements relating to permanent pasture

12.2 Please provide details of any resulting reductions and/or exclusions from the SPS

12.3 Please provide details of any inspections that have been carried out under the scheme by the RPA or one of its designated specialist agencies

12.4 Please provide full details of any appeals, representations or complaints made in respect of proposed reductions and/or exclusions

12.5 Please confirm the following:-

12.6 Is any part of the Property a Special Protection Area (SMR1)?

---

[14] For enquiries in the Scheme year 2009.

12.7 Will any substances listed in Schedule of List 1 or List 2 substances in the Environment Agency's Groundwater Regulation fact sheet have been disposed of on the Property during the current year (SMR2)?

12.8 Will you have used sewage sludge on the Property during the current year (SMR3)?

12.9 Is any of the Property within a Nitrate Vulnerable Zone (SMR4)?

12.10 Is any of the Property within a Special Area of Conservation and/or are there to the Seller's knowledge any European protected species on the Property (SMR5)?

12.11 Are there any uncultivated or semi-natural areas on the Property (GAEC5)?

12.12 Is any of the Property a designated site of Special Scientific Interest (GAEC6)?

12.13 Are there any Scheduled Monuments on the Property (GAEC7)?

12.14 Are there any visible public rights of way on the Property (GAEC8)?

12.15 Will any eligible agricultural land on the Property not have been used in agricultural production in the current year (GAEC12)? If so, please give details

12.16 Are there any stone walls on the Property (GAEC13)?

12.17 Are there any hedgerows on the Property (GAEC14 & 15)?

12.18 Are there any watercourses on the Property (GAEC14)?

12.19 Please confirm that all qualifying fields on the Property have/will have the required 2 metre buffer zone around the hedges and watercourses

12.20 If the Buyer is to receive the payment arising from the Seller's SPS application:-

12.20.1 please provide details of any final or provisional notification of the value of Entitlements and any payments received to date

12.20.2 please confirm whether or not the Seller has yet received payment

12.20.3 has the Seller opted in writing to receive payment in Euros?

13. Milk quota

13.1 Does the Property have the benefit of any milk quota? If so, please:-

13.1.1 supply full details

13.1.2 confirm that the Seller will co-operate for the Buyer's accession to the quota concerned

14. Sugar beet contract tonnage

14.1 Does the Property have the benefit of any sugar beet contracted tonnage entitlement or milk quota? If so, please:-

14.1.1 confirm that the Seller will co-operate to have the contract transferred to the Buyer

14.1.2 is any sugar beet compensation payment outstanding, whether expected to be received under the Single Farm Payment Scheme or otherwise? If so, please provide details and confirm that it will be available to the Buyer

15. Water and drainage

15.1 Is any water used on or for the Property obtained otherwise than from a main supply? If so, please:-

15.1.1 supply a copy of the requisite water abstraction or other licence and any supplemental documents

15.1.2 confirm that the conditions of the licence(s) have been fully complied with, or supply details of any breach

15.2 If any non-mains water for which no licence is held by the Seller is used on the Property, please supply full details of:-

15.2.1 the source of such water

15.2.2 the amount consumed or used

15.2.3 the use to which the water is put

15.3 Is there any dam or weir on the Property which requires an impounding licence?

15.4 If payments are made pursuant to any licence relating to water, are they made at the full rate? If not, please supply details of the payments made and a copy of any relevant document

15.5 Are there any water meters or water abstraction meters serving the Property?

15.6 Is the Seller aware of any unsuccessful application for any licence relating to water? If so, please supply details

15.7 If the Property abuts upon a river, or if the Property is in any way affected by any watercourses, please:-

15.7.1 confirm that the ownership of the entire watercourse passes with the Property so far as co-extensive

15.7.2 confirm that there are no public or private rights along the co-extensive banks of the river or watercourse

15.7.3 confirm that the fishing is in hand

15.7.4 give the names and address of the appropriate River Authority and Drainage Board and say which of the river and watercourses they maintain

give details of the rights of drainage into the river watercourse

15.7.5 confirm that on completion plans of the land drains on the property will be handed over

16. Environment

16.1 Please provide a copy of any environmental audit, assessment or investigation of the Property that has been carried out by the Seller or anyone else

16.2 Has the Seller, or anyone to the Seller's knowledge, caused or allowed the Property or any part of it to be or become contaminated land as defined in section 78A of the Environmental Protection Act 1990? If so, please provide full particulars

16.3 Is the Seller aware of any past or present storage or deposit on the Property of:-

16.3.1 any hazardous substance in notifiable quantities as defined by the relevant Regulations?

16.3.2 any substance or material which may cause pollution to the environment (including pollution of controlled waters) or harm (as defined in section 78A of the Environmental Protection Act 1990)

16.3.3 any controlled waste as statutorily defined having been stored or deposited on the Property?

16.4 In relation to the Property, or any process or activity carried on there by the Seller or anyone else now or in the past, is the Seller aware of any potential or actual civil or criminal liability? If so, please provide full details

16.5 Is the Seller aware of any circumstances which might lead to the Environment Agency, any Waste Regulation Authority the Health and Safety Executive, or any other authority exercising any of its powers? If so, please provide full details

16.6 Have any notices been served by any competent authority under any legislation relating to public health, the control of pollution or environmental matters?

16.7 Is the Property affected by any matter under the Control of Pollution Act 1974? If so, please give details

16.8 Is any part of the Property designated as Environmentally Sensitive Area, a Nitrate Sensitive Area, a Nitrate Vulnerable Zone, a Site of Special Scientific Interest or any other special area or zone? If so, please provide copies of all relevant documents

16.9 Is the land subject to any environmental land management scheme (ELS, HLS, Countryside Stewardship or other)? If so, please supply a copy, with any other relevant documents

16.10 Is the Seller aware of any disused pipes or cables, footing or other structures under the Property?

16.11 Has there been grown on the Property to the Seller's knowledge any plant which has been genetically modified within the meaning of Part VI of the Environmental Protection Act 1990? If so, please give details

16.12 Please supply a copy of any authorisation in relation to the Property under the Ground Water Regulations

16.13 Please confirm that the Seller will serve notice within twenty one days of completion transferring any authorisation under the Ground Water Regulations to the Buyer

16.14 Please supply details of any disposal and tipping within the last five years

16.15 If the Property is in a Nitrate Vulnerable Zone, please supply details of Fertiliser and manure applied to the Property in the last six years on a field by field basis

17. Woodlands

17.1 Please supply full details of:-

17.1.1 any Woodland Grant Scheme

17.1.2 any other forestry grant

17.1.3 any tree planting scheme carried out during the last five years

17.1.4 Please supply full details of all felling licences and felling direction affecting the Property and confirm that all their terms have been complied with

18. Grants

Except as already disclosed, has the Property or any part of it been included in any scheme under which moneys have been paid by way of grant or otherwise?

19. Taxation

19.1 Is any part of the purchase price subject to VAT? If so, please:-

19.1.1 confirm that the purchase price stated in the draft contract is inclusive of VAT

19.1.2 supply evidence of option to tax

19.2 If no option to tax has been made, please confirm that the Seller will not opt to tax in relation to the Property on or before completion

General

20. Services

Please supply a copy of any wayleave agreements relating to any electricity supply which crosses the Property

21. Boundaries

21.1 Please specify (by reference to a plan or sketch) the boundary features and the ownership of each of them

21.2 Of those not owned by the Seller, which (if any) has the Seller maintained or regarded as his responsibility to maintain?

21.3 Please confirm that there are no discrepancies in the title deeds relating to the boundaries of the Property and that the physical boundaries of the Property correspond exactly to those shown in the title deeds

21.4 Is the Seller aware of any alterations to the physical boundaries of the Property whether before or during his period of ownership?

21.5 Is it necessary to enter upon any adjoining or neighbouring premises to maintain or repair any boundary feature or any building on the property? If so, please give details and state:-

21.5.1 by what right entry is obtained

21.5.2 whether any neighbour has objected? If so, please give details

22. Disputes

22.1 Please give details of the following of which the Seller is aware:-

22.2 any dispute (whether resolved or not) relating to the Property

22.3 any trespass or encroachment on the Property within the last three years

22.4 any complaints or causes of complaint within the last three years relating to any adjoining or neighbouring property

23. Notices

Except where already supplied in reply to a previous enquiry, please give particulars of all notices, whether given or received, relating to the Property of which the Seller is aware

24. Guarantees etc

Are there any guarantees, warranties, or other arrangements relating to any improvement to, or treatment or repair of, the Property or any items included in the sale? If so, please confirm that the Seller will assign (if appropriate) the benefit of any such guarantees etc

25. Services

25.1 Is any water supply to the Property metered? If not, has the Seller received notification that compulsory metering is to be imposed?

25.2 Does any water supply pass through or over property not included in the sale? If so:-

25.2.1 please give details of the routes

25.2.2 by what right?

25.2.3 please confirm that the Seller is not aware of any defect or want of repair in any of the services

25.2.4 please supply a copy of any wayleave agreements relating to any electricity supply which crosses the Property

26. Outgoings

26.1 Please specify all outgoings payable in respect of the Property

26.2 Please confirm that all outgoings payable before completion have been or will be paid by then and receipts for those payments will be produced on completion

27. Investment property

27.1 Is any part of the Property occupied by anyone who may by virtue of that occupancy ('the letting') be entitled to claim a tenancy under the Agricultural Holdings Act 1986 ('AHA 1986')? If so, please reply to the following enquiries:-

27.1.1 Please supply:-

full details (including age) of anyone occupying the whole or any part of the Property under the letting

27.1.2 a copy of any document evidencing all or any of the terms of the letting

27.1.3 if the letting is of only part of the Property, then insofar as they are not shown by any copy document supplied, please:-

27.1.4 supply a plan of the part of the Property comprised in the letting

27.1.5 supply full details of any rights enjoyed or suffered by that part

27.1.6 specify the commencement date of the letting

27.1.7 specify the period of the letting

27.1.8 specify the respective repairing and other obligations undertaken by the owner and the occupier

27.2 In relation to the rent or any other payment by the occupier:-

27.2.1 how much is it?

27.2.2 on what dates is it payable?

27.2.3 is it payable in advance or in arrear?

27.2.4 please say when and how it was last fixed and supply a copy of any relevant document

27.2.5 have there been any arrears during the last two years? If so, please give details

27.3 Have the owner and the occupier concluded any agreement as to the amount of the rent or any other sum which is to be payable for any future period? If so, please give particulars

27.4 Has any notice to quit relating to the Property or any part been served? If so, please supply a copy of the notice and any relevant details

27.5 Are there any outstanding matters arising from any notice served by the landlord:-

27.5.1 under the Law of Property Act 1925 section 146?

27.5.2 under AHA 1986 sections 26-28 and Schedule III Cases D, E and F

27.6 Please supply details of all agricultural tenancy successions

27.7 Please supply a list of, and all matter agreed as to:-

27.7.1 tenant's fixtures on the Property

27.7.2 tenant's improvements on the Property

27.8 Please supply copies of:-

27.8.1 all licences or consents granted for any tenant's improvements that have been carried out or are to be carried out

27.8.2 all notices to the Agricultural Land Tribunal or approval by the Tribunal (Agricultural Holdings Act 1986 section 67)

27.9 Have there been any other tenant's improvements for which consent was NOT granted? If so, please give full details

27.10 Has the owner refused any application for consent from the occupier? If so, please give details

PRECEDENT 1.3 AGRICULTURAL – SUPPLEMENTAL TO COMMERCIAL PROPERTY ENQUIRIES (PURCHASE OF BARE LAND)

ENQUIRIES BEFORE CONTRACT
(AGRICULTURAL – BARE LAND)
(SUPPLEMENTAL TO CPSE 1)

PROPERTY:

PARTIES:

Reference	:	Reference	:
Buyer's Solicitors		Seller's Solicitors	
Dated	:		:

Bare land

1. Please confirm there are no buildings on the Property

Occupancy

2. Please confirm that:-

2.1 vacant possession of the whole will be given on completion

2.2 no right of holdover will be required. If so, please state when and in what condition the Seller proposes to leave the Property after any crop now growing has been harvested

2.3 If vacant possession of the whole will not be given:-

2.4 please supply details of all persons other than the Seller now occupying or farming any part of the Property (other than any dwelling)

2.5 on what basis do those persons occupy or farm the part concerned?

2.6 Please specify any sporting rights which are not in hand and supply full details of, and a copy of any relevant document relating to, the exercise of any such rights

Rights of way

3.1 Is there any public footpath or other public right of way crossing the Property?

3.2 Has any footpath or other public right of way on the Property been ploughed up? If so, please supply details and a copy of the relevant notice

In-filling

4.1 Has there been any in-filling on the Property? If so, please:-

4.2 supply full details, including the dates and the nature of the in-filling

4.3 state the use to which the land had previously been put

Flooding/subsidence

5.1 Please give details of any flooding that has affected the Property

Disease

6.1 Have any cattle which have been on the Property within the last six years suffered from bovine spongiform encephalopathy? If so, please give details

6.2 Have any animals on the Property been affected by foot and mouth disease, blue tongue or any other notifiable disease? If so, please give details

6.3 Is any part of the Property affected by disease, pests or persistent weeds, eg wireworm, rhizomania, blackgrass? If so, please supply details

6.4 Are there any animal carcasses buried on the Property? If so, where?

Adverse matters

7.1 Is the Seller aware that any part of the Property is subject to any liability for chancel repair, corn rent or other unusual liability or payment? If so, please give full details

7.2 Is the Seller aware of any past existing or proposed public or private rights of way claimed or exercised over the Property? If so, please supply full details, including copies of all orders and notices

7.3 Please supply full details of any of the following of which the Seller is aware:-

7.3.1 any other rights or informal arrangements affecting the Property

7.3.2 any restrictions affecting the Property other than any referred to in the draft contract

7.3.3 any interests set out in Schedules 1, 3 and 12 to the Land Registration Act 2002

7.4 Does the Seller own or occupy any adjoining or nearby property? If so, please:-

7.4.1 specify the property concerned

7.4.2 supply details of all quasi-easements or analogous rights or privileges enjoyed by such property over or affecting the Property, with details of the part of that property which has the benefit of them

7.5 Has the land been used by the public or the local inhabitants for open air recreation? If yes, give details

Copyhold

8. Was any part of the Property formerly copyhold? If so, please supply details

Planning

9.1 Please supply a copy of each planning permission which relates to the Property or any part of it

9.2 If any part of the Property is used for a non-agricultural use, please supply full details

9.3 Is there an established use certificate relating to any part of the Property or has there ever been an unsuccessful application for such a certificate? If so, please supply a copy with its plan (if any)

9.4 Is there any ancient monument on any part of the Property? If so, please supply particulars

9.5 Is the Seller aware that any part of the Property is, or is likely to be, designated as an area of archaeological importance? If so, please give details

IACS – Single Payment Scheme ('SPS')

10. Entities

10.1 Please confirm the identity of the SPS claimant in respect of the Property

10.2 If the SPS claimant is a separate entity, please confirm that that entity will be party to and sign the contract to transfer entitlements ('Entitlements') to the Buyer or his nominee

11. Entitlements

11.1 How many Entitlements (to 2 decimal places) are to be transferred?

11.2 Please confirm that the hectarage of the Property comprising SPS eligible land is to be transferred under the proposed transaction

11.3 Of the Entitlements to be transferred please provide details of:-

(a) Normal; formerly Set-aside Entitlements[15]

(b) Normal Entitlements

(c) Special Entitlements

11.4 Please confirm that all Entitlements to be transferred were validly applied for

11.5 Please provide details of the Entitlements for which payments are being claimed this year

11.6 What part of the purchase price is being apportioned to the Entitlements?

11.7 If the Property being transferred forms only part of the Seller's Holding is it intended that the Seller remain in occupation so as to validly claim payment for this year?

11.8 If the Property being transferred is the Seller's whole Holding what is the intention regarding this year's payment?

11.9 Please confirm that a transfer application will be submitted as soon as possible

---

[15] During 2009 Scheme year, the Set-aside Entitlements registered with RPA will be converted to Normal Entitlements.

11.10 Does any of the Property lie within English Upland SDA or English Moorland within the Upland SDA? If so please provide details

11.11 Please provide details of the SPS eligible agricultural areas of the Property

11.12 Please identify which areas of the Property are not eligible for payments under SPS (such as non grazable woodland, forests (but not commercial forests tree nurseries in woodland or land in eligible forestry scheme), peat production, non agricultural use)

11.13 Please identify any areas of the Property which are permanent pasture

11.14 Please identify any areas of orchards or grazed woodlands

11.15 Have any of the Entitlements to be transferred been applied for on the basis of commons rights of the Seller (Annex 5c of the SP5a)? If so, please provide details

11.16 Do any of the Entitlements have an historic element which derives in full or in part from dairy payments, and for which an uplift in value was awarded in 2006?

11.17 Please provide details of any non agricultural use on eligible SPS land including, where relevant, specific dates and areas

11.18 Please provide copies of the Rural Land Register maps for the property. If these have not yet been finalised please provide copies of correspondence, forms and plans submitted to the RPA

11.19 Please provide a copy of the latest SP5a Form and any supplemental forms and documentation submitted to the RPA together with confirmation as to when this was received

11.20 Please provide full details of any disputes or anomalies relating to the Seller's claim to establish Entitlements and claim payments

11.21 Please provide full details of any licence agreements affecting the Property

11.22 Please provide details of any contracting arrangements affecting the Property

11.23 Please supply a copy of the latest statement of payments and allocation statement from the RPA

12. Other IACS and ERDP payments

12.1 Are any of the following relevant to the Property:-

(a) area payment for nuts

(b) aid for energy crops

(c) protein crop premium

(d) Hill Farm Allowance[16]

12.2 (a) Please provide details of any agri-environment, Environmental Stewardship or similar national schemes relevant to the Property

(b) If such schemes relate to both the Property and other land not being transferred, what is the Seller's intention regarding such schemes?

13. Non-food contracts

---

[16] From 2010 SPS year, the Hill Farm Allowance will be replaced by Uplands Entry Level Stewardship (UELS). It is understood at the time of writing that UELS agreements can start from July 2010.

13.1 Please provide details of any non-food crop contracts (whether in formerly set-aside land or otherwise[17])? What is the Seller's intention in relation to any such contracts (transfer to Buyer/agency arrangement)?

13.2 Is any part of the Property used to grow hemp for fibre? If so, what is the Seller's intention in relation to the supply contract and licence?

14. Cross-compliance and eligibility

14.1 Please provide full details of any breach of Statutory Management Requirements ('SMR'), standards relating to good agricultural and environmental condition ('GAEC'), or requirements relating to permanent pasture

14.2 Please provide details of any resulting reductions and/or exclusions from the SPS

14.3 Please provide details of any inspections that have been carried out under the scheme by the RPA or one of its designated specialist agencies

14.4 Please provide full details of any appeals, representations or complaints made in respect of proposed reductions and/or exclusions

14.5 Please confirm the following:-

(a) is any part of the Property a Special Protection Area (SMR1)?

[17] For enquiries in the Scheme year 2009.

(b) will any substances listed in Schedule of List 1 or List 2 substances in the Environment Agency's Groundwater Regulation fact sheet have been disposed of on the Property during the current year (SMR2)?

(c) will you have used sewage sludge on the Property during the current year (SMR3)?

(d) is any of the Property within a Nitrate Vulnerable Zone (SMR4)?

(e) is any of the Property within a Special Area of Conservation and/or are there to the Seller's knowledge any European protected species the Property (SMR5)?

(f) are there any uncultivated or semi-natural areas on the Property (GAEC5)?

(g) is any of the Property a designated Site of Special Scientific Interest (GAEC6)?

(h) are there any Scheduled Monuments on the Property (GAEC7)?

(i) are there any visible public rights of way on the Property (GAEC8)?

(j) will any eligible agricultural land on the Property not have been used in agricultural production in the current year (GAEC12)? If so, please give details

(k) are there any stone walls on the Property (GAEC13)?

(l) are there any hedgerows on the Property (GAEC14 & 15)?

(m) are there any watercourses on the Property (GAEC14)?

(n) please confirm that all qualifying fields on the Property have/will have the required 2 metre buffer zone around the hedges and watercourses

14.6 If the Buyer is to receive the payment arising from the Seller's SPS application:-

(a) please provide details of any final or provisional notification of the value of Entitlements and any payments received to date

(b) please confirm whether or not the Seller has yet received payment

(c) has the Seller opted in writing to receive payment in Euros?

Milk quota

15.1 Does the Property have the benefit of any milk quota? If so, please:-

(a) supply full details

(b) confirm that the Seller will co-operate for the Buyer's accession to the quota concerned

Sugar beet contract tonnage

16.1 Does the Property have the benefit of any sugar beet contract tonnage? If so, please:-

(a) supply full details

(b) confirm that the Seller will co-operate to have the contract transferred to the Buyer

16.2 Is any sugar beet compensation payment outstanding, whether expected to be received under the Single Farm Payment Scheme or otherwise? If so, please provide details and confirm that it will be available to the Buyer

Water and drainage

17.1 Is any water used on or for the Property obtained otherwise than from a main supply? If so, please:-

17.1.1 supply a copy of the requisite water abstraction or other licence and any supplemental documents

17.1.2 confirm that the conditions of the licence(s) have been fully complied with, or supply details of any breach

17.2 If any non-mains water for which no licence is held by the Seller is used on the Property, please supply full details of:-

17.2.1 the source of such water

17.2.2 the amount consumed or used

17.2.3 the use to which the water is put

17.3 Is there any dam or weir on the Property which requires an impounding licence? If so, please:-

17.3.1 supply details and a copy of any licence held

17.3.2 confirm that the conditions of the licence have been fully complied with, or supply full details of any breach

17.4 If payments are made pursuant to any licence relating to water, are they made at the full rate? If not, please supply details of the payments made and a copy of any relevant document

17.5 Are there any water meters or water abstraction meters serving the Property? If so, please:-

17.5.1 supply a plan showing their positions

17.5.2 confirm that each of them is in good working order and condition

17.6 Is the Seller aware of any unsuccessful application for any licence relating to water? If so, please supply details

17.7 If the Property abuts upon a river, or if the Property is in any way affected by any watercourses, please:-

17.7.1 confirm that the ownership of the entire watercourse passes with the Property so far as co-extensive

17.7.2 confirm that there are no public or private rights along the co-extensive banks of the river or watercourse

17.7.3 confirm that the fishing is in hand

17.7.4 give the names and address of the appropriate Environment Agency office and Drainage Board and say which of the river and watercourses they maintain

17.7.5 give details of the rights of drainage into the river watercourse

17.8 Please confirm that on completion plans of the land drains on the Property will be handed over

Environment

18.1 Please provide a copy of any environmental audit, assessment or investigation of the Property that has been carried out by the Seller or anyone else

18.2 Is any part of the Property designated as Environmentally Sensitive Area, a Nitrate Sensitive Area, a Nitrate Vulnerable Zone, a Site of Special Scientific Interest or any other special area or zone? If so, please provide copies of all relevant documents

18.3 Is the land subject to any environmental land management scheme? If so, please supply a copy, with any other relevant documents

18.4 Has there been grown on the Property to the Seller's knowledge any plant which has been genetically modified within the meaning of Part VI of the Environmental Protection Act 1990? If so, please give details

18.5 Please supply a copy of any authorisation in relation to the Property under the Ground Water Regulations

18.6 Please confirm that the Seller will serve notice within twenty one days of completion transferring any authorisation under the Ground Water Regulations to the Buyer

18.7 Please supply details of any disposal and tipping within the last five years

18.8 If the Property is in a Nitrate Vulnerable Zone, please supply details of Fertiliser and manure applied to the Property in the last six years on a field by field basis

18.9 Is the Seller aware of any disused pipes or cables, footings or other structures under the Property?

Woodlands

19.1 Please supply full details of:-

19.1.1 any Woodland Grant Scheme

19.1.2 any other forestry grant

19.1.3 any tree planting scheme carried out during the last five years

20. Please supply full details of all felling licences and felling direction affecting the Property and confirm that all their terms have been complied with

Taxation

21.1 Is any part of the purchase price subject to VAT? If so, please:-

21.1.1 confirm that the purchase price stated in the draft contract is inclusive of VAT

21.1.2 supply evidence of option to tax

21.2 If no option to tax has been made, please confirm that the Seller will not opt to tax in relation to the Property on or before completion

General

22. Guarantees etc

Please confirm that the Seller will assign (if appropriate) the benefit of any guarantees etc in relation to the physical condition of the Property of which he has the benefit

23. Services

Please supply a copy of any wayleave agreements relating to any electricity supply which crosses the Property

24. Investment property

24.1 Is any part of the Property occupied by anyone who may by virtue of that occupancy ('the letting') be entitled to claim a tenancy under the Agricultural Holdings Act 1986 ('AHA 1986')? If so please reply to the following enquiries

Please supply:-

24.1.1 full details (including age) of anyone occupying the whole or any part of the Property under the letting

24.1.2 a copy of any document evidencing all or any of the terms of the letting

24.2 If the letting is of only part of the Property, then insofar as they are not shown by any copy document supplied, please:-

24.2.1 supply a plan of the part of the Property comprised in the letting

24.2.2 supply full details of any rights enjoyed or suffered by that part

24.2.3 specify the commencement date of the letting

24.2.4 specify the period of the letting

24.2.5 specify the respective repairing and other obligations undertaken by the owner and the occupier

24.3 In relation to the rent or any other payment by the occupier:-

24.3.1 how much is it?

24.3.2 on what date is it payable

24.3.3 is it payable in advance or in arrear?

24.3.4 please say when and how it was last fixed, and supply a copy of any relevant document

24.3.5 have there been any arrears during the last two years? If so, please give details

24.4 Have the owner and the occupier concluded any agreement as to the amount of the rent or any other sum which is to be payable for any future period ? If so, please give particulars

24.5 Has any notice to quit relating to the Property or any part been served? If so, please supply a copy of the notice and any relevant details

24.6 Are there any outstanding matters arising from any notice served by the landlord:-

24.6.1 under the Law of Property Act 1925 section 146?

24.6.2 under AHA 1986 sections 26–28 and Schedule III Cases D, E and F

24.7 Please supply details of all agricultural tenancy successions

24.8 Please supply a list of, and all matter agreed as to:-

24.8.1 tenant's fixtures on the Property

24.8.2 tenant's improvements on the Property

24.9 Please supply copies of:-

24.9.1 all licences or consents granted for any tenant's improvements that have been carried out or are to be carried out

24.9.2 all notices to the Agricultural Land Tribunal or approval by the Tribunal (AHA 1986 section 67)

24.9.3 Have there been any other tenant's improvements for which consent was NOT granted? If so, please give full details

24.10 Has the owner refused any application for consent from the occupier? If so, please give details

PRECEDENT 2 AGREEMENT FOR SALE AND PURCHASE (LONG FORM)

[*The precedent below is in a long form so as to offer a choice of clauses suitable for various circumstances. There are standard clauses relating to title, completion, transfer etc, which will be necessary for every transaction, but the clauses relating to the agricultural aspect of any transaction will obviously depend on particular circumstances. It would be a rare situation indeed if all or most clauses needed to be included. Please note that the precedent is drafted on the basis of the sale of land with SPS Entitlements.*]

THIS AGREEMENT is made on [*date*]

PARTIES:-

(1) [*name*] of [*address*] ('the Seller')

(2) [*name*] of [*address*] ('the Buyer')

and

[(3) [*name*] of [*address*] ('the Second Seller')]

AGREED TERMS:-

1. **Definitions**

In this Agreement the following expressions have the following meanings unless inconsistent with the context:-

[Assets	:	assets of the Seller's business (other than the Property) agreed to be sold to the Buyer and listed in the Schedule of assets]
[Buildings	:	Buildings on the Property]
Buyers Solicitors	:	
[Buyer's Agents	:	]
Contract Rate	:	[*insert interest rate or method of calculating the interest*]
Completion	:	the completion of the sale and purchase of the Property in accordance with the provisions of this Agreement
Completion Date	:	
Deposit	:	

Entitlements	:	have the meaning given to it in clause [ ]
LRA 2002	:	the Land Registration Act 2002
Part 1 Conditions	:	the conditions in Part 1 of the Standard Commercial Property Conditions (Second Edition) and Condition means any one of them
Part 2 Conditions	:	the conditions in Part 2 of the Standard Commercial Property Conditions (Second Edition)
Plan	:	the plan [attached hereto] – [attached to the form of the draft Transfer]
[Property	:	the freehold land [and buildings] at [*address*] and registered at HM Land Registry [with title absolute] under title number(s) [*number/s*)]
[Property	:	the freehold land [and buildings] shown edged red on the Plan and being part of the property registered at HM Land Registry with title [absolute] under title number(s)[*numbers/s*]]
[Property	:	the freehold land [and buildings] known as [*name*] comprised within a conveyance dated [*date*] made between [*name*] and [*name*] [and shown edged red on the Plan]]
[Property	:	the freehold land [and buildings] shown edged red on the Plan and being part of the land comprised in a conveyance dated [*date*] made between [*name*] and [*name*]]
Purchase Price	:	(exclusive of any amount arising in respect of VAT) which shall be apportioned as follows:- for the Property [('Property Price')] for the Entitlements [('Entitlements Price')]
[Retention	:	]
[Tenancies	:	tenancies affecting the Property [*date*] as per the Schedule of Tenancies]
[Seller's Agents	:	]
Seller's Solicitors	:	

VATA	:	Value Added Tax Act 1994
VAT	:	value added tax chargeable under the VATA and any similar replacement and any additional replacement tax

2. **Interpretation**

In this Agreement unless otherwise specifically stated:-

2.1 a reference to a clause or Schedule is a reference to a clause of or schedule to this Agreement

2.2 a reference to a party to this Agreement includes all the persons comprising that party (or such of them as are relevant in the particular context) and the obligations of the persons so included (if more than one) can be enforced against each of them individually or against some or all of them jointly

2.3 words importing only one gender include all genders; words importing persons include corporations and vice versa

2.4 a reference to a paragraph is a reference to a paragraph of the relevant schedule to this Agreement

2.5 a **person** includes a corporate or unincorporated body

2.6 unless otherwise specified, a reference to a particular law is a reference to it as it is in force for the time being, taking account of any amendment, extension, application or re-enactment and includes any subordinate legislation for the time being in force made under it

2.7 writing or written includes faxes but not e-mail

2.8 the headings are for convenient reference only and shall not be taken into account in the construction or interpretation of this Agreement

3. **Standard Conditions and variations**

3.1 The Part 1 Conditions are incorporated in this Agreement so far as they:-

(a) apply to a sale by private treaty

(b) relate to freehold property

(c) are not inconsistent with the other clauses in this Agreement and

(d) have not been modified or excluded by any of the other clauses in this Agreement

3.2 Where the context so admits terms used or defined in this Agreement have the same meaning when used in Part 1 Conditions

[3.3 Part 2 Conditions are not incorporated in this Agreement]

3.4 In Part 1 Conditions:-

3.4.1 Conditions 1.1.4(a), 1.3.3(b), 1.3.5(c), [1.4,] 3.1.3, 3.3, 6.3.1, 6.4.2, 6.6.5 [7.2, 7.13, 7.14] and 8.1.1 shall not apply[18]

3.4.2 there shall be added at the end of condition 1.3.1 the words 'and may be given by the solicitors of the party concerned'

3.4.3 there shall be added at the end of condition 1.3.1 the words 'and may be given by the solicitors of the party concerned'

3.4.4 1.00 pm shall be substituted for 2.00 pm in conditions 8.1.2 and 8.1.3

3.4.5 the words 'gives notice under' shall be substituted for the words 'exercises his option' in condition 8.3.2

3.4.6 there shall be added at the end of condition 9.5.2 (a) (iii) 'on which the Seller shall be entitled to interest at the contract rate from the completion date'

[3.4.7 the following Part 2 Conditions are included in this Agreement: { }][19]

4. **Sale and Purchase Price**

The Seller sells and the Buyer buys the Property [and the Entitlements] for the Purchase Price [and the Second Seller sells and the Buyer buys the Entitlements for the Entitlements Price] upon the terms and subject to the conditions herein contained

5. **Deposit**

A deposit equal to [ten] per cent of the Purchase Price shall be paid on the date of this Agreement to the Seller's Solicitors as [agents for the Seller] [stakeholders][20]

[*Note – All VAT clauses are drafted on the assumption that you are acting for the Seller*]

6. **VAT**

[***Exempt sale clause:-***

Please use this clause if the Property:

---

18 Part 1 Conditions should be considered in the context of a particular transaction. If there is a sub-sale or where deposit is greater than 10% of the purchase price or held by the Seller's solicitors as agent – Conditions 1.5 and 2.2 may have to be excluded or modified. Condition 1.4 should be excluded if the sale will be regarded as TOGC for the purpose of VAT legislation, and VAT clause should reflect TOGC status.

19 Care should be taken when the transaction is to be regarded as TOGC. It may be preferable to draft a detailed VAT clause rather than rely on Condition A2 of Part 2 Conditions.

20 Condition 2.2.2 should be excluded if the deposit is to be held as agents. Where the sale is subject to VAT, the tax point is normally the date of the transfer but if the deposit is held by the Seller's solicitors as agents, the tax point arises at the date of exchange of contracts. If the buyer is to opt to tax (for example, for the purpose of TOGC) the tax point is the 'relevant date' for notifying HMRC in writing of the option.

- *is exempt, ie the Seller has not opted to tax and the Property does not comprise a freehold new building or a building less than three years old*
- *is sold with vacant possession, ie there are no leases or agreements for leases in place and*
- *the sale is not a transfer of business as a going concern (TOGC)*

Condition 1.4 applies to the transaction and the Seller must be certain that it can give warranties included in that Condition. Otherwise, the Condition should be excluded.

NOTE:

The Seller should consider whether or not it should opt to tax before Completion. The result of not opting is that the Seller is unable to recover any VAT on its costs. Furthermore, if the Property is subject to the capital goods adjustment rules the Seller should seek specialist VAT advice as there may be a clawback of input tax previously recovered unless the Seller opts to tax before Completion [VAT provisions re Entitlements – NB distinguish between First and Second Seller if required].]

6.1 All amounts expressed as payable by the Buyer to the Seller under or pursuant to this Agreement are expressed exclusive of VAT which may be chargeable

6.2 If any VAT is chargeable on any supply made by the Seller under or pursuant to this Agreement the Buyer will on receipt of a valid VAT invoice pay to the Seller an amount equal to that VAT as additional consideration on Completion

[***Standard rated sale – not TOGC***

Please use this clause if the Property is taxable and the sale cannot be a TOGC]

6.1 The Seller confirms that it has opted to tax in relation to the Property (or the sale is for any other reason a taxable supply on which VAT is chargeable

6.2 All amounts expressed as payable by the Buyer to the Seller under or pursuant to this Agreement are expressed exclusive of VAT and the Buyer will pay to the Seller on receipt of a valid VAT invoice an amount equal to that VAT as additional consideration on Completion]

[***Sale is regarded as TOGC***[21]

This clause may be suitable if the sale is of property subject to existing tenancies, or where there is a transfer of a business which includes the Property as the main or one of the assets. The parties should seek a specialist tax advice in view of the warranties given]

---

21 VAT provisions relation to transfers as going concern are complex and the parties to the transaction should seek specialist tax advice.

6.1 All amounts expressed as payable by the Buyer to the Seller under or pursuant to this Agreement are expressed exclusive of VAT which may be chargeable

[6.2 The parties agree and intend that section 49(1) of VATA and Article 5 of the Value Added Tax (Special Provisions) Order 1995 will apply to the transfer of the Property [and Assets] and the Seller and the Buyer will each use reasonable endeavours to secure that pursuant to these provisions the transfer of the Property [and the Assets][22] is treated as neither a supply of goods nor a supply of services for the purposes of VAT but as the transfer of [part of] a business as a going concern]

OR

[The parties agree that the business of letting the Property for a consideration is capable of being operated separately as a business and intend that section 49(1) of VATA and Article 5 of the Value Added Tax (Special Provisions) Order 1995 ('**1995 Order**') will apply to the transaction provided by this Agreement and the Seller and the Buyer will each use reasonable endeavours to secure that pursuant to these provisions the transfer of [a part of] the Seller's business is treated as neither a supply of goods nor a supply of services for the purposes of VAT but as a going concern]

6.3 The Seller warrants that:-

6.3.1 it (or a representative member of VAT group) is registered for VAT and will provide the Buyer with a copy of its VAT registration certificate before the Completion Date and

6.3.2 it has used the Property for carrying out [the business of letting the Property for consideration] [the business of] and

6.3.3 it (or a relevant associate for the purposes of paragraph 2 or relevant group member for the purposes of paragraph 21 to Schedule 10 to VATA) made option to tax on which applies to the Property and

6.3.4 the option to tax was effective from [*date*] and was duly notified to HM Revenue and Customs on [*date*][23]

6.3.5 the option to tax has not been revoked

6.4 The Buyer warrants that:-

---

22 Transfer as a going concern situation may arise not only if there is a letting business but also if the Seller is transferring business assets to the buyer who, following the transfer will carry on a business of the same kind as the Seller.

23 If the Seller has not opted to tax in respect of the property, or the Seller has not been granted permission from HMRC by the relevant date, but TOGC provisions apply then there will be no VAT payable. In such situation the buyer does not have to opt to tax before the relevant date. Where the Seller did not opt to tax and does not intend to make an option in respect of the property, a warranty may be required that the Seller did not and will not make option to tax on or before completion (as the buyer may not then be able to make valid option to tax by the relevant date).

6.4.1 it (or a representative member of its VAT group) is duly registered for VAT and it will provide the Seller with a copy of its VAT registration certificate before the Completion Date and

6.4.2 it is not certified under Agricultural Flat Rate Scheme and

6.4.3 it has made or immediately after the date hereof will make an option to tax in relation to the Property pursuant to paragraph 2 or 21 of Schedule 10 to VATA with effect from on or before the relevant date (as defined in Article 5 of the 1995 Order) and

6.4.4 it has already notified or will duly notify HM Revenue and Customs of the option to tax and will supply the Seller before the relevant date (as defined in Article 5 of the 1995 Order) with a copy of the option to tax and of the acknowledgement by HM Revenue and Customs of receipt of the Buyer's notification and

6.4.5 it has not revoked and will not revoke the option to tax referred to in clause 6.4.3 and

6.4.6 following Completion it will use [the Property for carrying out the business of [ ] [letting it for consideration] [the Property and the Assets] for carrying out the business of [ ] (whether or not as part of any existing business of the Buyer) and

6.4.7 [(in a case where the Property is let to a single tenant) it is not part of the same VAT group as that single tenant]

6.5 The Buyer hereby notifies the Seller for the purposes of Article 5(2A) of the 1995 Order that the provisions of Article 5(2B) of the 1995 Order do not apply to him and the Buyer covenants that it will not apply on or following Completion

NOTE: Amendments to this clause 6.5 should not be accepted. Please take specialist advice if the Buyer amends this

6.6 If HM Revenue and Customs at any time assess the Seller or any member of the Seller's VAT group to VAT in relation to the sale of the Property [and Assets] or anything else contained in this Agreement or determine that the transaction provided for by this Agreement constitutes or includes a supply of goods or services for the purpose of VAT the Buyer will on the Completion Date (or within 5 working days of the Seller notifying the Buyer of the assessment or determination if such notification takes place after the Completion Date) pay to the Seller the VAT which is attributable to the sale and on receipt of the amount in respect of VAT the Seller will provide to the Buyer a valid VAT invoice

[6.7 The Seller and the Buyer do not intend to make a joint application under regulation 6(1) of the 1995 Order to HM Revenue and Customs for the Buyer to be registered for VAT under the Seller's VAT registration number]

OR

[6.7 The Seller and the Buyer intend to make a joint application under regulation 6(1) of the 1995 Order to HM Revenue and Custom for the Buyer to be registered for VAT under the Seller's VAT registration number

6.8 If the HM Revenue and Custom shall register the Buyer under the Seller's VAT registration number then the Seller shall deliver to the Buyer on Completion (or as soon as reasonably practicable afterwards):-

6.8.1 such VAT records relating to the Property [and Assets] as are required to be kept for VAT purposes by the provisions of paragraph 6 of Schedule 11 to VATA or

6.8.2 a written declaration from HM Revenue and Custom confirming that all VAT records relating to the Property [and Assets] that are required to be kept for VAT purposes whether the provisions of paragraph 6 to Schedule 11 as aforesaid, may be retained by the Seller]

6.9 If the VAT records relating to the Property [and the Assets] are delivered to the Buyer by the Seller:-

6.9.1 the Buyer shall preserve the records for the period of six years from Completion (or for such longer period as may be required by legislation relating to VAT) and allow the Seller and its servants and agents to inspect the records during normal business hours upon a reasonable written notice and to take such copies and extracts (at the Seller's cost) as the Seller may reasonably require]

6.10 The Buyer shall indemnify the Seller [on a full indemnity basis] in respect of any liability of the Seller (or the representative member of its VAT group) to pay HM Revenue and Customs any interest, penalty or surcharge (together with any costs incurred by the Seller in connection with such interest, penalty or surcharge) by reason of the late payment of any VAT charged on the sale of the Property [and the Assets] (or any part of it) [or any part of them] or breach of any obligation or warranty of the Buyer pursuant to the provisions of this Agreement

7. **Title guarantee**

Subject to the terms of this Agreement the Seller shall transfer the Property with [full] [limited] title guarantee

8. **Deducing title**

[The Seller's title to the Property has been deduced to the Buyer's Solicitors before the date of this Agreement and the Buyer is deemed to have full knowledge of the title number(s) [*insert the registered number*] dated [*date*] at [*time*] and is not entitled to raise any enquiry, objection, requisition or claim in relation to it save for the matters arising out of or as a result of the Buyer's pre-completion searches]

OR

[The Seller's title to the property has been deduced to the Buyer's Solicitors before the date of this agreement and commences with [*description of root document*]. The Buyer is deemed to have full knowledge of the title and is not entitled to raise any objection, enquiry or requisition in relation to it, save for any matters arising out of or as a result of the Buyer's pre-completion searches]

9. **Vacant possession**

9.1 The Property will be sold with vacant possession on Completion [subject to clause [*number*] and the rights of [*name*] (their agents employees or contractors) with or without machinery to enter on to the Property from time to time up to and including [*date*] for the purpose only to [tend to and harvest the crops presently on field [*number*]]

9.2 [The Seller shall make good to the Buyer's reasonable satisfaction any damage done to the Property by reason of the exercise of the rights reserved by this clause [*number*] and shall compensate the Buyer in respect of any damage done not capable of being made good]

10. **Matters affecting the Property**

10.1 The Property is sold free from incumbrances other than:-

10.1.1 any matters other than monetary charges contained or referred to in [the entries or records made in registers maintained by HM Land Registry as at [*date and* time] under title number(s) [*numbers*]] [*the root of the title*]

10.1.2 any matters other than financial charges disclosed or which would have been disclosed by the searches and enquiries which a prudent purchaser would have made before entering into this Agreement

10.1.3 wayleaves consents permissions or privileges in respect of any service pipes wires cables conduits poles stays pylons or other apparatus

10.1.4 liability to repair walls fences roadways drains sewers pipes wires cables conduits dykes ditches or the like

10.1.5 local land charges (whether registered before or after the date of this Agreement or not) and matters capable of registration as local land charges whether or not so registered

10.1.6 notices served and orders demands proposals or requirements made by any local or other authority or company whether before or after the date of this Agreement

10.1.7 actual or proposed orders directions notices charges restrictions conditions agreements and other matters arising under the Town and Country Planning legislation

10.1.8 ecclesiastical or civil duties and payments charged upon or payable out of the Property

10.1.9 any matters which [are unregistered interests which override first registration under Schedule 1 to the LRA 2002 and/or unregistered interests which override registered dispositions under Schedule 3 to the LRA 2002] [which would be unregistered interest which override first registration under Schedule 1 to the Land Registration Act 2002]

10.1.10 drainage rates and charges and other outgoings

10.1.11 incidents of tenure rights of way (whether public or private) light air and drainage and other easements quasi-easements rights liberties and privileges

10.1.12 any matters which the Seller does not and could not reasonably know about

10.1.13 any matters discoverable by inspection of the Property by the date of this Agreement

10.1.14 public requirements

[10.1.15 [*any specific matters that affect the Property*]]

10.2 The Seller shall not be required to define any of the matters or things referred to in this clause or ascertain the ownership of any walls fences ditches or such other matters and things as aforesaid beyond such information (if any) as is afforded by the documents of title

10.3 The Buyer is deemed to have full knowledge of the matters referred to in this clause and will raise no enquiry, requisition, objection or claim in respect of them

11. **Exclusion of requisitions – description**

11.1 The Buyer shall raise no enquiry or requisition in respect of:-

11.1.1 any minor discrepancy (whether as to Ordnance Survey enclosure numbers or areas or otherwise) between any of the plans on or attached to the documents of title and any of the descriptions contained or referred to in such documents or between such plans or descriptions and the Ordnance Survey plan

11.1.2 any minor discrepancy between any such description or plan on the one hand and any description or plan contained or referred to in this Agreement on the other hand

11.1.3 any reference to an incorrect edition or the absence of any reference to the relevant edition of the Ordnance Survey plan in any of the documents of title

11.1.4 any variation between different editions of the Ordnance Survey plan[24]

---

[24] Where appropriate also refer to the official copy entries in the description of the Property.

12. **Transfer**

[12.1 The transfer to the Buyer shall be in the form of the Transfer annexed to this Agreement at Annexure [*number*]][25]

OR

[12.1 the Transfer to the Buyer will contain:-

12.1.1 a declaration as to the title guarantee with which the transfer is made as stated in this Agreement

12.1.2 a provision that all matters recorded at the date of the transfer in registers open to public inspection are deemed to be within the actual knowledge of the Buyer for the purposes of section 6(2)(a) of the Law of Property (Miscellaneous Provisions) Act 1994 notwithstanding section 6(3) of that Act

12.1.3 a covenant by the Buyer by way of indemnity only on its own behalf and on behalf of its successors in title to observe and perform the charges, incumbrances, covenants and restrictions contained or referred to in the [(*the root of the title document*)] [(*registers of the title*)] insofar as they are subsisting and capable of taking effect and will keep the Seller indemnified against all proceedings costs claims and expenses arising from any future failure to do so

12.1.4 a provision that the Property will not by virtue of the transfer have any rights, easements or the benefit of any other matters over the land retained by the Seller other than those (if any) which are expressly mentioned or granted by the transfer and a provision of section 62 of the Law of Property Act 1925 will not apply to the Transfer

[12.1.5 (*provisions for specific rights and obligations, including restrictive and positive covenants applicable to the transaction*)]

12.2 The Seller and the Buyer will execute the Transfer in duplicate

13. **Insurance**

[13.1 With effect from the date of this Agreement the Property is at the Buyer's risk and the Seller is under no obligation to insure the Property][26]

13.2 No damage to or destruction of the Property nor any deterioration in its condition however caused will entitle the Buyer either to any reduction of the Purchase Price or to refuse to complete or delay Completion

---

[25] On transfer of parts, or where complex provisions are required it is usual to agree the transfer deed and attach the agreed draft to the contract.

[26] If the Seller is to maintain the policy, Condition 7 may need to be excluded or modified unless its provisions are acceptable to the parties.

14. **Buyer's admissions**

The Buyer hereby acknowledges that:-

14.1 it has inspected the Property before the date of this Agreement and buys it with full knowledge of its actual state and condition

14.2 it enters into this Agreement solely as a result of its own inspection and not in reliance upon any representation or warranty (whether written oral or implied) made by or on behalf of the Seller save for any representation or warranty contained in written replies given by the Seller's Solicitors to any preliminary enquiries or points in correspondence raised by the Buyer's Solicitors prior to the date of this Agreement (and then only subject to any terms and conditions upon which such replies were expressed to be given and to the extent that such replies were not capable of verification by inspection survey or searches or enquiries capable of being raised of any local or other authority or body whether or not such inspection survey or searches shall have been made or such enquiries shall have been raised by or on behalf of the Buyer)

14.3 it has been given the opportunity to inspect survey measure carry out tests upon and generally satisfy itself as to the matters affecting the Property and its fitness for any purpose for which the Buyer wishes or intends to use the same and the Seller shall not be liable for any latent or other physical defect relating to or affecting the Property or any part or parts thereof and without prejudice to the generality of the foregoing no warranty is given by the Seller as to the fitness or suitability of the Property or any part or parts thereof for any purpose whatsoever

[15. **Quotas – no transfer**

It is not intended that the benefit of any quota or contract tonnage now held with or applicable to the Property or any part thereof should pass to the Buyer or with the Property and the Buyer shall do all such things and sign all such documents as the Seller may reasonably require to achieve that position]

16. **Entitlements**

16.1 In this clause the following definitions shall apply:-

16.1.1 'the Regulations' means Council Regulation (EC) No 73/2009 and any Commission Regulations supplementary thereto, and any relevant subsequent EC and/or UK Government regulations and/or administrative provisions applicable in England and implemented by the UK Government

16.1.2 'Entitlements' means the following entitlements to the single income payment held by the Seller [the Second Seller] in its name pursuant to the Regulations namely:-

Type	Number	Value
[ ]	[ ]	[ ]

and reference to the Type of entitlements has the same meaning as that contained in the Regulations

16.1.3 'DEFRA' means the Department for the Environment Food and Rural Affairs or any successor department

16.1.4 'RPA' means the Rural Payments Agency or any successor body carrying out similar functions

16.1.5 'farmer' shall have the same meaning as that contained in Article 2(a) of the Council Regulations

16.2 The Property is sold TOGETHER WITH the Entitlements

16.3 The Seller [the Second Seller] and the Buyer agree to act in fairness and co-operate in good faith with each other and to take all (if any) reasonable steps to ensure the transfer of the Entitlements from the Seller [the Second Seller] to the Buyer [or the Buyer's the Buyer's nominee] from the earliest date possible under the Council Regulation and the Commission Regulations so that they are available to be transferred to the Buyer [or The Buyer's nominee] to enable the Buyer [or The Buyer's nominee] to make a valid claim for the single payment in 20[ ] and will accordingly respond promptly to requests made by the other party, DEFRA and/or the RPA for documentation approvals information or assistance

16.4 The Seller [the Second Seller] and the Buyer warrant each with the other that they will do all such acts and all such things and enter into such documents (if any) as shall be necessary to enable the Entitlements to be transferred to the Buyer [or the Buyer's nominee] and in particular and without prejudice to the generality of the foregoing shall comply with [(and where the transfer is to the Buyer's nominee the Buyer will procure compliance with)] the requirements of and the procedures laid down in or to be laid down pursuant to the Council Regulation and the Commission Regulations to allow the transfer to Agreement

16.5 The Seller [the Second Seller] [and the Second Seller] also warrant that:-

16.5.1 it [they] will not seek to claim any of the Entitlements that are intended to be transferred to the Buyer [or The Buyer's nominee] pursuant to this Agreement nor will it [they] by any act or omission on its [their] part permit any third party to do so

16.5.2 it has [they have] not received any communication from the RPA to indicate that there are any problems with the establishment and/or validation of Entitlements to it [them] or that would in any way prevent it [them] being transferred to the Buyer [or The Buyer's nominee] in accordance with the terms of this Agreement

16.5.3 it has [they have] not done nor will it [they] do anything which may invalidate their claim to establish the Entitlements

16.5.4 on receiving notice from the RPA of the validation of the Entitlements it [they] will promptly send to the RPA notification of a transfer of the Entitlements to the Buyer [or The Buyer's nominee] on Form RLE1 (or such other form as the RPA may from time to time require) and will use its [their] best endeavours to effect the transfer to the Buyer [or The Buyer's nominee] for its use in conjunction with the Buyer's or [The Buyer's nominee's] claim for single payment in 20[ ]

16.5.5 it [they] will indemnify the Buyer in respect of any shortfall in the number of the Entitlements that may actually be transferred to the Buyer [or the Buyer's nominee] as opposed to those warranted to be transferred in accordance with the terms of this Agreement whether such shortfall is occasioned by any act of negligence omission or default on its [their] part and also against any loss of single payment arising from any failure on its [their] part or the part of others on its [their] behalf to notify the RPA of the transfer by the 2 April 20[ ]

16.6 The Buyer also warrants with the Seller [[and] with the Second Seller] that it will observe and fully comply [(and where the transfer is to the Buyer's nominee will procure compliance)] with the cross compliance requirements of the Single Payment Scheme from the Completion Date until the end of the calendar year following the Completion Date to ensure that it does not jeopardise the Seller's [the Second Seller's] claim for single farm payment in respect of all other property belonging to and/or in the occupation of the Seller [[and] the Second Seller] in respect of which it is claiming such payment AND in the event of any failure on the part of the Buyer [or the Buyer's nominee] to comply with this warranty it will compensate the Seller [and the Second Seller] for all losses occasioned by such failure on its part [or the Buyer's nominee's part] or the part of others acting on its behalf arising whether directly or indirectly from the Buyer's [or The Buyer's nominee's] failure to observe the terms of this sub-clause to include but without prejudice to the generality hereof all losses of payment and loss of entitlements (if any) together with all legal or other costs incurred by the Seller [and/or the Second Seller] in putting themselves back into the position they would have been in had the Buyer [or the Buyer's nominee] not failed to observe the terms of this sub-clause

16.7 If after the date of this Agreement there are changes in the Council Regulation or the Commission Regulations or in the interpretation of one or other or all of them by DEFRA or the RPA or otherwise before [the Second Seller and] the Seller [have] has fully complied with the terms of this clause in such a way that the objects of this clause are not achieved either party may give notice to the others to appoint an arbitrator in accordance with the following paragraph to re-draft the terms of this clause in such a way as may fairly achieve the objects of this clause in the event that the parties are not able to agree such re-draft between them and the costs of such arbitration shall be borne equally by the Seller[, the Second Seller] and the Buyer

16.8 If any dispute arises under this clause either party may give notice to the other[s] in writing specifying the dispute and requesting that agreement be reached on the identity of a single arbitrator to be appointed to determine the dispute in accordance with the provisions of the Arbitration Act 1996 and if no arbitrator has been appointed by agreement within one month of the notice either party may make a written request to the President for the time being of the Royal Institution of Chartered Surveyors for him to appoint an arbitrator to determine the matter in dispute

16.9 If on Completion the Entitlements are not transferred to the Buyer [or to the Buyer's nominee] the Buyer will deposit on Completion the Retention with the Buyer's Solicitors which will be held by the Buyer's Solicitors and payable to the parties with accrued interest in the following circumstances:-

16.9.1 in the event that the Seller [the Second Seller] is successful in transferring to the Buyer [or the Buyer's nominee] the Entitlements by [*date*] the Retention shall be paid to the Seller [the second Seller] within seven days of the Buyer [or the Buyer's nominee] receiving notification from the RPA of the Entitlements having been registered in the Buyer's [or The Buyer's nominee's] name

16.9.2 in the event that the Seller [the Second Seller] is successful in transferring to the Buyer [or its The Buyer's nominee] some but not all of the Entitlements a proportion of the Retention being £[ ] times the number of entitlements actually transferred to the Buyer [or the Buyer's nominee] shall be paid to the Seller [the Second Seller] within seven days of the Buyer [or the Buyer's nominee] receiving notification from RPA of the registration and the remainder of the Retention shall be released to the Buyer

16.9.3 In the event that no Entitlements are transferred to the Buyer [or the Buyer's nominee] by the [*date*] the Retention shall be released forthwith to the Buyer

16.9.4 If following registration of the transfer of the Entitlements to the Buyer [or the Buyer's nominee] the Retention or part of it (as appropriate) remains unpaid for more than seven days the Buyer shall pay to the Seller [the Second Seller] interest on the outstanding balance at the Contract Rate from the date of the payment becoming due until the date of actual payment

[16.10 The Seller [the Second Seller] shall be entitled to retain any payment received under the Entitlements in respect of the [*year*] calendar year]

17. **Sugar beet contract – transferred**

17.1 If the Buyer is a registered grower the Seller will give up [*insert*] tonnes of A and B sugar beet contract tonnage to the British Sugar Corporation immediately upon Completion so as to facilitate the grant of a replacement contract to the Buyer

17.2 The Seller shall at the expense of the Buyer use such reasonable endeavours and execute such further documents as may reasonably be required to enable the Buyer to obtain the benefit of the said contract tonnage (less the appropriate deduction)

17.3 If the Buyer is not a registered grower it shall at its own expense use such reasonable endeavours and execute such further documents as may reasonably be required to enable the Seller to retain the said contract tonnage

18. **Tenantright**

18.1 In addition to the Purchase Price the Buyer shall on Completion pay to the Seller the sum of £[ ] in respect of all items in the nature of tenantright

OR

19. **Tenantright – valuation by Seller's agent**

19.1 In addition to the Purchase Price the Buyer shall pay to the Seller a sum equal to the valuation of the items specified in the next sub-clause in accordance with the Agriculture (Calculation of Value for Compensation) Regulations 1978 to 1983 and the prices and recommendations of the Central/Lincolnshire Association of Agricultural Valuers and for the purposes of such valuation all appropriate notices shall be deemed to have been served the Seller shall be deemed to be the outgoing tenant and the Buyer shall be deemed to be the landlord of the Property under a yearly tenancy to which the Agricultural Holdings Act 1986 applies and of which the last year had commenced one year before the Completion Date

19.2 The items to be so valued are:-

19.2.1 seeds sown cost of cultivations (including beneficial cultivation to set-aside land) fertilisers and sprays applied growing crops residual manurial values and the unexpired value of lime and slag applied

19.2.2 all sprays fertilizers seeds fuel and oil in store as at the date for valuation enhancement where appropriate

19.3 The valuation shall be made as at the date on which the Buyer takes occupation of the Property or the part to which the valuation relates

19.4 The amount of such valuation shall be assessed by the Seller's Agents whose valuation shall be final and binding on the parties hereto

19.5 The amount payable under this clause shall be paid on Completion or within seven days of being ascertained (whichever is the later)/within seven days of being ascertained together with interest thereon at the contract rate from the date as at which the valuation is to be made until payment but neither party shall be entitled to delay Completion on the ground that the amount of such valuation has not by then been ascertained

OR

20. **Tenantright – valuation by respective agents**

20.1 In addition to the Purchase Price the Buyer shall pay to the Seller a sum equal to the valuation of the items specified in the next sub-clause in accordance with The Agriculture (Calculation of Value for Compensation) Regulations 1978 to 1983 and the prices and recommendations of the Central/Lincolnshire Association of Agricultural Valuers and for the purposes of such valuation all appropriate notices shall be deemed to have been served the Seller shall be deemed to be the outgoing tenant and the Buyer shall be deemed to be the landlord of the Property under a yearly tenancy to which the Agricultural Holdings Act 1986 applies and of which the last year had commenced one year before the Completion Date

20.2 The items to be so valued are:-

20.2.1 seeds sown cost of cultivations

20.2.2 fertilisers and sprays applied growing crops residual manurial values and the unexpired value of lime and slag applied

20.2.3 all sprays fertilizers seeds fuel and oil in store as at the date for valuation

20.2.4 enhancement where appropriate

20.3 The valuation shall be made as at the date on which the Buyer takes occupation of the Property or the part to which the valuation relates

20.4 Subject to the provisions of the next two following paragraphs of this sub-clause such valuation shall be made and agreed by the Seller's Agents and the Buyer's Agents

20.5 Unless the Buyer's Agents are named in this Agreement the Buyer shall notify their name and address to the Seller's Agents in writing within five working days from the date hereof (as to which time shall be of the essence) and in default thereof the valuation shall be carried out by the Seller's Agents alone

20.6 If the respective agents have not agreed the valuation within [ ] weeks of the Seller's Agents submitting the claim to the Buyer's Agents the matter shall be referred to arbitration by a single arbitrator under the provisions of the Arbitration Act 1996 the arbitrator to be agreed between the respective agents or in default of such agreement to be nominated by the President for the time being of the Royal Institution of Chartered Surveyors upon the application of either party but at the joint and equal expense of the Seller and the Buyer

20.7 The valuation made agreed or determined in accordance with this clause shall be final and binding on the parties hereto

20.8 The amount payable under this clause shall be paid on Completion or within seven days of being ascertained (whichever is the later)/within seven days of being ascertained but neither party shall be entitled to delay Completion on the ground that the amount of such valuation has not by then been ascertained

OR

21. **No tenantright**

21.1 No sum shall be payable by the Buyer in respect of any item in the nature of tenantright or otherwise beyond the Purchase Price

22. **Included items**

22.1 Only the items listed in the Schedule of Equipment are included in the sale

[22.2 The sum of [*sum*] (being part of the Purchase Price) shall be apportioned to the Plant and equipment and machinery included in the sale]

23. **Excluded items**

23.1 All items listed in the Schedule of Excluded Items are excluded from the sale

23.2 The items excluded from the sale in so far as they belong to the Seller may be removed by the Seller by [*date*] but any such items not so removed by the Seller shall thereupon become the property of the Buyer without payment the Seller making good as soon as reasonable practicable any damage caused to the Property as a result of such removal

23.3 In so far as such items do not belong to the Seller the Seller shall not be required to remove them and the provisions of the preceding sub-clause of this clause shall not apply to them

24. **Holdover – crops and storage**

24.1 There are reserved out of the sale all crops growing on the Property at the date hereof together with the right for all appropriate persons with or without vehicles and equipment to come upon the Property after Completion from time to time up to and including [*date*] with or without equipment for any of the purposes of cultivating harvesting storing and removing such crops

24.2 The Property will be left in the condition in which it is after the crops have been properly harvested

24.3 There is also reserved the right to store and ventilate [*insert crop*] in the buildings on the Property up to and including [*date*] together with full rights of access for all appropriate persons with or without vehicles and equipment for any of the purposes of depositing protecting inspecting conditioning and loading for delivery

24.4 The Buyer shall provide such facilities labour fuel electricity and use of machinery as the Seller reasonably requires in connection with any of the rights reserved by this clause but shall be entitled to recover from the Seller the reasonable cost of such provision on written request with appropriate supporting documentation

24.5 The crops shall be stored at the risk of the Seller and the Buyer shall not be liable for any loss or deterioration of the crops whether before or after harvest

25. **Holdover – use and removal of Buildings**

25.1 For the period of [ ] months following Completion (or for such lesser period as the Seller may specify by notice in writing to the Buyer expiring not earlier than the date of that notice) the Seller reserves the right:-

25.1.1 to use the Buildings or any of them in a similar way as heretofore for the purpose of

25.1.2 to dismantle and remove the Buildings or any of them

in each case together with all appropriate rights of access for himself his employees and contractors with or without vehicles and equipment at all times and for all reasonable purposes

25.2 During the subsistence of the rights specified in sub-clause 1 of this clause:-

25.2.1 the Seller shall discharge all outgoings in respect of the Buildings and shall indemnify the Buyer in respect thereof

25.2.2 the Buildings and their contents shall be at the entire risk of the Seller who may (but shall not be under any obligation to) insure the Buildings or any of them

25.3 Any of the Buildings not removed by the end of the period referred to in sub-clause 1 of this clause shall vest in the Buyer without payment at the expiration of that period

26. **Stock/equipment sale**

26.1 There is reserved the right to hold an auction sale of live and dead stock equipment and machinery in the enclosure numbered [*number*] on the Ordnance Survey plan on or before [*date*] (in connection with which auction sale the Buyer undertakes to leave an agreed area within the said enclosure in stubble and unploughed until after the sale) and the right for all appropriate persons (including the purchasers of such stock equipment and machinery) to enter upon the Property with employees or contractors and any necessary equipment for the purpose of storing and removing the same at any time up to and including [*date*]

26.2 There is also reserved the right to store the said dead stock equipment and machinery in the buildings on the Property with all reasonable rights of access in connection therewith until [*date*] or if earlier seven days after the auction sale provided for in this clause

26.3 Neither the Seller nor any other person shall incur any liability whatsoever to the Buyer for any damage done to the Property or any buildings thereon in the exercise or purported exercise of the rights reserved by this clause

27. **Early entry – cultivation**

27.1 On payment to the Seller of a further deposit equal to ten per cent of the Purchase Price the Buyer shall be permitted as licensee and not as tenant to enter on the Property/each cropping area forming part of the Property after it has been cleared of the current crop for the purpose of cultivating the crops on the Property/preparing the area concerned for and planting the following crop but for no other purpose

27.2 The Buyer shall:-

27.2.1 ensure that in the exercise of the permission granted by this clause no damage or loss is occasioned to the Property or the Seller and as little inconvenience as practicable is occasioned to the Seller

27.2.2 pay or indemnify the Seller against all outgoings and other expenses in respect of the Property or the part concerned (as the case may be) for the period during which that right is exercisable

27.3 No interest shall be payable by the Buyer solely by reference to the permission granted by this clause

27.4 The permission granted by this clause:-

27.4.1 is personal to the Buyer save that he may exercise it by any employee or agricultural contractor of the Buyer

27.4.2 ends on the earlier of the date for Completion and the rescission of this Agreement

28. **Early entry – Buildings**

28.1 On payment to the Seller of a further deposit equal to ten per cent of the Purchase Price the Buyer shall be permitted as licensee and not as tenant to store in the Buildings

28.2 The Buyer shall:-

28.2.1 be permitted all appropriate rights of entry on the Property with or without vehicles and equipment to enable him to exercise the permission granted by this clause but for no other purpose

28.2.2 not alter the Buildings in any way

28.2.3 ensure that in the exercise of the permission granted by this clause no damage or loss is occasioned to the Property or the Seller and as little inconvenience as practicable is occasioned to the Seller

28.2.4 pay or indemnify the Seller against all outgoings and other expenses (including the cost of electricity used) in respect of the Buildings or their use for the period during which the permission is exercisable

28.3 No interest shall be payable by the Buyer solely by reference to the permission granted by this clause

28.4 The permission granted by this clause:-

28.4.1 is personal to the Buyer save that he may exercise it by any employee or contractor of the Buyer

28.4.2 ends on the early of the date for Completion and the rescission of this Agreement

28.5 As from the date hereof the risk of damage to the buildings passes to the Buyer[27]

29. **Drainage rates**

The Buyer shall pay and indemnify the Seller in respect of the drainage rates payable in relation to those parts of the Property upon which the Buyer enters from and including the respective dates of entry and on the whole of the Property from and including the Completion Date

[30. **Tenancies**

30.1 The Property is sold and will be transferred subject to but with the benefit of the tenancies the details of which are set out in the Schedule of Tenancies

30.2 Details of the Tenancies having been supplied to the Buyer on or before the date of this Agreement the Buyer shall be deemed to buy with full knowledge of its provisions and shall raise no enquiry or requisition

30.3 The Buyer shall in the Transfer covenant with the Seller that the Buyer and the persons deriving the title under him will at all times observe and perform the obligations on the part of the landlord and the conditions applicable to the Tenancies and will indemnify the Seller

---

[27] Not required if there is a general provision that risk passes to the Buyer immediately.

against all actions claims and liability arising out of any breach non-observance or non-performance thereof]

31. **Completion**

31.1 Completion shall take place on the Completion Date

32. **Non-merger**

The provisions of this Agreement shall not merge on Completion of the Transfer so far as they remain to be performed[28]

33. **Entire agreement**

33.1 This contract constitutes the entire agreement and understanding of the parties

33.2 The Buyer acknowledges and agrees that in entering into this Agreement, it does not rely on and shall have no remedy in respect of any statement, representation, warranty, collateral agreement or other assurance (whether negligently or innocently made) of any person (whether party to this Agreement or not) other than as expressly set out in this Agreement or the documents annexed to it or in any written replies which the Seller's Solicitor has given to any written enquiries raised by the Buyer's Solicitor before the date of this Agreement. Nothing in this clause shall, however, operate to limit or exclude any liability for fraud

34. **Contracts (Rights of Third Parties) Act 1999**

Except where otherwise expressly provided a person who is not a party to this Agreement has no rights under the Contracts (Rights of Third Parties) Act 1999 to enforce its terms

35. **Provision of official copies**

Within two weeks of completion of registration of the Buyer's title to the Property at HM Land Registry the Buyer shall apply for and deliver to the Seller's solicitors official copies of the register and plan for [both] the Property [and the Seller's retained land as remaining comprised in title number [*insert number*][29] on the sale of part]

36. **Water abstraction licence**

36.1 The water abstraction licences specified in the Water Licence Schedule and any other water abstraction licence for the Property ('the Licences') shall be transferred from the Seller to the Buyer and the Seller shall assist in signing any documentation required to facilitate such a transfer

---

28 See Standard Condition 9.4.

29 On sale of part.

36.2 For the avoidance of doubt the full extent of the Licences shall be transferred to the Buyer

37. **Environmental scheme(s)**

37.1 The Buyer agrees with the Seller that on completion he will take over the obligations of the Seller in respect of the Scheme(s) specified in the Environmental Scheme Schedule

37.2 The Buyer will indemnify the Seller against all claims demands actions proceedings damages penalties (including interest) and liability arising in respect of such Schemes arising from the Buyer's failure to comply with sub-clause 1 of this clause or any breach arising in respect of such Schemes save any breach committed by the Seller

38. **Employees**

The Seller warrants to the Buyer that there are no employees[30] whose contract of employment will be transferred to the Buyer whether under the terms of the Transfer of Undertakings (Protection of Employment) Rights 2006 or otherwise and the Seller will indemnify the Buyer against all costs claims and demands of whatever nature which may be made against the Buyer arising out of the employment of employees previously working on the Property

39. **Notices**

39.1 Any notice given under this Agreement must be in writing and signed by or on behalf of the party giving it

39.2 Any notice or document to be given or delivered must be given by delivering it personally or sending it by pre-paid recorded delivery to the address and for the attention of the relevant party as follows:-

(a) to the Seller at:-

or at the Seller's Solicitors, quoting Ref.:-

(b) to the Buyer at:-

marked for the attention of:-

or at the Buyer's Solicitors, quoting Ref.:-

39.3 Giving or delivering a notice or a document to a party's solicitors has the same effect as giving it to that party

39.4 Any such notice or document will be deemed to have been received:-

---

[30] The Seller should only give this warranty if either there were no persons employed in relation to the Seller's business on the Property or if a valid compromise agreement has been signed by (each) employee. Employment legislation is very complex and specialist advice should be sought if there is any doubt as to the employment status of any persons involved in the Seller's business, and as to the compromise agreement(s).

(a) if delivered personally, at the time of delivery provided that if delivery occurs before 9.00 am on a working day, the notice will be deemed to have been received at 9.00 am on that day, and if delivery occurs after 5.00 pm on a working day, or on a day which is not a working day, the notice will be deemed to have been received at 9.00 am on the next working day

(b) in the case of pre-paid first class or recorded delivery post on the second working day after posting

(c) in the case of fax message one hour after despatch provided that if delivery occurs before 9.00 am on a working day, the notice shall be deemed to have been received at 9.00 am on that day, and if delivery occurs after 5.00 pm on a working day, or on a day which is not a working day, the notice will be deemed to have been received at 9.00 am on the next working day

39.5 In proving delivery, it will be sufficient to prove that delivery was made or that the envelope containing the notice or document was properly addressed and posted as a prepaid first class or recorded delivery letter or that the fax message was properly addressed and transmitted, as the case may be

39.6 A notice or document delivered under this Agreement will not be validly given or delivered if sent by e-mail

40. Rights of third parties

A person who is not a party to this Agreement will not have any rights under or in connection with it by virtue of the Contracts (Rights of Third Parties) Act 1999 [provided that the expression 'the party' shall include the permitted assignee]

41. Governing law and jurisdiction

This Agreement will be governed by and construed in accordance with the law of England and Wales

This Agreement has been entered into on the date stated at the beginning of it

[SCHEDULE OF ASSETS]

[SCHEDULE OF TENANCIES]

[SCHEDULE OF PLANT AND EQUIPMENT]

[SCHEDULE OF EXCLUDED ITEMS]

[WATER LICENCE SCHEDULE]

[ENVIRONMENTAL SCHEME SCHEDULE]

SIGNED by)

[for and on behalf of] the Seller)

SIGNED by)
[for and on behalf of] the Buyer)

[SIGNED by)
[for and on behalf of] the Second Seller])

3 SINGLE PAYMENT SCHEME CONTRACT CLAUSES

[*Below are some alternative clauses dealing with the single payment scheme which you may wish to consider as an alternative to those set out in the contract included here.*]

PRECEDENT 3.1 A SALE OF LAND WITH ENTITLEMENTS

[*These clauses are suggested for inclusion in contracts for the sale and purchase of agricultural property together with entitlements. They will continue to require amendment as matters progress and RPA and professional interpretation changes.*

They can be included either in the body of the contract or in a separate schedule to the contract for sale as follows.]

In this Agreement/Schedule

1.1 Definitions

1.1.1 'the Council Regulation' means Regulation (EC) No 73/2009

1.1.2 'the Commission Regulations' means the Council Regulation and any Commission Regulations supplementary thereto and all relevant subsequent regulations and administrative provisions implemented by the EC

1.1.3 'DEFRA' means the Department of the Environment Food and Rural Affairs or any successor department

1.1.4 'RPA' means the Rural Payments Agency or any successor body carrying out its function

1.1.5 'the Single Payment Scheme' means the single payment scheme established under the Council Regulation and the Commission Regulations and any relevant subsequent EC or UK Government regulations and/or administrative provisions applicable in England ('the Regulations'), and the statutory guidance or codes of practice applicable in England and implemented by the EC and/or the UK Government

1.1.6 'Entitlements' means the entitlements held by the Seller in his name in accordance with the Regulations being the entitlement to the single income payment and any other entitlements to financial support of whatever kind relating to the Property and deriving from the Regulations comprising:-

[[ ] being [ ]% of the total number of entitlements to the single payment allocated to the Seller]

Or,

[Type(s) Number(s) Value(s)

and reference to Type(s) of entitlements has the same meaning(s) as that contained in the Regulations]

[*If all of the Seller's Entitlements are being sold then the second alternative needs to be used; if only some of the Entitlements are being transferred, for example on a sale of part of the Seller's property, either of the alternative sets of words in square brackets can be used. In the event that only some of the Entitlements are being transferred it would also be advisable to ensure the Buyer gives a warranty to observe the cross compliance requirements contained in the Regulations so that the Seller's remaining entitlements are not jeopardised during the remainder of the calendar year following the completion date.*]

1.2 Sale of Entitlements

The Property is sold TOGETHER WITH the Entitlements

1.3 Parties' obligations and agreements

1.3.1 The Seller and the Buyer agree to act in fairness and co-operate in good faith with each other and to take all (if any) reasonable steps to ensure the transfer of the Entitlements from the Seller to the Buyer from the earliest date possible under the Regulations so that they are available to be transferred to the Buyer to enable the Buyer to make a valid claim for the single farm payment in 20[ ] and will accordingly respond promptly to requests made by the other party, DEFRA and/or the RPA for documentation approvals information or assistance

1.3.2 The Seller and the Buyer warrant each with the other that they will do all such acts and all such things and enter into such documents (if any) as shall be necessary to enable the Entitlements to be transferred to the Buyer and in particular and without prejudice to the generality of the foregoing shall comply with the requirements of and the procedures laid down in or to be laid down pursuant to the Regulations to allow the transfer to the Buyer to be made in accordance with the provisions of this Agreement

1.3.3 The Seller also warrants that:-

1.3.3.1 he will not seek to claim any of the Entitlements that are intended to be transferred to the Buyer pursuant to this Agreement nor will he by any act or omission on his part permit any third party so to do

1.3.3.2 he has not received any communication from the RPA to indicate there are any problems with the Entitlements that would in any way prevent them being transferred to the Buyer in accordance with the terms of this Agreement

1.3.3.3 he has not done nor will he do anything which may lead to any loss of the Entitlements or any of them or the right to receive the single farm payment

1.3.3.4 he will indemnify the Buyer in respect of any shortfall in the number of the Entitlements that may actually be transferred to the Buyer as opposed to those warranted to be transferred in accordance with the terms of this Agreement whether such shortfall is occasioned by any act of negligence omission or default on his part and also against any loss of single payment arising from any failure on his part or the part of others on his behalf to notify the RPA of the transfer by the 2 April 20[ ], or otherwise howsoever

1.3.5 The Buyer also warrants with the Seller that he will observe and fully comply with the cross compliance requirements of the Single Payment Scheme from the Completion Date until the end of the calendar year next following the Completion Date to ensure that he does not jeopardise the Seller's claim for single farm payment in respect of any other property belonging to and/or in the occupation of the Seller in respect of which he is claiming such payment AND in the event of any failure on the part of the Buyer to comply with this warranty he will compensate the Seller for all losses occasioned by such failure on his part or the part of others acting on his behalf arising whether directly or indirectly from the Buyer's failure to observe the terms of this sub-clause to include but without prejudice to the generality hereof all losses of payment and loss of entitlements (if any) together with all legal or other costs incurred by the Seller in putting himself back into the position he would have been in had the Buyer not failed to observe the terms of this sub-clause

1.3.6 If after the date of this Agreement there are changes in the Regulations or in the interpretation of one or other or all of them by DEFRA or the RPA or otherwise before the Seller has fully complied with the terms of this Schedule in such a way that the objects of this Schedule are not achieved either party may give notice to the other to appoint an arbitrator in accordance with the following paragraph to re-draft the terms of this Schedule in such a way as may fairly achieve its objects in the event that the parties are not able to agree such re-draft between them and the costs of such arbitration shall be borne equally by the Seller and the Buyer

1.3.7 If any dispute arises under this Schedule either party may give notice to the other in writing specifying the dispute and requesting that agreement be reached on the identity of a single arbitrator to be appointed to determine the dispute in accordance with the provisions of the Arbitration Act 1996 and if no arbitrator has been appointed by agreement within one month of the notice either party may make a written request to the President for the time being of the Royal Institution of Chartered Surveyors for him to appoint an arbitrator to determine the matter in dispute

[1.4 Retention

A retention of [ ] thousand pounds (£) ('the Retention') shall be made on Completion which will be held by the Buyer's solicitors and payable to the parties with accrued interest in the following circumstances:-

1.4.1 in the event that the Seller is successful in transferring to the Buyer the Entitlements by the [*date*] the Retention shall be paid to the Seller on or before the [*date*]

1.4.2 in the event that the Seller is successful in transferring to the Buyer some but not all of the Entitlements a proportion of the Retention corresponding to the number of the Entitlements transferred divided by the Entitlements shall be paid to the Seller and the remainder of the Retention shall be released to the Buyer

1.4.3 in the event that no Entitlements are transferred to the Buyer by the [*date*] the Retention shall be released forthwith to the Buyer]

PRECEDENT 3.2 SALE OF LAND WITHOUT ENTITLEMENTS

[*The following clauses can be used where the Seller is retaining his entitlements and the sale to the Buyer is of bare land (aka naked acres!) without entitlements*

Again these can be included in the body of the contract or in a separate schedule to the contract for sale as follows.]

In this Agreement/Schedule

1. Definitions

1.1 'the Council Regulation' means Regulation (EC) No 73/2009

1.2 'the Commission Regulations' means the Council Regulation and any Commission Regulations supplementary thereto

1.3 'DEFRA' means the Department of the Environment Food and Rural Affairs or any successor department

1.4 'RPA' means the Rural Payments Agency or any successor body carrying out its function

1.5 'the Single Payment Scheme' means the single payment scheme established under the Council Regulation the Commission Regulations and any relevant subsequent EC or UK Government regulations and/or administrative provisions applicable in England ('the Regulations'), and the statutory guidance or codes of practice applicable in England and implemented by the EC and/or the UK Government

1.6 'Entitlements' means the entitlements held by the Seller in his name in accordance with the Regulations being the entitlement to the single income payment and any other entitlements to financial support of whatever kind relating to the Property and deriving from the Regulations

2. Sale of Property without Entitlements

The Property is sold WITHOUT any Entitlements

3. Parties' obligations and agreements

3.1 The Seller and the Buyer agree to act in fairness and co-operate in good faith with each other and to take all (if any) reasonable steps to ensure the Entitlements remain with the Seller

3.2 The Buyer warrants that he will do all such acts and things and enter into such documents (if any) as shall be necessary to ensure the Entitlements remain with the Seller

3.3 The Buyer further warrants that he will comply with all of the requirements of the Regulations and any subsequent regulations made by the EC and/or UK Government from the date hereof until the date that DEFRA or the RPA on its behalf confirm the Entitlements have been retained by the Seller

3.4 The Seller shall notify the Buyer of any shortfall in the Entitlements on being notified of such (if any) by DEFRA or the RPA and the Buyer shall indemnify the Seller in respect of any such shortfall in the Entitlements within twenty eight days of receiving notification from the Seller where such shortfall has resulted from the Buyer's negligence omission or default under this Agreement but not otherwise

3.5 The Buyer further warrants with the Seller that he will observe and fully comply with the cross compliance requirements of the Single Payment Scheme from the Completion Date until the end of the calendar year next following the Completion Date to ensure that he does not jeopardise the Seller's claim for single farm payment in respect of any other property belonging to and/or in the occupation of the Seller in respect of which he is claiming such payment AND in the event of any failure on the part of the Buyer to comply with this warranty he will compensate the Seller for all losses occasioned by such failure on the Buyer's part or on the part of others acting on his behalf arising whether directly or indirectly from the Buyer's failure to observe the terms of this sub-clause to include but without prejudice to the generality hereof all (if any) losses of payment and losses of entitlements together with all legal or other costs incurred by the Seller in putting himself back into the same position that he would have been in had the Buyer not failed to observe the terms of this sub-clause

3.6 If after the date of this Agreement there are changes in the Regulations or in the interpretation of one or other or all of them by DEFRA or the RPA or otherwise before the objects of this Schedule have been achieved the Seller may give notice to the Buyer to appoint an arbitrator in accordance with the following paragraph to re-draft the terms of this Schedule in such a way as may fairly achieve its objects in the event that the Seller and the Buyer cannot agree such re-draft between them and the costs of such arbitration shall be borne equally by the Seller and the Buyer

3.7 If any dispute arises under this Schedule either party may give notice to the other in writing specifying the dispute and requesting that agreement be reached on the identity of a single arbitrator to be appointed to determine the dispute in accordance with the Arbitration Act 1996 and if no arbitrator has been appointed by agreement within one month of such notice then either party may make a written request to the President for the time being of the Royal Institution of Chartered Surveyors for him to appoint an arbitrator to determine the matter in dispute

Chapter 3

TENANCIES UNDER THE AGRICULTURAL HOLDINGS ACT 1986

INTRODUCTION

PRECEDENTS

1 Tenancies under the Agricultural Holdings Act 1986 ('AHA 1986')
  - 1.1 Succession tenancy agreement
  - 1.2 Succession tenancy – short form by reference
  - 1.3 Assignment of tenancy
  - 1.4 Assignment of oral tenancy
  - 1.5 Agreement recording terms of tenancy
  - 1.6 Deed of variation
  - 1.7 Surrender agreement
  - 1.8 Deed of surrender
  - 1.9 Agreement to surrender where part of holding has been sub-let
  - 1.10 Deed of surrender

2 Notice pursuant to AHA 1986, section 6 being a request to enter into a written tenancy agreement

3 Notice under LTA 1987, section 48

4 Permission to erect fixture under AHA 1986, section 10

5 Demand for arbitration of rent under AHA 1986, section 12

6 Memorandum of rent review

7 Memorandum of rent increase after landlord's improvement

8 Landlord's consent to tenant's improvement

9 Notice of intention to claim compensation under AHA 1986, section 72

10 Notice of intention to claim compensation on termination of tenancy pursuant to AHA 1986, section 83

11 Notice of intention to re-enter or forfeit

12 Notice to quit to personal representatives (the Case G Notice)

13 Notice to tenant to pay rent due in the prescribed form under SI 1987/711

14 Notice to tenant to remedy breach of tenancy by doing work of repair maintenance or replacement

15 Notice to tenant to remedy breach of tenancy (not being a notice requiring the doing of any work for repair, maintenance or replacement)

16 Suggested designation for inclusion in a will

17 Application for direction giving entitlement to a tenancy

18 Notification of the application to the landlord
19 Reply to application for direction giving entitlement to tenancy
20 Application for direction giving entitlement to tenancy (retirement)
21 Notice to landlord

INTRODUCTION

Since the last edition of this book the Regulatory Reform (Agricultural Tenancies) (England and Wales) Order 2006[1] (RRO 2006) has made amendments to the Agricultural Holdings Act 1986 (AHA 1986) and the Agricultural Tenancies Act 1995 (ATA 1995). The amendments which affect Agricultural Holdings Act tenancies relate to the application of AHA 1986 to new tenancies, rent reviews, succession tenancies, end of term compensation and arbitration.

ATA 1995 will generally apply to all new tenancies. However, ATA 1995 provides that AHA 1986 will continue to apply to (1) tenancies beginning before 1 September 1995, (2) succession tenancies,[2] (3) tenancies created by an implied surrender and regrant,[3] (4) tenancies granted pursuant to written contracts made before 1 September 1995[4] and (5) tenancies where the Evesham custom applies.[5] The RRO 2006[6]has added another category, namely where a tenancy:

> '... is granted to a person who, immediately before the grant of the tenancy, was the tenant of the holding, or of any agricultural holding which comprised the whole or a substantial part of the land comprised in the holding, under a tenancy in relation to which the 1986 Act applied, and is so granted by a written contract of tenancy indicating (in whatever terms) that the 1986 Act is to apply in relation to the tenancy.'[7]

The references to 'substantial part' means a substantial part by reference to either area or value.[8]

---

1 SI 2006/2805.
2 ATA 1995, s 4((1)(b), (c), (d) and s 4(2), (2A), as amended by RRO 2006, art 12.
3 ATA 1995, s 4(1)(f). This originally applied where the new tenancy was 'so granted merely because a *purported variation of the previous tenancy* (not being an agreement expressed to take effect as a new tenancy between the parties) ha[d] effect as an implied surrender followed by the grant of the tenancy'. RRO 2006, art 12(4) has replaced these words with the following 'is so granted because an agreement between the parties (not being an agreement expressed to take effect as a new tenancy between the parties) has effect as an implied surrender followed by the grant of the tenancy'.
4 ATA 1995, s 4(1)(a).
5 ATA 1995, s 4(1)(e).
6 RRO 2006, art 12(5).
7 ATA 1995, s 4(1)(g).
8 See amendments made to ss 34, 35, and 49 of AHA 1986 by RRO 2006, art 4, and the insertion of ss 69(A), 73(2) and Sch 9, para (5)(1A) by art 6.

Where AHA 1986 applies to a new tenancy by virtue of s 4(1)(g) and the rent payable is unchanged from the rent payable under the previous tenancy, disregarding any changes resulting from adjustments to the boundary of the holding, the 3-year rent review cycle is uninterrupted.[9]

Succession tenancies

Succession tenancies may be granted either by the Agricultural Land Tribunal or by agreement.

Sections 36 (succession tenancies on death) and 50[10] (succession tenancies on retirement) of AHA 1986 provide that the principal source of livelihood of the potential successor must be derived from his 'agricultural work on the holding or on an agricultural unit of which the holding forms part', for at least 5 out of the 7 years before the death or retirement of the previous tenant. The RRO 2006[11] adds a new sub-s 36(6):

> '(6) The reference in subsection (3)(a) above to agricultural work carried out by a person on the holding or on an agricultural unit of which the holding forms part includes –
>
> (a) agricultural work carried out by him from the holding or an agricultural unit of which the holding forms part, and
> (b) other work carried out by him on or from the holding or an agricultural unit of which the holding forms part,
>
> which is of a description approved in writing by the landlord after the commencement of this subsection.'

A similar amendment is made to s 41(1) where the eligible person condition although not fully satisfied is satisfied to a material extent.

The terms of the new tenancy may depend to an extent on the terms of the original tenancy, in view of the provisions of ss 47 and 48 of AHA 1986. However, as part of the negotiations in connection with the grant of a succession tenancy, it is often agreed that the terms of the tenancy will be updated and, for example, provisions relating to milk quota and entitlements included, if not already provided for, and obligations in respect of repairs varied. It is imperative that the provisions of s 4 of ATA 1995 are met, otherwise the new tenancy will be a farm business tenancy. In some cases, the form of the old tenancy will be acceptable to the parties. A short form of new tenancy by reference to the terms of the old one is accordingly included.

9 RRO 2006, art 8 adds para 7 to Sch 2 of AHA 1986.
10 Sections 36(3)(a) and 50(2)(a).
11 RRO 2006, art 5.

Successions by agreement[12] may be carried out either by surrender and grant of a new tenancy or by assignment. In the event of a surrender and grant the tenant must ensure that the formalities required by ATA 1995, s 4(1)(d) are met. If not, he will acquire only a farm business tenancy. This chapter contains a full length form of succession tenancy. However, a new tenancy under art 12(5) of the RRO 2006 is now an option.

On a surrender and the grant of a new tenancy, the landlord may become entitled to 100% relief for inheritance tax purposes if he has owned the holding for 7 years. The tenant, on the other hand, may have problems with capital gains tax and inheritance tax. There may be stamp duty land tax on the new tenancy agreement.

In the case of an assignment, the landlord must ensure that the event is indeed a succession. If not, there will be a transfer of the tenancy without using one of the statutory successions referred to in s 37 of AHA 1986. There will be no inheritance tax advantage (if appropriate) for the landlord. The tenant, on the other hand, will have capital gains tax and inheritance tax advantages and will pay no stamp duty land tax.

For a tenancy granted before 12 September 1984, there must be a grant or assignment of the whole or a substantial part of the original holding to one eligible person. For a tenancy granted on or after 12 September 1984, the transaction can be a grant or assignment of all or part of the holding and can be to an eligible person or to an eligible person and another or others.[13]

Where there is more than one tenant and the potential successor is related to one of them, the retirement route must be followed.

One of the items to be addressed in planning for succession will be the existing tenant's will. The tenant may designate in his will a named individual as the person he wishes to succeed under AHA 1986, s 39(4).The Tribunal must be satisfied as to the eligibility and suitability of the designated successor if the tenancy is to remain with the family.

If the family owns land in addition to renting it, the prospective applicant for a succession tenancy should not acquire land which would constitute a commercial unit. It may, therefore, be appropriate for a discretionary will to be drawn up, so that the prospective applicant will be only one of the potential

---

12 Section 4(2) has been amended by art 12 of the RRO 2006. An agreed succession is one granted to a close relative of the previous tenant and either is granted (alone or jointly with other persons) to a person who was (or had become) the sole surviving applicant for a direction of an agricultural land tribunal for a tenancy or is granted and begins, before the giving of any retirement notice by the previous tenant. If no retirement notice is given, it must be granted and begin before the death of the previous tenant.

13 AHA 1986, s 37(2), (4), (5), (8).

beneficiaries. Under current law, if matters can be resolved within 2 years of the death, the discretionary will trust can be wound up in a tax-efficient manner.

Obtaining a written tenancy agreement

Existing oral tenancy agreements may be reduced to writing. Great care must be taken with the form of the agreement to ensure that it does not create a new tenancy which would be a farm business tenancy but merely confirms the terms of the existing one.

Section 6 of AHA 1986 applies where there is no written agreement embodying all the terms of the tenancy or there is an agreement but one or more of the matters specified in Sch 1 to the Act is not included (s 6(1)). Either the landlord or the tenant may apply for the tenancy to be reduced to writing or for the missing term(s) to be recorded.

The matters to be included are set out in Sch 1, but they are not all-embracing and other terms might have been agreed and could therefore be properly included. Schedule 1 provides the starting point. Although not included in Sch 1, the model clauses as to repairs and insurance contained in the Agricultural (Maintenance, Repair and Insurance of Fixed Equipment) Regulations 1973[14] will apply unless some other repair obligation has been agreed and is recorded.

The precedent given is restricted to the matters contained in Sch 1 but there may be other terms which the parties can agree.

As an unwritten tenancy will be assignable, it is essential that a landlord of such a tenancy serves a s 6 notice forthwith.

A tenant under an oral tenancy may consider assignment to others or to a company. Any such assignment made after a s 6 notice is served will be void. It is not necessary to notify the landlord immediately.

Even if the tenancy is oral the assignment must be by deed.

Notices and prescribed forms

Apart from a selection of conveyancing-related documents, we have also included some commonly used notices and a number of prescribed forms. It is hoped that this will be convenient for the practitioner. It cannot be stressed too highly, however, that AHA 1986 is full of technicalities and no action should be taken, or omitted, without a proper understanding of the implications. In addition, there are tight time-limits with no opportunity of extension.

---

14 SI 1973/1473.

Reference must therefore be made to the leading text-books, Scammell and Densham's *Law of Agricultural Holdings*,[15] and Muir, Watt and Moss *Agricultural Holdings*,[16] or if necessary to an expert skilled in the area of law. This should be done before the precedents in this book are used, rather than afterwards when a problem has arisen.

While every effort should be made to ensure that notices are correct, not every defect will invalidate the notice: see *Mannai Investment Co Limited v Eagle Star Life Assurance Co Limited*[17] and subsequent cases A notice to quit an annual tenancy can specify either the anniversary date of the tenancy or the date before.[18] However, errors in prescribed forms, especially in respect of the warnings, are likely to be more strictly treated by the judiciary: see, for example, *Sabella v Montgomery*.[19]

Recovery of possession

When considering the diversification uses envisaged later in this book the tenant must be aware both of the restrictions in his tenancy agreement and the possibility of the landlord recovering possession under Case B if planning permission is obtained for non-agricultural use.

Some comfort can be derived in respect of the former from *Pennell v Payne*[20] where an unlawful sub-letting was held not to constitute an irremediable breach under Case E.

Rent reviews

Section 12(2) of AHA 1986 provided that, where the rent payable in respect of a holding was referred to arbitration, the arbitrator should determine the rent properly payable at the date of the reference. The amendment made by art 3 of the RRO 2006 provides that the arbitrator should determine the rent properly payable at the next termination date (as defined in s 12(4)) following the date of the demand for arbitration.

End of term compensation

Section 69(1) provides that where a tenant has remained on the holding for two or more tenancies, he does not lose the right to end of tenancy compensation for an improvement which was carried out during an earlier tenancy of the same holding. There are similar provisions relating to a landlord's right to end of tenancy compensation in respect of dilapidations (s 73) and to old

---

15 Butterworths, 9th edn, 2007.
16 Sweet & Maxwell, 14th edn, 1998.
17 [1995] 1 WLR 1508.
18 *Yeandle v Reigate BC* [1996] 1 EGLR 20.
19 (1998) 09 EG 153.
20 (1995) 06 EG 152.

improvements (para 5 of Sch 9). The RRO 2006[21] amends these provisions in relation to compensation payable on termination of a tenancy, where that tenancy was granted after that article came into force, so that they also apply where the earlier tenancy was of a holding which comprised the whole or a substantial part of the land comprised in the holding.

Arbitration

Section 84 of AHA 1986 provided that matters to be determined by arbitrations under AHA 1986 were to be determined in accordance with the provisions of Sch 11 to AHA 1986, and not in accordance with AHA 1996. Schedule 11 has been repealed by the RRO 2006.[22] Arbitrations now take place under the Arbitration Act 1996.

Section 84 has been amended so that it now reads:

> '(1) Any matter which by or by virtue of this Act or regulations made under this Act is required to be determined by arbitration under this Act shall, notwithstanding any agreement(under a contract of tenancy or otherwise) providing for a different method of arbitration, be determined by a single arbitrator
>
> (2) The arbitrator shall be a person appointed by agreement between the parties or, in default of agreement, a person appointed on the application of either of the parties by the President of the RICS.
>
> (3) If the arbitrator dies, or is incapable of acting, a new arbitrator may be appointed as if no arbitrator had been appointed.
>
> (4) No application may be made to the President of the RICS for an arbitrator to be appointed by him under this section unless the application is accompanied by such fee as may be prescribed as the fee for such an application; but once the fee has been paid in connection with any such application no further fee shall be payable in connection with any subsequent application for the appointment by him of a new arbitrator in relation to that arbitration.
>
> (5) Where by virtue of this Act compensation under an agreement is to be substituted for compensation under this Act for improvements or for any such matters as are specified in Part II of Schedule 8 to this Act, the arbitrator shall award compensation in accordance with the agreement instead of in accordance with this Act.
>
> (6) In this section "the RICS" means the Royal Institution of Chartered Surveyors.'

21 RRO 2006, art 6.
22 RRO 2006, art 7.

PRECEDENTS

1 TENANCIES UNDER THE AGRICULTURAL HOLDINGS ACT 1986 ('AHA 1986')

PRECEDENT 1.1 SUCCESSION TENANCY AGREEMENT

PARTICULARS

Date of Agreement	:	
The Landlord	:	
The Tenant	:	
The Holding	:	Farm extending to some [ ] acres/hectares or thereabouts as more particularly described in the First Schedule hereto and shown edged red on the plan ('the Plan') annexed hereto
Commencement Date	:	
Term Date	:	[*date*]
Commencing Rental	:	[ ] per year payable in arrear/advance [*delete as appropriate*] on the Rent Days in each year
Rent Days	:	
Landlord's Address for service of notices (Landlord and Tenant Act 1987, section 48(1))	:	
Redundant Items	:	those specified or subsequently deemed to be specified in the Second Schedule hereto
Tenant's Fixtures	:	those specified in the Third Schedule hereto
Milk Quota – litres	:	with a butterfat base of [ ]%
Single Payment Scheme	:	the single payment scheme governed by Council Regulation (EC) No 73/2009 and all the relevant subsequent Commission and other regulations, administrative provisions, statutory guidance or codes of practice implemented by the EC and/or the UK Government ('the Regulations')

Entitlements : such entitlements to the single income payment and any other entitlement to financial support of whatever kind deriving in any way from the Regulations and made available to the Tenant [pursuant to the Regulations] [from the estate of the late (*deceased tenant*)]

1. Interpretation

In this Agreement:-

1.1 the attached Particulars are incorporated and unless the context otherwise requires expressions in the Particulars have the respective meanings set against them in the Particulars

1.2 words importing one gender only include the other genders words importing the singular include the plural number and vice versa

1.3 words importing persons include corporations and vice versa

1.4 the expression 'the Landlord' includes the person for the time being entitled to the reversion immediately expectant on the Tenancy

1.5 the expression 'the Tenant' includes the person for the time being entitled to the Tenancy

1.6 where there are two or more persons included in the expression 'the Landlord' or 'the Tenant' the obligations of such persons under this Agreement shall be joint and several

1.7 the expression 'the Tenancy' means the tenancy hereby created

1.8 the expression 'AHA 1986' means the Agricultural Holdings Act 1986

1.9 the expression 'Relevant Quotas' means such quotas contracts premium rights or similar arrangements (as permitted by the context) as are from time to time attaching to or available in respect of the Holding or any crops produced thereon or to which the Tenant or any other person is entitled in respect of production on the Holding whether existing at the commencement of the Tenancy or introduced thereafter and 'Relevant Quota' shall be construed accordingly

1.10 any reference to 'Holding' shall where the context admits include any part or parts thereof

1.11 any provision not to do an act or thing shall import an obligation not to cause or permit such act or thing to be done

1.12 any reference to a statutory provision regulation or code of practice shall be deemed to include any later provision substituted for or amending the same and any subsidiary provision made under powers thereby conferred

1.13 the clause headings are for convenient reference only and shall not affect its construction

2. Letting

2.1 Subject as hereinafter mentioned and in consideration of the rent agreements and conditions hereinafter following the Landlord agrees to let and the Tenant agrees to take the Holding

2.2 Part IV of AHA 1986 shall apply to the Tenancy hereby created

[2.3 *If Entitlements are being transferred to the Tenant from the estate of the deceased former tenant that transfer needs to be referred to here with the definition of Entitlements containing details of those transferred*]

2.4 The parties hereby agree that the Tenancy is a first/second [*delete as appropriate*] succession and the Tenant is both eligible and suitable within the meaning of Part IV of AHA 1986

3. Matters included in the Holding

There are included with the Holding all appurtenances thereof and all fixtures and fittings thereon except any which the outgoing tenant is entitled to remove and any which are to be regarded as fixtures belonging to the Tenant (in respect of which section 10 of AHA 1986 as hereinafter modified is to apply)

4. Matters affecting the Holding

The Holding is let subject to:-

4.1 the exceptions and reservations herein mentioned

4.2 such public or private rights liabilities provisions agreements and other matters as affect the Holding

4.3 the rights of the outgoing tenant (if any)

4.4 any existing tenancies or occupancies of any dwellings on the Holding

5. Period of tenancy

The Tenancy shall commence on and include the Commencement Date and continue until the next following Term Date and (unless determined on the first anniversary of the Term Date) thereafter from year to year until determined

6. Rent

The Holding is let at the Commencing Rental payable on the Rent Days in each year the first (proportionate if appropriate) payment to be made [on the Commencement Date if the Commencing Rental is payable in advance] [on the first Rent Day next following the Commencement Date if the Commencing Rental is payable in arrears] and where the rent is otherwise payable in arrear the last payment shall if demanded be paid in advance on the Rent Day next preceding the termination of the Tenancy

7. Exceptions and reservations

There are excepted and reserved out of the Holding:-

7.1 *Game*

subject to the provisions of the Ground Game Act 1880 (as amended by section 12 of the Wildlife and Countryside Act 1981) and the Ground Game (Amendment Act) 1906 all deer foxes game wildfowl woodcock snipe fish hares and rabbits and all birds listed in Part 1 of the Second Schedule to the Wildlife and Countryside Act 1981 and the nests and eggs of all birds in this clause referred to with the right for the purposes of shooting fishing hawking sporting taking preserving and rearing such wildlife and destroying pigeons and other pests to enter onto the Holding with such equipment and persons as the Landlord authorises [and also the right to hunt or permit hunting on foot or with horse and hound]

7.2 *Water*

all springs of water or other wells ponds streams and water courses (including any underground water) with the right for the Landlord and all persons authorised by the Landlord to have access thereto and to take water therefrom by means of pumps and other constructions and equipment and pipes laid or to be laid on across or under the Holding or otherwise (subject to sufficient water being left for the Tenant for domestic and agricultural purposes other than irrigation)

7.3 *Timber*

all timber timber-like trees fruit trees tellers pollards saplings and underwood with the right for the Landlord and all persons duly authorised to make fell cut remove the same and to plant upon any coppice or woodland any quantity of young trees as the Landlord thinks fit including the right to enter the Holding for such purposes with or without motor vehicles plant and equipment but doing as little damage as practicable and paying the Tenant reasonable compensation for any damage done

7.4 *Minerals*

all mines minerals quarries stones sand brick-earth clay gravel petroleum and its relative hydrocarbons and all other gases and substances in upon or under the Holding including the air space void created from time to time by the winning working getting and carrying away of any of such substances whether opened or unopened within upon or under the Holding together with the right for the Landlord and all persons authorised by the Landlord and his lessees and agents and all persons authorised by them or any of them with workmen and others from time to time and at all times hereafter whether by entry on the surface or by underground working to win work get and carry away the said substances and any like substances within upon or under any adjacent or other lands with full powers for those purposes:-

7.4.1 to withdraw vertical and lateral support from the surface of the Holding and from any buildings or works now erected or hereafter to be erected thereon or from any crops standing thereon

7.4.2 to sink pits or shafts open quarries and erect buildings and construct place and maintain any plant machinery equipment apparatus or materials in on or under the Holding

7.4.3 to constitute any requisite access tracks roads or railways fences pipes drains sewers wires cables conduits or other apparatus in under or over the Holding

7.4.4 to use the surface of the Holding for the temporary storage or deposit of any of the said substances or any waste

7.4.5 to do all other things necessary or convenient for such purposes or any of them notwithstanding any subsidence or other injury or damage that may thereby be occasioned to the Holding or any building or works or crops as aforesaid or any other injury or damage or loss whatsoever arising whether directly or indirectly from any such working or operation as aforesaid which may be sustained by the Tenant save that the Landlord shall pay to the Tenant reasonable compensation for damage thereby done to the surface buildings or crops and shall allow an abatement of rent in respect of the surface of the Holding of which the Tenant is deprived by reason of the exercise of any such rights and liberties as aforesaid

7.5 *Roads*

the right at all times and for all purposes for the Landlord and the occupiers and tenants of any adjoining or neighbouring land of the Landlord and all other persons authorised by the Landlord or by such occupiers or tenants to use the roads tracks and footpaths within the Holding together with all other ways hitherto used or enjoyed across the Holding and in the case of each such road track or way so used (other than footpaths) either with or without animals vehicles implements machinery or plant

7.6 *Services*

7.6.1 the right for the Landlord and the tenants or other occupiers of any adjoining or neighbouring land of the Landlord to the free passage of drainage water gas oil telecommunications and electricity from and to such other land through the watercourses channels drains sewers pipes cables and wires belonging to or running through the Holding

7.6.2 the right for the Landlord and all persons authorised by the Landlord to erect lay maintain inspect repair renew and use within upon under or over the land comprised in the Holding telecommunications and electricity lines and cables and gas oil water drainage and sewage pipes and any necessary apparatus in connection therewith respectively with full power for the Landlord and such other persons as aforesaid and the agents and workmen of the Landlord and such other persons to enter onto the Holding for the purpose of inspecting repairing replacing and renewing such lines cables pipes and apparatus respectively doing as little damage as practicable and paying to the Tenant reasonable compensation for any damage done

7.7 *Repairs*

the right to enter and execute repairs for which the Tenant is liable under any provision hereof and which he has failed to start within two months or complete within three months of being required so to do by written notice given by the Landlord and the whole or the proportionate part for which the Tenant is liable (as the case may be) of the expense thereof shall be payable by the Tenant forthwith after such repairs have been executed and shall be recoverable as rent in arrears PROVIDED THAT nothing in this clause shall relieve the Tenant from his liability under any provision hereof to execute repairs

7.8 *Cultivation*

the right after seven days' written notice to enter and execute acts of husbandry or cultivation which the Tenant after giving or receiving notice to quit is in the opinion of the Landlord or his agent neglecting or performing in a dilatory or otherwise unsatisfactory manner and to take such manure as the Landlord shall require for such purpose and the Tenant shall provide accommodation for any tractors machinery fertiliser seed or other items brought onto the Holding by the Landlord in exercise of this right

7.9 *Entry*

7.9.1 the right for the Landlord and all persons authorised by the Landlord (where appropriate with or without workmen animals carts motor vehicles plant machinery implements guns and other articles) to enter onto the Holding:-

7.9.1.1 for the due and proper exercise and enjoyment of his rights hereunder

7.9.1.2 for the due and proper performance of his obligations hereunder

7.9.1.3 for any other reasonable purpose

7.9.2 the right to permit any party involved in an arbitration or any other legal proceedings for which an inspection of the Holding is or may be relevant to inspect the Holding whether or not the Landlord is one of the parties to the arbitration or proceedings

7.10 *Wayleaves*

7.10.1 the right to grant any wayleave contract easement or licence to any public or local authority or public utility company or other company or person with the right to authorise agents and servants of the grantee with or without motor vehicles plant and equipment to enter upon the Holding and carry out all appropriate works and in particular but without prejudice to the generality of the foregoing the right to run pipes drains conduits cables wires or other works (either already existing or any new ones) for the benefit of any adjoining or neighbouring land of the Landlord and the right to carry out works for the benefit of such adjoining or neighbouring land the Landlord paying reasonable compensation for any damage done to the surface

7.10.2 the benefit of all such contracts easements or licences existing at the date hereof or granted or made under this clause 7.10 and all rents and other payments reserved thereby (except annual payments made for disturbance to the occupier) and the benefit of any other similar agreements or contracts now or at any time affecting the Holding

7.11 *Holdover*

if the Commencement Date falls between 1 July and 31 December the right of the Landlord and all persons authorised by the Landlord with or without motor vehicles plant and equipment to enter at any time or times upon the Holding for the purpose of storing any produce of the Holding on hard standings or in clamps or pies until 1 January immediately following the Commencement Date or in covered storage of adequate capacity until the following 25 March the Landlord making an allowance against the Commencing Rental in respect of such hard standings clamps or pies in a sum to be agreed with the Tenant and failing agreement to be decided by a single arbitrator in accordance with the provisions of clause 11.18

7.12 *Antiquities*

the right to all antiquities on the Holding together with the right for persons authorised by the Landlord to carry out scientific and research studies of whatsoever nature on the Holding the Landlord paying reasonable compensation for all damage and interference occasioned thereby to the Tenant's business

7.13 *Recordings*

the right for persons authorised by the Landlord to enter on the Holding with or without motor vehicles plant equipment cameras and visual and sound recording equipment and to use the Holding (other than the interiors of dwellings) for the purpose of making sound or visual recordings of whatever nature the Landlord paying reasonable compensation for any interference with the Tenant's business and damage done

8. Tenant's agreements

The Tenant hereby agrees with the Landlord as follows:-

8.1 *Rent*

8.1.1 to pay the Commencing Rental at the times and in the manner aforesaid

8.1.2 to pay any additional yearly rents determined in accordance with section 13 of AHA 1986 such additional rents to be paid (on the same basis as to period and timing as the Commencing Rental) on the Rent Days and with effect from the date of completion of the improvement to which such increase of rental values is attributable such date of completion for the purposes hereof being the date on which the said improvement is first capable of use as to which date the certificate of the Landlord's agents shall be conclusive

8.1.3 to pay the rents referred to in clauses 8.1.1 and 8.1.2 hereof until a different rent is substituted therefor either by agreement or by arbitration under the provisions of AHA 1986 which thereafter shall be the rent payable by the Tenant. In the event of a new rent for the Holding either being agreed or being determined by arbitration the same shall unless otherwise specifically agreed in writing be deemed to be made under this clause and not to create a new tenancy

8.1.4 all such rents shall be paid to the Landlord free from all deductions whatsoever and without set off

8.2 *Interest*

to pay interest on any rent or other moneys payable to the Landlord hereunder which shall not have been paid by the due date for payment at the rate of three per cent above the base rate from time to time of such clearing bank as may be specified from time to time by the Landlord or if no such base rate is available at such other reasonably equivalent rate as the Landlord may from time to time specify

8.3 *Outgoings*

to pay all present or future taxes rates assessments impositions and outgoings whatsoever of an annual or other recurring nature (other than Landlord's drainage rates)

8.4 *Incoming*

to pay to the Landlord on demand the amount payable or to become payable by the Landlord to the outgoing tenant (if any) in respect of the Holding

8.5 *Water expenses*

to reimburse to the Landlord on demand all charges costs and expenses reasonably incurred by the Landlord in connection with the provision or the improved provision of water to the Holding

8.6 *Residence*

personally to reside in the farmhouse and make the same his usual place of residence and personally to farm the Holding

8.7 *Occupation*

8.7.1 not to assign underlet or part with the possession of or share the occupation of the Holding PROVIDED THAT the Tenant may permit the occupation of cottages and their gardens by a person or persons in the full time employment of the Tenant on the Holding but only if such occupation shall be as assured shorthold tenants without security of tenure

8.7.2 except as expressly required to do so by or under the provisions hereof not without the Landlord's written consent (such consent not to be unreasonably withheld) to enter into any scheme imposing limitations on cropping or stocking or requirements for public access

8.7.3 not without the Landlord's written consent (such consent not to be unreasonably withheld) to place any tent or caravan upon the Holding nor to take in lodgers in the farmhouse or cottages on the Holding

8.7.4 not without the Landlord's written consent (such consent not to be unreasonably withheld) to let or sell any grass keep or growing crops on the Holding nor to take on to the Holding livestock belonging wholly or in part to any other person and not to enter into a contract of agistment with any other person in relation to the Holding

8.8 *Charges*

to give written notice to the Landlord forthwith upon giving any charge over his farming stock under the Agricultural Credits Act 1928 or any other similar enactment

8.9 *Acknowledgement*

not to give an acknowledgement as to the entitlement to any right of light or other easement over or in respect of the Holding

8.10 *Water disposal*

not to enter into any agreement or licence for the supply of water from any spring or well pond stream water course or underground water supply on the Holding to any adjoining or neighbouring premises

8.11 *Recordings*

8.11.1 not without the written approval of the Landlord to permit or authorise any person to take for hire or reward any film or other recording (in whatever form and whether of sight or sound) of or on the Holding or any natural or manmade feature thereon or anything done or occurring thereon but so that this shall not restrict the liberty of the Tenant to take such film or other recording as he shall think fit for the personal and domestic use and enjoyment of himself or his family and friends or in direct connection with his agricultural business

8.11.2 forthwith to pay to the Landlord any money received by the Tenant in consequence of any breach of this provision such money to be deemed for the purpose of calculating interest thereon to have become due on the date of the breach

8.12 *Death of the tenant*

to take all such steps as are necessary to ensure that:-

8.12.1 if the Tenant or any of them dies during the Tenancy written notice of the death and the date thereof shall be given to the Landlord within one month of such death

8.12.2 office or duly certified copies of all probates of wills letters of administration orders of court and other instruments affecting these presents or the Tenancy shall be served on the Landlord within one month from the respective dates thereof

8.13 *Proceedings*

to permit the Landlord to take proceedings or do any other act in the name of the Tenant which the Landlord considers appropriate to preserve or protect his interest in the Holding and to cooperate with the Landlord in relation thereto the Landlord indemnifying the Tenant against any costs charges and expenses he may incur in connection with any such matter

8.14 *Repairs*

[*Note: not required if clause 10 applies.*]

[8.14.1 to keep in good and substantial repair order and condition at all times during the Tenancy using the best and most suitable materials for the same all dwellings and other buildings on the Holding together with all fixtures and fittings drains sewers water supplies pumps fences live and dead hedges gates external boundary walls post stiles bridges culverts ponds watercourses ditches roads and yards in and upon the Holding or on other property but serving the Holding or which during the Tenancy may be erected or provided on the Holding or serve the Holding and to keep clean and in good working order all roofs valleys eaves guttering and down pipes gulleys and grease traps and without prejudice to the generality of the foregoing to do and perform all such acts and matters (except as to insurance against fire and other risks) which are the liability of a landlord or a tenant in accordance with Part I or Part II of the Schedule to the Agriculture (Maintenance Repair and Insurance of Fixed Equipment) Regulations 1973, SI 1973/1473] [*This will need checking carefully in each case*]

8.14.2 to replace and upon replacement adequately to paint as may be proper all items of fixed equipment which need replacing from time to time

8.15 *Drainage*

without prejudice to the generality of the provisions of clause 8.14 to maintain the drainage on all parts of the Holding requiring drainage in the most approved manner practised on lands of a similar nature in the district and in the case of field drains and outfalls to open up the land as necessary and to clean the drains and tiles as required

8.16 *Common items*

to pay to the Landlord on demand an appropriate proportion of the cost from time to time incurred by the Landlord in the maintenance and repair of all items used in common with adjoining or neighbouring property

8.17 *Alterations*

not to carry out any alteration or make any addition to the farmhouse buildings or cottages (if any) or erect any new buildings or make any other improvement within the meaning of section 67 and the Seventh Schedule to AHA 1986 on the Holding except with the written consent of the Landlord (such consent not to be unreasonably withheld) or the approval of the Minister as thereby provided

8.18 *Waste*

not to commit any wilful or voluntary waste spoil or destruction on the Holding or do thereon anything which in the opinion of the Landlord may be or become a nuisance or annoyance to the Landlord or to the owners or occupiers of adjoining or neighbouring land

8.19 *Ancient monuments*

not in any way to interfere with or injure any tumulus stone circle or other ancient remain structure or monument (including known buried archaeological remains) or any actual or proposed Site of Special Scientific Interest (whether statutorily protected or not) and not to permit any person to enter on the Holding (with or without metal detectors tools hammers or other equipment) for the purpose of searching for or removing any object or specimen of antiquarian archaeological or scientific interest (save under the rights hereinbefore reserved to the Landlord) but if it shall come to the knowledge of the Tenant that any such object or specimen has been found or observed on the Holding forthwith to report the same to the Landlord in writing and to give such assistance as the Tenant reasonably can to enable possession therof to be given to the Landlord or as he shall direct

8.20 *Record of condition*

to bear the cost of making any record of condition under section 22 of AHA 1986 which the Tenant requires

8.21 *Fire precautions*

to maintain adequate fire fighting equipment throughout the Holding and to comply in all respects with any requirements or recommendations made by the local fire service

8.22 *Notices*

forthwith to notify the Landlord in writing of any notice proposal or order served on him under or by virtue of any statutory provision and if so required to produce to the Landlord such notice proposal or order and allow the Landlord to make copies thereof

8.23 *Legislation*

to observe and comply fully with the provisions of any statute order regulation rule direction code of practice or other thing which extends to or affects the Holding or anything done or omitted or to be done or omitted thereon

8.25 *Insurance of crops and stock*

8.25.1 at all times during the Tenancy to insure against destruction or damage by fire lightning aircraft explosion riot civil commotion malicious persons or theft (and including consequential loss relating to such loss destruction or damage) all live and dead farming stock and crops for the time being on the Holding to the full market value thereof (delivered to the Holding) or the cost of replacement (if greater) and so that every policy of insurance shall:-

8.25.1.1 be effected with an insurance company approved in writing by the Landlord and be endorsed with a note of the Landlord's interest as Landlord of the Holding

8.25.1.2 not be avoidable on account of the storing handling or keeping of petrol or any other inflammable substance on the Holding or for any other reason connected with the user of the Holding

8.25.1.3 provide cover for the return to the Holding of the full equivalent manurial value of any crops destroyed or damaged insofar as the return thereof is required for the fulfilment by the Tenant of his obligations under this Agreement

8.25.2 To lay out the insurance monies received by virtue of such insurance in replacement of live and dead farming stock and crops and the return of the appropriate manurial value to the Holding (as the case may be) and to make good any deficiency out of his own money

8.25.3 to produce to the Landlord on demand the policy or policies of insurance and the receipt for the premium in respect of the current period of insurance

8.25.4 if such insurance shall not be effected or maintained or if such policy or receipt shall not be produced in accordance with the terms hereof and the Landlord shall effect such insurance cover as the Landlord may determine to be appropriate to make good such default (which he is hereby authorised to do) to repay to the Landlord on demand all premiums paid for such purpose and all other expenses incurred by the Landlord in connection therewith

8.26 *Landlord's insurance*

8.26.1 not to do or suffer to be done on the Holding any act or thing which shall or may invalidate the Landlord's policy of insurance in respect of the Holding and/or any other property or cause the premium thereon to be increased or increase the risk of any insured peril to any of the buildings or fences for the time being on the Holding and/or any other property but to take full and proper precautions to protect the Holding from risk of any insured peril and in particular to take all responsible precautions when installing petrol oil gas or electric engines and when storing petrol paraffin oil or similar fuel or lubricants which shall be kept in proper containers and wherever possible in a detached building and in such cases to observe and perform every requirement of any statutory or other relevant provision

[8.26.2 To pay to the Landlord on demand a sum equal to the gross premium or premiums paid or payable by the Landlord for effecting or maintaining the insurance which he is hereby required to effect and maintain] [*if applicable*]

8.27 *Liability insurance*

8.27.1 at all times during the Tenancy to maintain adequate insurance in respect of all employer's risks for which the Tenant may be liable

8.27.2 at all times during the Tenancy to maintain insurance in respect of all third party liability risks in respect of which liability may fall on the Landlord in relation to the Holding the policy or policies of such insurance to be with an insurance company approved in writing by the Landlord and be endorsed with a note of the Landlord's interest as landlord of the Holding and to provide cover in respect of each and every claim of not less than five million pounds or such higher sum as the Landlord may from time to time direct in writing

8.27.3 to apply all monies received through such policies for the purpose in respect of which the same has been paid and to make good any deficiency from his own resources

8.28 *Environmental practice*

8.28.1 not to contravene any legislation intended to prevent or control pollution or to protect the environment which shall from time to time be in force and to comply with all such statutory or regulatory requirements or codes of practice from time to time and to use the Tenant's best endeavours to prevent the pollution of rivers water courses and water supplies and the air and land on the Holding

8.28.2 when using spray chemicals to take all reasonable care to ensure that adjoining hedges trees and crops are not adversely affected and to make good or pay full compensation for any damage done

8.28.3 in performing his obligations hereunder and his obligation to comply with the rules of good husbandry so far as possible to select chemicals and to adopt farming practices which cause the least harm to game birds wildlife (other then vermin and pests) and the environment

8.29 *Trees*

to use his best endeavours to protect from damage any trees coppice wood or underwood growing upon the Holding and in particular:-

8.29.1 not to cut down cut top or lop or drive nails into or otherwise injure any trees growing on the Holding

8.29.2 not to cut any tellars whether growing from stools or otherwise without the written consent of the Landlord

8.29.3 to protect all trees from damage by livestock machinery or otherwise howsoever

8.29.4 to prevent any waste occurring in relation to any trees on the Holding

8.30 *Game*

8.30.1 to use his best endeavours to protect from damage all [foxes] game (other than ground game) wildfowl woodcock snipe fish and protected birds as hereinbefore referred to and their nests and eggs and not to interfere with measures taken by the Landlord or persons authorised by him to rear and preserve game

8.30.2 to warn off all poachers and trespassers and allow his name to be used in any proceedings which the Landlord may take against any poachers or trespassers (the Landlord indemnifying the Tenant against any costs charges and expenses he may thereby incur) and so far as possible to stop all encroachments

8.30.3 not to do any act or thing that may contravene the provisions of the Wildlife and Countryside Act 1981 and to give full written particulars to the Landlord within seven days of the receipt of any notice relating to the Holding served under or by virtue of that Act and upon request to produce the same to the Landlord

8.30.4 not to make or keep rabbit warrens on the Holding and to destroy any that may exist

8.30.5 the period of twelve months mentioned in proviso (b) to sub-section (3) of section 20 of AHA 1986 (within one month of the expiration of which any claim under the said section 20 for damage by game must be made) shall be the twelve calendar months ending on the sixth day of April in each year

8.31 *Deleterious substances*

not without the written consent of the Landlord to bring on to the Holding any material or substance inconsistent with the Tenant's obligations under this Agreement

8.32 *Refuse*

8.32.1 not to deposit any refuse waste redundant material or redundant machinery of any kind on the Holding

8.32.2 within seven days of receipt thereof to comply with any notice issued by the Landlord requiring removal from the Holding of any such material or item as is referred to in this clause in accordance with all statutory provisions relating to such removal then in force

8.33 *User*

to use the Holding for the purposes of agriculture only [and not to use the Holding for any other trade or for advertising purposes or for the holding of any show race match game or contest whether of animals birds vehicles or persons]

8.34 *Cultivation*

8.34.1 to cultivate and manage the Holding according to the best standard and system of cultivation used in the district and so as to comply with the rules of good husbandry as defined by section 11 of the Agriculture Act 1947 and so as not to impoverish or deteriorate the land and to leave the same clean and in good condition

8.34.2 except as permitted by or under any statutory provision not to change the system of cropping or type of farming without the written consent of the Landlord (such consent not to be unreasonably withheld)

8.34.3 to take all reasonable precautions to prevent the introduction of eelworm rhizomania and other pests and diseases on the Holding and in particular not to grow on the same part of the Holding during any period of four consecutive years more than one crop of potatoes and one crop of sugar beet including sugar beet grown for seed

8.34.4 in the last year of the Tenancy to leave any arable land in a course normally practised in the district or as the Landlord shall otherwise direct in writing

8.35 *Grass Land*

8.35.1 to manage the Grass Land in the most approved manner and also to destroy and prevent from seeding and spreading all rushes thistles nettles thorns ragwort gorse and other weeds and to keep down moles and rats and other vermin and spread all molehills and ant hills

8.35.2 not to remove from the Holding any turf or topsoil without the previous written consent of the Landlord and forthwith to pay to the Landlord any money received by the Tenant in consequence of a breach of this provision such money to be deemed for the purpose of calculating interest thereon to have become due on the date of the breach

8.35.3 not to break up or convert into tillage any part of the Holding which shall be considered permanent pasture and for this purpose (and for the purposes of paragraph 2 of the Second Schedule of the Agriculture Act 1947 and section 14 of AHA 1986 or either of them) the whole of that part of the Holding described in the Cultivation column of the First Schedule hereto as pasture meadow grass or ley and also any land subsequently laid to grass for which the Landlord has prescribed seeds shall be so considered

8.35.4 he shall be at liberty not to break up and convert into tillage any of the remainder of the Grass Land on the Holding unless:-

8.35.4.1 he shall at least one year before the termination of the Tenancy sow down to temporary pasture in a husband like manner with a seed mixture approved by the Landlord in writing the whole of that part of the Holding so broken up so that such pasture is fully established before the termination of the Tenancy

8.35.4.2 if the Tenant obtains the written consent of the Landlord he may sow down to temporary pasture in substitution for any grass land ploughed up an equal area of arable land elsewhere on the Holding upon such terms as the Landlord may direct in writing but such consent shall not imply any obligation upon the part of the Landlord to provide fencing or a water supply for such substituted land nor shall it imply the Landlord's consent to the provision of fencing or water supplied by the Tenant as an improvement under the terms of the Seventh Schedule of AHA 1986

[8.35.4.3 no compensation shall be payable to the Tenant in respect of any temporary pasture laid down by him in accordance with this clause]

8.35.5 after giving or receiving notice to quit not to plough up any temporary grass land on the Holding sown more than five years prior to the end of the Tenancy without giving to the Landlord in writing the option to purchase the same at valuation

8.35.6 if the Tenancy shall terminate between 1 January and 30 June in any year not to stock the new seeds after the 1 October immediately preceding such termination

8.36 *Orchards*

to keep all orchards and gardens in a proper state of cultivation well manured and in good heart to protect from damage by cattle or otherwise howsoever and to prune at proper seasons all fruit trees and bushes and at his own expense to replace dead or worn out fruit trees with other fruit trees suitable to the district PROVIDED THAT the Tenant may at his option in lieu of replacing individual trees grub up and replace the whole or a substantial part of an orchard in accordance with a scheme approved in writing by the Landlord or in default of such approval found to be a reasonable scheme by arbitration hereunder

8.37 *Manurial value crops*

not to sell off or remove from the Holding any hay straw roots or green crops but annually to consume such produce on the Holding PROVIDED THAT the Tenant may except in the case of such produce grown in the last year of the Tenancy sell off and remove such produce on returning to the Holding the equivalent manurial value of the produce sold off

8.38 *Records*

8.38.1 to keep upon the Holding and maintain a field book showing how every field or parcel of land has been stocked cropped and cultivated in each year of the Tenancy and to make the same available to the Landlord at all reasonable times

8.38.2 to supply a copy to the Landlord of any return or other form which the Tenant is required to make or complete pursuant to any United Kingdom or European Union requirements or directives and if so required by the Landlord to consult with the Landlord in regard to the completion of such return or form so as to preserve and safeguard in all respects the interest of the Landlord

8.38.3 to keep true accounts of all crops or produce sold off and of the equivalent manurial value returned to the Holding and to produce such accounts to the Landlord or his agent forthwith upon receiving at any time a written request to do so

8.39 *Boundaries*

save in accordance with the rules of good husbandry not to alter or destroy any hedge fence or boundary on the Holding nor do anything whereby the size or shape of any field shall be rendered different from its present size or shape

8.40 *Manure*

to spread annually on the Holding all manure made on it and at the termination of the Tenancy to leave the same on the Holding for the Landlord or his incoming tenant and at no time to remove or sell off from the Holding any farmyard manure

8.41 *Disease*

to give the Landlord immediate notification (to be confirmed in writing forthwith) of any outbreak or suspected outbreak of any notifiable disease of crops or livestock on the Holding

8.42 *Burning*

not to burn any crop residues on land from which a crop has been harvested except in accordance with the provisions of any by-laws Regulations and Codes of Practice from time to time in force

8.43 *Genetically modified organisms*

not to grow or permit to be grown on the Holding any plant which has been 'genetically modified' within the meaning of Part VI of the Environmental Protection Act 1990

8.44 *Groundwater Regulations*

to comply with the provisions of the Groundwater Regulations 1998, SI 1998/2746 ('the Groundwater Regulations') and in particular:-

8.44.1 to supply the Landlord with details of disposal and tipping on the Holding in each year to 31 March within one month of the expiration thereof

8.44.2 to maintain an authorisation in respect of the Holding under the Groundwater Regulations

8.44.3 to take all necessary steps to transfer such authorisation to the Landlord or the incoming occupier on the termination of the tenancy and to give notice of such transfer to the Environment Agency within twenty one days thereof

8.44.4 to indemnify the Landlord against all costs claims proceedings and demands of whatsoever nature and howsoever arising from any breach of the Groundwater Regulations

8.45 *Nitrate Vulnerable Zones*

if the Holding is situated within a Nitrate Vulnerable Zone:-

8.45.1 to retain all records required by the Nitrate Pollution Prevention Regulations 2008, SI 2008/2549 and on the termination of the Tenancy to hand to the Landlord or the incoming occupier such records for the period of five years preceding the end of the Tenancy

8.45.2 to comply with the provisions of the said Regulations and to indemnify the Landlord against all claims proceedings and demands of whatever nature and howsoever arising from any failure to comply therewith

8.46 *Farm Assurance Schemes*

8.46.1 at all times at his own expense to ensure that the requirements of the Farm Assurance Schemes specified in Schedule 7 or any schemes replacing the same are met in respect of the Holding and the crops produced thereon

8.46.2 if the Tenant wishes to register the Holding for the purposes of any Farm Assurance Scheme to comply at his own expense with the requirements thereof including any works required in respect of any fixed equipment on the Holding

8.46.3 at his own expense to ensure that the requirements of any Farm Assurance Scheme available in respect of crops grown on the Holding are met

8.46.4 On the expiration of the Tenancy to hand to the Landlord all records required as evidence that the Tenant has complied with the requirements of all such Schemes

8.47 *Sewage sludge*

8.47.1 not to use or permit to be used sludge as defined in the Sludge (Use in Agriculture) Regulations 1989 on the Holding or any part thereof

OR

8.47.2 not to use or permit to be used sludge as defined in the Sludge (Use in Agriculture) Regulations 1989 on the Holding or any part thereof otherwise than:-

8.47.2.1 in accordance with the said Regulations

8.47.2.2 in accordance with any Code of Practice or Guidelines relating thereto from time to time published by the Government or any Government Ministry or Agency

8.47.2.3 after obtaining the Landlord's consent in writing and in compliance with any conditions specified in writing by the Landlord

8.48 *Health and safety*

8.48.1 to comply in respect of the Holding and the Tenant's activities thereon with the Regulations relating to Health and Safety at Work and all Codes of Practice in respect thereof for the time being in force

8.48.2 forthwith to notify the Landlord of any defects under the Health and Safety Regulations in respect of the Holding or any areas used by the Tenant in common with other persons

8.48.3 to indemnify the Landlord in respect of any liability arising in respect of the Holding under the Health and Safety Regulations and in respect of any areas used in common to reimburse the Landlord a fair payment according to user of the costs of any works required to be undertaken by the Landlord to comply with such Regulations

8.49 *Water abstraction licence*

8.49.1 where a previous occupier of the Holding held a licence to abstract water to give notice to the Water Authority of the change of occupancy of the Holding within one month from the date on which the Tenant becomes the occupier in order to preserve the licence to abstract water

8.49.2 not to allow to lapse any licence taken over from a previous occupier of the Holding or granted during the Tenancy

8.49.3 on the termination of the Tenancy to make such application for the continuance or renewal of the licence as may be requisite for the time being under regulations then in force to preserve the licence for the benefit of the Landlord or his nominee without payment

8.50 *IACS*

in each year to register the Holding or the Tenant (as the case may be) under all relevant schemes or arrangements for any compensatory payments established by the United Kingdom or the European Union and to supply to the Landlord on demand copies of all applications for such registration and any other documents relating thereto

8.51 *Milk quota*

if milk quota shall have been transferred from the Landlord or outgoing tenant to the Tenant on the commencement of the Tenancy (details of such quota being set out in the Particulars):-

8.51.1 to fulfil and maintain such quota with no less litres (subject only to generalised cuts) and the same butterfat by milk produced on the Holding

[8.51.2 not to use any land not comprised in the Holding for the production of milk or ancillary purposes]

8.51.3 to procure on the termination of the Tenancy the transfer of such quota without payment to the Landlord or his nominee and to sign all requisite forms for that purpose

8.51.4 to pay to the Landlord the value of any such quota not transferred to the Landlord or his nominee as aforesaid such value to be the open market value thereof on the date of termination

8.51.5 not to sign any contract for the supply of milk from the Holding without ensuring appropriate provisions for the continuation of the acquisition of milk from the Holding in the event of the Tenancy being terminated and on the termination of the Tenancy to use his best endeavours to transfer any contract for the supply of milk to the Landlord or his nominee if the Landlord so requires

8.51.6 in relation to the period which falls both prior to the termination of the Tenancy and within the milk quota year current at that termination not to produce a quantity of milk which is greater than the due proportion for that period as indicated on the monthly quota statement relating to the Holding

8.51.7 and if he produced a quantity of milk in excess of the proportion calculated in accordance with clause 8.51.6 to indemnify the Landlord and any other occupier of the Holding against any levy liability payable in respect of such excess

8.52 *Single Payment Scheme*

to comply with the provisions of the Fifth Schedule in relation to the Single Payment Scheme

8.53 *Relevant Quotas*

8.53.1 at all times during the Tenancy to do all such acts and things as shall be required (including producing a sufficient quantity of any relevant product and signing and lodging within any applicable statutory or other time-limits such documents as are required) to maintain all Relevant Quotas

8.53.2 to notify the Landlord in writing of any changes in the level of any of the Relevant Quotas within seven days of the Tenant becoming aware of any such change

8.53.3 to use his best endeavours to obtain for the Holding suitable quotas contracts premium rights or similar arrangements under any new scheme introduced after the commencement of the Tenancy

8.53.4 at all times during the Tenancy to carry out all obligations imposed on the Tenant as occupier or producer or quota or contract owner in connection with notices compliance with directions payments of levies maintenance of records or otherwise in connection with every Relevant Quota

8.53.5 forthwith to furnish to the Landlord all information which the Landlord may reasonably request concerning allocation of any Relevant Quota or concerning other farming activities of the Tenant which may affect or tend to affect the allocation of any Relevant Quota whether such farming activities are carried out by the Tenant directly or indirectly and whether they relate to the Holding or to other land

8.53.6 if the Landlord at any time or times wishes to agree or record any figures facts or calculations concerning any Relevant Quota and the relative rights of the parties thereto under any United Kingdom or European scheme legislation or directive from time to time in force then to endeavour to agree the same with the Landlord and in default of agreement the matter shall be referred to arbitration either in accordance with the arbitration provisions of such scheme legislation or directive or hereunder

8.53.7 not to dispose of transfer lend lease or otherwise deal with the whole or any part of a Relevant Quota as to result in the Relevant Quota concerned lapsing or reducing during the Tenancy or the Relevant Quota concerned being transferred off the Holding or otherwise becoming unavailable for use on the Holding on the termination of the Tenancy

8.53.8 not by any direct or indirect act or omission in respect of any land or property not included in the Holding to allow any Relevant Quota to pass to any other person whatsoever or become attached to land not comprised in the Holding

8.53.9 on quitting the Holding to take all such steps as may be required to procure the transfer to the Landlord or such person as the Landlord shall in writing require as his successor in respect of each of the Relevant Quotas [on payment of the then current market value of the same/payment of the sum of [ ] pounds/without charge or payment]

8.54 *General indemnity*

without prejudice to any right or remedy of the Landlord in relation to the Tenant's obligations hereunder at all times to keep the Landlord indemnified against all actions claims and liabilities arising out of or in connection with any of those obligations

8.55 *Crops on termination*

at the termination of the Tenancy to leave upon the Holding all the unconsumed hay straw roots or green crops (other than peas beans potatoes sugar beet carrots or other vegetable crops) produced on the Holding during the last year of the Tenancy

8.56 *Vacant possession*

to take every available measure (including the giving of any necessary notice) to enable vacant possession of the Holding (including any cottages or houses comprised therein) to be given to the Landlord upon the termination of the Tenancy whether by notice forfeiture or otherwise

9. Landlord's Agreements

The Landlord hereby agrees with the Tenant as follows:-

9.1 *Tenantright*

to pay to the Tenant on quitting the value to an incoming tenant of all items in the nature of tenant right such value to be calculated in accordance with the Agriculture (Calculation of Value for Compensation) Regulations 1978 save that the Landlord may set off:-

9.1.1 any sum due to the Landlord for rent breaches of agreement waste or otherwise

9.1.2 if the year of quitting shall not coincide with a year during which the external painting or internal painting and decoration must be executed then a due proportion of the cost of such painting or decoration the deduction for external painting being one-fifth and that for internal painting papering and decorating one-seventh of the estimated total cost in respect of each year that has elapsed since the last repainting papering or decorating or since the commencement of the Tenancy (whichever shall be the later)

9.1.3 the estimated cost of such repairs and replacements to the Landlord's fixed equipment as may in the opinion of the Landlord be required at the time of quitting

9.2 *Insurance*

to insure the farmhouse cottages (if any) and farm buildings (except the buildings and fixtures belonging to the Tenant and any Redundant Items) against destruction or damage by fire and such other risks as the Landlord shall choose and in the event of destruction or damage by an insured risk (except to the extent that such insurance shall have been vitiated by some act or default of the Tenant or any person for whose acts or defaults the Tenant may be responsible) to reinstate or replace the same if its reinstatement or replacement is required for the fulfilment of his responsibilities to manage the Holding in accordance with the rules of good estate management (and in the event of any shortfall in the insurance monies, to make good the same from his own resources)

9.3 *Quiet enjoyment*

to permit the Tenant on paying the rent and performing and observing the agreements by the Tenant herein contained peaceably to hold the Holding without any lawful interruption by the Landlord or any person deriving title under him

9.4 *Single Payment Scheme*

to comply with the provisions of the Fifth Schedule in relation to the Single Payment Scheme

10. Repairs and insurance

[*Note: not required if clause 8.14 applies.*]

The Landlord and the Tenant hereby agree that the Fourth Schedule hereto shall apply to the Tenancy and hereby agree with each other to comply with their respective obligations therein contained [*the wording to tie in with that at clause 8.14 above*]

11. Provisos and conditions

Provided always and it is hereby agreed and declared as follows:-

11.1 *Re-entry*

if at any time:-

11.1.1 the rent or any other payment due to the Landlord from the Tenant hereunder is in arrear for more than twenty one days after any of the days appointed for payment thereof (whether formally demanded or not) or

11.1.2 there shall be a breach of any of the agreements by the Tenant herein contained or implied or

11.1.3 the Tenant or any of them shall become insolvent for the purposes of AHA 1986 or if the interest of the Tenant or any of them under this Agreement shall be taken in execution or if any execution shall be levied on the goods or chattels of the Tenant or any of them or if any stock or crops on the Holding shall be taken under a bill of sale or sold under the authority of a charge made under the Agricultural Credits Act 1928 or the Tenant shall be convicted of stealing sheep or cattle or

11.1.4 a receiver or administrator is appointed in respect of the Tenant's goods property or undertaking or any part thereof

and if the Landlord shall give to the Tenant at least two months' written notice of such non-payment breach or event stating in such notice that the Landlord accordingly relies upon this sub-clause then in any or every such case it shall be lawful for the Landlord or any person authorised by him in that behalf at any time thereafter to re-enter upon the Holding or upon any part thereof in the name of the whole and thereupon the Tenancy shall absolutely determine without prejudice to the right of action or other remedies of the Landlord in respect of any breach of any of the agreements by the Tenant or any of them herein contained

11.2 *Insolvency*

if the Tenant shall be insolvent for the purposes of AHA 1986 or a receiver or administrator shall be appointed as aforesaid it shall be lawful for the Landlord at any time upon giving to the Tenant not less than three months' previous notice in writing to expire at any time to resume possession and determine the Tenancy

11.3 *Distress*

upon seizure by the Landlord under distress for rent the Landlord shall not be obliged to sell any hay straw silage or crops (except the crops which the Tenant is permitted to sell under the terms hereof without returning an equivalent manurial value to the Holding) upon the terms that the same may be removed from the Holding but may exercise the power of sale on distrained goods conferred by statute by selling the same subject to the condition that such produce shall be consumed on the Holding or subject to some other condition which shall secure that the manurial value of such hay straw or crops shall be returned to the Holding. Upon any such seizure and sale it shall be lawful for the Landlord to grant to any purchaser of such hay straw or crops and for such period or periods as the Landlord may think fit the use of such part or parts of the Holding as the Landlord may think necessary or proper for the purpose of harvesting storing consuming or otherwise dealing with such hay straw or crops and without making any compensation to the Tenant in respect thereof

11.4 *Tenant's fixtures and improvements*

11.4.1 the Tenant's right to and in respect of fixtures erected by him or acquired on entry shall be limited to those given to him by section 10 of AHA 1986

11.4.2 if the Tenant shall not within two calendar months after the termination of the Tenancy have removed from the Holding any fixtures or fittings to which section 10 of AHA 1986 applies and in relation to which the Landlord has not given written notice of his intention to purchase under that section such fixtures or fittings shall on the expiration of that period immediately become the property of the Landlord without payment

11.4.3 the items specified as such in Schedule 3 are agreed to be the Tenant's improvements

11.5 *Redundant items*

11.5.1 the items specified in the Second Schedule hereto are agreed as obsolete or redundant to the farming of the Holding

11.5.2 if at any time or times during the continuance of the Tenancy either party shall wish to treat any buildings or other items of fixed equipment on the Holding (other than those included in the Second Schedule hereto) as obsolete or redundant to the proper requirements of the Holding then in default of agreement such party shall be entitled on giving two months' written notice to the other to have the question whether such buildings or other items of fixed equipment or any of them are obsolete or redundant referred to arbitration under AHA 1986 and if it shall be agreed or if such arbitration shall award that such buildings or other items of fixed equipment or any of them are obsolete or redundant then as from the date of such agreement or award (as the case may be) the buildings or other items of fixed equipment so agreed or awarded to be obsolete or redundant shall be deemed to be included in the said Second Schedule

11.5.3 in relation to the items specified or deemed to be included in the Second Schedule both parties shall be relieved from all liability to repair or insure the same save that the Tenant shall keep them in a safe condition

11.5.4 the Tenant may use any redundant buildings or other items of fixed equipment at his own risk and without any obligation (save as aforesaid) to carry out any repairs to them but he must keep them in a clean and tidy condition

11.5.5 the Landlord shall have the right at any time or times (whether or not the Tenant is using them in accordance with the immediately preceding sub-clause) if he shall so desire to enter and remove any obsolete or redundant fixed equipment at his own expense

11.6 *Cottages*

if any of the cottages or other houses or premises comprised in the Tenancy shall at any time have been unoccupied for a continuous period of not less than six months (other than by reason of the temporary absence of an employee or former employee of the Tenant who can demonstrate an intention to return) or occupied in breach of the provisions hereof then it shall be lawful for the Landlord to give to the Tenant not less than three months' written notice to expire at any time to resume possession and determine the Tenancy as to the part of the Holding comprising the said cottage house or premises and its curtilage and appurtenances. In respect of any such property the Landlord may require the Tenant (at the Tenant's expense) to apply to the Court for the purpose of obtaining such orders as may be necessary to secure vacant possession thereof and relief ancillary thereto

11.7 *Termination for non agricultural use*

it shall be lawful for the Landlord at any time or times upon giving to the Tenant not less than three months' previous written notice to expire at any time to resume possession and determine the Tenancy as to the whole or any part of the Holding for any purpose not being the use of the Holding or such part for agriculture and whether the use thereof for any of such purposes is to be made by the Landlord or by a purchaser lessee assignee or other person deriving title from or through the Landlord

11.8 *Determination as to part*

if there shall be a determination of the Tenancy as to any part of the Holding (whether under the foregoing provisions hereof or by surrender or otherwise) then subject to any contrary agreement the Tenant shall have the same rights as to compensation and to a proportionate reduction of rent as would have been applicable had such determination been by notice to quit pursuant to section 31 of AHA 1986 but without the need to serve any notice required by that section and in the event of the parties hereto failing to agree as to the amount of such compensation or proportionate reduction of rent the same shall be determined by arbitration hereunder

11.9 *Rent reviews*

11.9.1 as between the parties hereto paragraph 4(2) of Schedule 2 to AHA 1986 shall be read and construed as if there were added at the end thereof:-

> 'An increase or reduction of rent by reason of the determination of the Tenancy as to part of the Holding (whether by notice to quit surrender or otherwise) or the addition of land to the Holding in either case whether with or without a variation in the amount actually payable'

11.9.2 on any rent review the question of the rent payable in respect of the Holding shall be referred to arbitration hereunder and it shall be determined as nearly as may be on the same basis as it would have been determined under s12 and Schedule 2 of AHA 1986

11.10 *Severed reversion*

if at any time the reversionary estate expectant on the Tenancy shall be severed then for the purposes of rent reviews section 12 of AHA 1986 shall be read and construed as if the words 'or a severed part of the reversionary estate therein' were added after the word 'holding' where it first occurs in that section and as if the words 'or a severed part of the reversionary estate therein (as the case may be)' were added after the word 'holding' where it subsequently occurs in that section

11.11 *Set-off*

the Landlord shall be entitled to set off against and deduct from any moneys which may at any time be payable by the Landlord to the Tenant in respect of the Holding any moneys which may be payable to the Landlord by the Tenant in respect of the Holding whether such sum so payable by or to the Landlord shall be of a liquidated character or not

11.12 *Misdescription*

in the event of the description of the Holding in the First Schedule hereto being by reference to an Ordnance Survey plan or areas the fact that Ordnance Survey areas are measured to the centre of the hedge shall not in the case of the boundary of the Holding be taken to indicate that the Holding ends at the centre of the hedge

11.13 *Custom*

the rights of the parties under this Agreement or otherwise in respect of the Tenancy shall not depend on or be affected by any custom of the country

11.14 *Perennial crops*

nothing herein contained shall be deemed to be a consent within the meaning of AHA 1986 to the making of any improvement nor shall the Holding be deemed to be let as a market garden but if the Tenant shall wish to remove (except from the gardens and orchards of the farmhouse and cottages) any asparagus rhubarb or other perennial crop or any fruit bushes planted or purchased by him before the termination of the Tenancy he shall first give to the Landlord written notice of such his wish three months before the end of the Tenancy and to the extent that the Landlord shall not have agreed in writing to purchase them by valuation within one month of receiving such notice the Tenant may remove such items but shall make good and replace the surface of the soil

11.15 *Value added tax*

where under the terms of this Agreement the Tenant is obliged to make any payment to the Landlord which attracts value added tax (or any tax replacing it) the Tenant shall be responsible for the payment of such value added tax (or any tax replacing it)

11.16 *Sale*

in the event of the sale of the Holding the responsibility for the Landlord's obligations hereunder shall pass to the purchaser and no action shall lie against the Landlord for their non-observance after the date on which the rent accrues to the purchaser

11.17 *Notices*

where any notice is required under this Agreement to be given by or to the Landlord such notice may be given by or to his duly authorised agent any notice (including notices to quit) to be given by either party under this Agreement may be served in the manner provided by section 93 of AHA 1986 or as regards any notice to the Tenant by being sent to him by registered post or recorded delivery to the Holding or by handing it to the Tenant or his agent or by affixing the notice to some conspicuous part of the Holding and any notice to the Tenant shall be sufficient though the same is not addressed to the Tenant by name but only addressed to him by the description of the Tenant of the Holding

11.18 *Arbitration*

all questions valuations and claims hereafter in dispute between the Landlord and the Tenant under this Agreement not compulsorily referred to arbitration by any statutory provision shall be referred to arbitration under the provisions of the Arbitration Act 1996 by a single arbitrator to be appointed in default of agreement by the President for the time being of the Royal Institution of Chartered Surveyors

AS WITNESS the hands of the parties hereto the day and year first before written

FIRST SCHEDULE
THE HOLDING

	Ordnance Survey				Area	
Parish	Sheet No	Enclosure No	Field Name	Cultivation	Hectares	Acres

SECOND SCHEDULE
REDUNDANT ITEMS

THIRD SCHEDULE
TENANT'S FIXTURES AND IMPROVEMENTS

FOURTH SCHEDULE

[*Note: not required if clause 8.14 applies.*]

General provisions and limitations

1. Nothing contained in this Agreement shall oblige either the Landlord or the Tenant:-

1.1 to maintain repair replace or insure any item of fixed equipment which is specified or deemed to be included in the Second Schedule

1.2 to execute any work if and so far as the execution of such work is rendered impossible (except at prohibitive or unreasonable expense) by reason of subsidence of any kind or by the blocking of outfalls which are not under the control of either the Landlord or the Tenant

2. The Landlord shall not be obliged to execute repairs or replacements or to insure buildings or fixtures which are the property of the Tenant or to execute repairs or replacements rendered necessary by the wilful or negligent act or omission of the Tenant or any member of his household or any of his employees licensees or invitees

The landlord's rights and obligations

3. To execute all repairs and replacements to the following parts of the farmhouse cottages and farm buildings on the Holding namely:-

3.1 the main walls and exterior walls (including walls of open and covered yards and garden walls) of whatever construction (but excluding the interior covering of such walls save to the extent that such interior covering has been adversely affected by any default by the Landlord in carrying out his obligations) and the chimney stacks including the pots BUT at his own option the Landlord may substitute a stock proof fence for any garden wall and thereafter paragraph 12.1.2 of this Schedule shall apply to such substituted fence

3.2 the roofs (subject to the provisions of paragraphs 10 and 14.1 of this Schedule)

3.3 the timbers carrying the roofs ceilings and floors

3.4 the floors doors windows and skylights including the frames (except glass glass substitute locks sashcords and fastenings)

3.5 the staircases of the farmhouse and cottages including banisters and hand rails (but not ladders)

3.6 the eaves guttering and downpipes

4. To execute all repairs and replacements to the water supply pipes laid underground on the Holding (subject to the provisions of paragraph 14.2 of this Schedule)

5. Save to the extent that replacement is rendered necessary by any failure by the Tenant to comply with his obligations to replace the following items when they or any of them are beyond repair:-

5.1 the electric wiring system of the premises (except the fittings)

5.2 the soil drains sewers gulleys manholes inspection chambers cesspools septic tanks filter beds and sewage disposal apparatus (excepting the covers tops and gratings of these items and the media of any filter bed)

5.3 the wells boreholes reservoirs pumphouses and pumping equipment

5.4 the bridges (but not culverts) and cattle grids

6. When in the opinion of the Landlord it shall be necessary and in a proper workmanlike manner to paint gas-tar creosote or otherwise effectively treat with a preservative material (at the option of the Landlord) the whole of the exterior wood and ironwork on the Holding including the inside of all external doors and windows opening outwards of the farm buildings usually so dealt with but subject to the provisions of paragraph 15 of this Schedule

7. If the Tenant does not start work on the repairs or replacements for which he is liable under paragraphs 9–17 of this Schedule within two months or if he fails to complete them within three months of receiving from the Landlord a written notice (not being a notice requiring him to remedy a breach of the terms of the Tenancy by doing work of repair maintenance or replacement in a form prescribed under the Agricultural Holdings (Forms of Notice to Pay Rent or to Remedy) Regulations 1987, SI 1987/711 specifying the necessary repairs or replacements and calling on him to execute them the Landlord may enter and execute such repairs or replacements and recover the reasonable cost from the Tenant forthwith

The tenant's rights and obligations

8. To pay to the Landlord on demand:-

8.1 any sum recoverable by the Landlord under paragraph 7 of this Schedule

8.2 a proportion not exceeding one-half of the reasonable cost of:-

8.2.1 repairs and replacements executed by the Landlord to the items specified in paragraph 4 of this Schedule

8.2.2 exterior painting tarring creosoting or treatment with other preservative material executed by the Landlord in accordance with paragraph 6 of this Schedule

9. To repair so long as practicable and to use his best endeavours to keep clean and in good repair until worn out the items specified in paragraph 5 of this Schedule

10. To provide free of charge suitable straw or reed and spars for the repair or renewal of thatch

11. To replace anything mentioned in paragraph 5 of this Schedule which has worn out or otherwise become incapable of repair if the need for replacement arises out of the Tenant's failure to repair it

12. Except insofar as such liabilities fall to be undertaken by the Landlord under the foregoing provisions of this Schedule:-

12.1 to keep and leave clean and in good tenantable repair order and condition:-

12.1.1 the farmhouse cottages and farm buildings on the Holding

12.1.2 all fences live and dead hedges gates gate posts and irons field walls posts stiles culverts covers and gratings (to inspection pits septic tanks cesspits filter beds wells and boreholes) the media in filter beds troughs taps tanks and all water pipes and fittings fixed above ground

12.2 To keep in good tenantable repair all roads yards (including the concrete or other surfaces thereof) watercourses sluices ditches ponds underdrains and outfalls

12.3 To keep clean and in good working order all roof valleys eaves guttering and downpipes and also to use carefully so as to protect from wilful reckless or negligent damage all items for the repair or replacement of which the Landlord is responsible under the provisions of this Schedule and report in writing immediately to the Landlord any damage however caused to items for the repair or replacement of which the Landlord is responsible

12.4 To replace and repair and, upon replacement or repair, adequately to paint or otherwise effectively treat with a proper and adequate preservative material all items of fixed equipment and to do any work where such replacement, repair or work is rendered necessary by the wilful or negligent act or omission of the Tenant or any member of his household or any of his employees licensees or invitees or by damage caused by the Tenant's stock

13. As often as may be requisite and in any case at intervals of not more than seven years properly to clean colour whiten paper and paint limewash or otherwise treat with materials of suitable quality the inside of the farmhouse cottages and farm buildings on the Holding which have been previously so treated including the internal structural steelwork of farm buildings (whether open-sided or not) and including the inside of all doors and windows of the farmhouse and cottages opening outwards and in the last year of the Tenancy to limewash the inside of all farm buildings which have previously been limewashed

14.1 To renew all broken or cracked tiles or slates and replace all slipped tiles or slates from time to time as the damage occurs but so that the Tenant shall not be required to carry out in any one year of the Tenancy such works to a cost greater than £[. . .]

14.2 To repair or renew any burst or damaged waterpipes underground as the damage occurs but so that the Tenant shall not be required to carry out in any one year of the Tenancy such works to a cost greater than £[. . .]

15. To cut trim or lay a proper proportion of the hedges in each year of the Tenancy so as to maintain them in good and sound condition

16. To dig out scour and cleanse all ponds water courses ditches and grips as may be necessary to maintain them at sufficient width and depth and to keep clear and free from obstructions all field drains and their outlets

FIFTH SCHEDULE
SINGLE PAYMENT SCHEME

1. In this Schedule:-

1.1 'DEFRA' means the Department for the Environment Food and Rural Affairs or any successor department

1.2 'RPA' means the Rural Payments Agency or any successor body carrying out similar functions

1.3 'Eligible Hectare' shall have the same meaning as in Article 34(2) of Council Regulation (EC) No 73/2009

2. The Tenant further covenants with the Landlord:-

2.1 to maintain the Entitlements upon the maximum eligible area of the Holding in accordance with the Regulations

2.2 at all times during the Term to do all acts and things which may be required to maintain the maximum number of Entitlements under the Regulations

2.3 during the Term to claim payment upon the Entitlements in accordance with the Regulations in such a way that neither the Entitlements nor any part of them are lost or charged or whereby they are removed from the Tenant whether permanently or temporarily

2.4 to use his best endeavours to obtain for the Holding any new or further entitlements or additional payment value for the Entitlements which become available under the Regulations or otherwise whether existing at or introduced after the commencement of the Tenancy and whether they become available by reference to the Entitlements, the Tenant's occupation of the Holding or otherwise

2.5 to keep and maintain records of all entitlements allocated to or transferred or leased in or out by the Tenant and their use together with any changes in their number type or payment value as may arise as a result of any official action and to permit the Landlord and any person nominated by him to inspect and take copies of that record in so far as it relates to the Entitlements

2.6 at all times during the Term to carry out all obligations imposed on the Tenant as occupier or producer with regard to all of the requirements of the Regulations

2.7 within 14 days of a written request being received to furnish the Landlord (in writing if so required) with all information which the Landlord may reasonably request concerning the Entitlements or concerning other farming activities of the Tenant which may affect or tend to affect the Entitlements whether such farming activities are carried out by the Tenant directly or indirectly and whether they relate to the Holding or to other land

2.8 prior to submitting any forms to DEFRA and/or the RPA in any year to agree with the Landlord the information required by the forms and the contents thereof and to obtain the Landlord's consent to that information and to the contents of the forms

2.9 not to dispose of transfer lend lease charge or otherwise deal with the Entitlements or any of them so as to result in the Entitlements lapsing or reducing in number or value during the Term or being transferred off the Holding or otherwise becoming unavailable for use in full on the Holding on the termination of the Tenancy

2.10 not by any direct or indirect act or omission in respect of any land or property not included in the Holding to allow the Entitlements to pass to any other person or become claimable in respect of land not comprised within the Holding

[2.11 not to make or permit anyone claiming on his behalf or through him to make any claim either during or at or following termination of the Tenancy which may result in the Entitlements or any of them ceasing to belong to or be available to the Landlord and if as a result of any such claim the Entitlements or any of them are lost to the Landlord to indemnify the Landlord against all losses which the Landlord suffers thereby including (but not necessarily limited to) all costs and expenses incurred in the acquisition of replacement entitlements which are comparable in all respects to those lost]

[2.12 on or after quitting the Holding to take all such steps and do all such acts as may be required to procure the transfer of the Entitlements to the Landlord or such person or persons as the Landlord nominates in writing]

delete as appropriate

*[at a consideration to be agreed between the parties being the open market value and in default of agreement to be determined by arbitration]

*[without payment]

*[the Landlord paying the Tenant compensation in the sum of one pound (£1)]

*[being a sum equivalent to the open market value of that part of the Entitlements that would represent the historic element of each of them at that time in the same way as if such were capable of being separately valued notwithstanding the fact that the Entitlements are not divisible AND in default of agreement to be determined by arbitration]

*[being [ ]% of the open market value of the Entitlements at the date of termination of the Tenancy]

[2.13 as the provisions of clause 2.12 also apply where the Tenancy is terminated in respect of any part of the Holding in that case to procure the transfer pro rata to the Landlord of the number of entitlements that the Entitlements bear to the number of Eligible Hectares in respect of which the Tenancy is terminated compared with the number of Eligible Hectares comprised in the Holding and the compensation payable shall also be calculated pro rata on the same basis]

2.14 during the final year of the Tenancy to obtain the Landlord's consent in respect of all matters concerning the management of the Entitlements to include cross compliance claiming the single payment and effecting the transfer of the Entitlements as the Landlord may direct

2.15 in so far as they remain to be observed and performed and are capable of taking effect after the termination of the Term howsoever determined the provisions of this Schedule shall remain in full force and effect notwithstanding the termination of the Tenancy

2.16 the Landlord covenants with the Tenant to comply with the terms of the Single Payment Scheme after the termination of the tenancy throughout the then subsisting calendar year

SIGNED by
in the presence of:-

SIGNED by
in the presence of:-

PRECEDENT 1.2 SUCCESSION TENANCY – SHORT FORM BY REFERENCE

THIS TENANCY AGREEMENT is made this [ ] day of [ ] 20[ ]

BETWEEN

(1) [*name*] of [*address*] ('the Landlord')

and

(2) [*name*] of [*address*] ('the Tenant')

WHEREBY IT IS AGREED as follows:-

1. This tenancy is an agreed succession to the tenancy which commenced on the [*date*] pursuant to a Tenancy Agreement made on the [*date*] between the Landlord (1) and [*name*] (2) ('the Principal Agreement') in respect of ALL THAT property situate at and known as [*Holding details*] ('the Holding') as the same is delineated and edged red on the plan annexed hereto

2. The Landlord hereby lets the Holding to the Tenant on a tenancy from year to year ('this Tenancy') commencing on the [*date*] at an initial rental of [*sum in words*] pounds (£ . . .) per annum payable on the days and in the manner set out in the Principal Agreement

[3. *If there are any entitlements to be transferred to the Tenant from the estate of the deceased former tenant that transfer needs to be referred to here with the definition of Entitlements containing details of those transferred. Reference should be made to the definitions and suggested clauses in Precedent 1.1*

If there are no entitlements then just the other five clauses alone will suffice]

4. Part IV of the Agricultural Holdings Act 1986 applies to this Tenancy

5. For the purposes of the Agricultural Holdings Act 1986 this agreed succession is the first/second [*delete as appropriate*] succession

6. This Tenancy is subject to the terms conditions covenants and provisions contained in the Principal Agreement as if the same had been repeated herein

AS WITNESS the hands of the parties hereto the day and year first before written

SIGNED by
in the presence of:-

SIGNED by
in the presence of:-

PRECEDENT 1.3 ASSIGNMENT OF TENANCY

THIS ASSIGNMENT is made the [ ] day of [ ] 20[ ]

BETWEEN:-

(1) [*name*] of [*address*] ('the Assignor')

(2) [*name*] of [*address*] ('the Assignee')

and

(3) [*name*] of [*address*] ('the Landlord')

NOW THIS DEED WITNESSES as follows:-

1. Definitions and interpretation

1.1 'the Property' means the property more particularly described in the Schedule

1.2 'the Tenancy Agreement' means an Agreement made this [*date*] between (1) [ ] and (2) [ ] and includes any deeds or documents supplemental thereto whether or not expressed to be so

1.3 words importing one gender shall be construed as importing any other gender

1.4 words importing the singular shall be construed as importing the plural and vice versa

1.5 where any party comprises more than one person the obligations and liabilities of that party under this assignment shall be joint and several obligations and liabilities of those persons

1.6 the clause headings do not form part of this assignment and shall not be taken into account in its construction and interpretation

2. Recitals

2.1 By the Tenancy Agreement [*name*] let the Property to [*name*] for a term beginning on [*date*] and continuing from year to year until determined as therein mentioned

2.2 The Assignor wishes to retire and assign all his interest in the Property and the Tenancy Agreement to the Assignee

2.3 The Assignee is the [son/daughter [*state relationship*]] of the Assignor and is a person who if the Assignor had died immediately before the date of this deed would have been a close relative of the Assignor for the purposes of the Agricultural Holdings Act 1986

2.4 The Landlord is [now] the landlord of the Property

2.5 The current annual rent is £[ ]

2.6 The Assignor has agreed with the Assignee for the assignment of the tenancy created by the Tenancy Agreement to the Assignee

2.7 The Tenancy Agreement contains a provision prohibiting the Assignor from assigning the tenancy thereby created

2.8 The Assignor has applied to the Landlord for consent to the assignment on the basis that such assignment will constitute a first/second succession for the purposes of Part IV of the Agricultural Holdings Act 1986 and on that basis notwithstanding but without prejudice to the provision in the Tenancy Agreement prohibiting assignment the Landlord has agreed to grant such consent in the manner hereinafter appearing to enable the Assignor to assign the whole of his estate and interest in the Property to the Assignee

[2.9 *If there are any entitlements to be transferred these will need to be referred to here with appropriate definitions, and details of entitlements to be included. Reference should be made to the definitions and suggested clauses in Precedent 1.1. If there are no entitlements then clauses 2.1 to 2.8 will suffice.*]

3. Licence to assign

The Landlord grants to the Assignor licence to assign the whole of the Assignor's estate and interest in the Property to the Assignee

4. Assignment

In pursuance of the said agreement and in consideration of the covenants by the Assignee contained in clause 5 the Assignor hereby assigns to the Assignee ALL THAT the Property and the Assignor with full title guarantee hereby releases to the Assignee all that his interest in the Property and in the said tenancy created by the Tenancy Agreement TO HOLD to the Assignee from the date hereof during the continuance of the tenancy created by the Tenancy Agreement as yearly tenant in the place of the Assignor SUBJECT TO the performance and observance of the terms and conditions of the said tenancy

5. Covenant for indemnity

With the object and intention of affording to the Assignor a full and sufficient indemnity but not further or otherwise the Assignee covenants with the Assignor that he and his successors in title will:-

5.1 during the tenancy pay the rents reserved under the Tenancy Agreement and perform all the terms and conditions on the part of the tenant therein contained and

5.2 keep the Assignor and his successors in title indemnified against all actions and claims demands losses costs damage and liabilities whatsoever arising by reason of any breach of the terms and conditions of the Tenancy Agreement

6. Covenant with the Landlord

The Assignee covenants with the Landlord that at all times hereafter during the remainder of the tenancy created by the Tenancy Agreement the Assignee shall pay the rent reserved by and observe and perform the covenants on the part of the tenant and the conditions contained in the Tenancy Agreement

7. Non-waiver

Nothing herein contained shall waive or be deemed to waive any breach of the obligations of the Assignor under the Tenancy Agreement which may have occurred prior to the date hereof or authorise or be deemed to authorise any further assignment and the covenants on the part of the Tenant and conditions contained in the Tenancy Agreement shall continue in full force and effect

8. First/second succession

The parties hereby agree that this Assignment shall be treated as a first/second succession for the purposes of Part IV of the Agricultural Holdings Act 1986 and that the Assignee is both eligible and suitable within the meaning of the said Act

IN WITNESS whereof the parties hereto have set their hands to this Deed the day and year first before written

SCHEDULE
THE PROPERTY

SIGNED as a Deed by
in the presence of:-

SIGNED as a Deed by
in the presence of:-

SIGNED as a Deed by
in the presence of:-

SIGNED as a Deed by
in the presence of:-

PRECEDENT 1.4 ASSIGNMENT OF ORAL TENANCY

THE ASSIGNMENT is made the [ ] day of [ ] 20[ ]

BETWEEN:-

(1) [*name*] of [*address*] ('The Assignor')

and

(2) [*name*] of [*address*] ('The Assignee')

NOW THIS DEED WITNESSES as follows:-

1. Definitions and interpretation

1.1 'the Property' means the property more particularly described in the Schedule

1.2 Words importing one gender shall be construed as importing any other gender

1.3 Words importing the singular shall be construed as importing the plural and vice versa

1.4 Where any party comprises more than one person the obligations and liabilities of that party under this assignment shall be joint and several obligations and liabilities of those persons

1.5 The clause headings do not form part of this assignment and shall not be taken into account in its construction and interpretation

[1.6 *If there are any entitlements to be transferred please refer to the definitions and suggested clauses in Precedent 1.1.*]

2. Recitals

2.1 The Assignor has since on or about the [*date*] been the tenant of the Property from year to year by virtue of an oral agreement made between (1) [*name*] and (2) the Assignor

2.2 The landlord of the property is now [*name*]

2.3 The current annual rent is £[ ]

2.4 The Assignor has agreed with the Assignee for the assignment of the Assignor's term and interest in the Property to the Assignee

3. Assignment

In pursuance of the said agreement and in consideration of the covenants by the Assignee contained in clause 4 the Assignor hereby assigns to the Assignee ALL THAT the Property and the Assignor with full title guarantee hereby releases to the Assignee all that his interest in the Property and in the said tenancy created by the said oral agreement TO HOLD to the Assignee for all of the yearly tenancy now vested in the Assignor SUBJECT TO the performance and observance of the terms and conditions of the said tenancy

4. Covenant for indemnity

With the object and intention of affording to the Assignor a full and sufficient indemnity but not further or otherwise the Assignee covenants with the Assignor that he and his successors in title will:-

4.1 during the tenancy pay the rents reserved under the tenancy created by the said oral agreement and perform all the terms and conditions of the tenancy and

4.2 keep the Assignor and his successors in title indemnified against all actions and claims demands losses costs damages and liabilities whatsoever arising by reason of any breach of the terms and conditions of the said tenancy

IN WITNESS whereof the parties hereto have set their hands to this Deed the day and year first before written

SCHEDULE
THE PROPERTY

ALL THAT land situate at [*address*] extending to [ ] acres or thereabouts shown edged red on the plan annexed hereto

SIGNED as a Deed by
in the presence of:-

SIGNED as a Deed by
in the presence of:-

PRECEDENT 1.5 AGREEMENT RECORDING TERMS OF TENANCY

THIS AGREEMENT is made the [ ] day of [ ] 20[ ]

BETWEEN:-

(1) [*name*] of [*address*] ('the Landlord')

and

(2) [*name*] of [*address*] ('the Tenant')

WHEREAS:-

1. The Tenant is the tenant from year to year of the agricultural holding known as [*name*] ('the Holding') situate at [*address*] and containing [ ] acres or thereabouts and being [the land registered with absolute title at HM Land Registry under the title number [*number*]] TOGETHER WITH all the Landlord's fixtures and fittings on the Holding

2. The Tenant has registered in his name entitlements to receive the single income payment ('the Entitlements') under the single payment scheme ('the Single Payment Scheme') now governed by Council Regulation (EC) No 73/2009 and all relevant subsequent Commission and other regulations administrative provisions statutory guidance or codes of practice implemented by the EC and/or the UK Government ('the Regulations') and has agreed with the Landlord as hereinafter appearing in connection with the same

3. The Landlord and the Tenant wish to record the terms of the tenancy in writing including the agreement between them regarding Entitlements

NOW IT IS AGREED as follows:-

1. Letting

The Holding is held on a tenancy from year to year until it shall be determined by either party giving to the other at least twelve months' previous notice in writing expiring on [ ] in any year

2. Rent

Annual rent of £[ ] is payable in advance/arrears on the [*date*] and the [*date*] in each year

3. Agreement by the Tenant

The Tenant agreed with the Landlord as follows:-

3.1 *To pay outgoings*

to pay all existing and future rates taxes assessments and outgoings of any kind charged or imposed upon the Holding or upon the owner or occupier thereof

3.2 *Return of manurial values in the event of the destruction by fire of harvested crops grown on the holding for consumption on it*

to return to the Holding the full equivalent manurial value of the crops destroyed insofar as the return is required for the fulfilment of the Tenant's responsibilities to farm in accordance with the rules of good husbandry

3.3 *Insurance*

to insure against damage by fire all dead stock on the Holding and such harvested crops as are grown on the Holding for consumption on it

3.4 *Assignment*

not to assign sublet or part with possession of the Holding or any part of it without the Landlord's consent in writing

4. Single Payment Scheme

The Tenant and the Landlord further agree to observe the provisions contained in the Schedule regarding the single payment scheme on their respective parts to be observed

5. Right of re-entry

5.1 If at any time:-

5.1.1 the rent or any part of it is in arrear for at least 21 days after becoming due or

5.1.2 there is any breach or non-observance by the Tenant of any of the covenants and conditions contained in this Agreement,

then the Landlord may without previously demanding any rents that may be due (after giving the Tenant two calendar months' notice) re-enter immediately upon the Holding or any part of it in the name of the whole and the tenancy is then to determine absolutely without prejudice to the rights of the Landlord in respect of any breach or non-observance of any of the agreements contained above and on the part of the Tenant to be performed and observed or any other rights and remedies of the Landlord

AS WITNESS the hands of the parties hereto the day and year first before written

SCHEDULE

1 In this Schedule:-

1.1 'the Regulations' means Council Regulation (EC) No 73/2009 and all relevant subsequent Commission Regulations administrative provisions statutory guidance or codes of practice implemented by the EC and/or the UK Government pursuant to the Regulations

1.2 'Entitlements' means such entitlements to the single income payment and any other entitlement to financial support of whatever kind deriving in any way from the Regulations as are held by the Tenant in his name

1.3 'DEFRA' means the Department for the Environment Food and Rural Affairs or any successor department

1.4 'RPA' means the Rural Payments Agency or any successor body carrying out similar functions

1.5 'Eligible Hectare' shall have the same meaning as in Article 34(2) of Council Regulation (EC) No 73/2009

2 The Tenant further covenants with the Landlord:-

2.1 to maintain the Entitlements upon the maximum eligible area of the Holding in accordance with the Regulations

2.2 at all times during the Term to do all acts and things which may be required to maintain the maximum number of Entitlements under the Regulations

2.3 during the Term to claim payment upon the Entitlements in accordance with the Regulations in such a way that neither the Entitlements nor any part of them are lost or charged or whereby they are removed from the Tenant whether permanently or temporarily

2.4 to use his best endeavours to obtain for the Holding any new or further entitlements or additional payment value for the Entitlements which become available under the Regulations or otherwise whether existing at or introduced after the commencement of the Tenancy and whether they become available by reference to the Entitlements, the Tenant's occupation of the Holding or otherwise

2.5 to keep and maintain records of all entitlements allocated to or transferred or leased in or out by the Tenant and their use together with any changes in their number type or payment value as may arise as a result of any official action and to permit the Landlord and any person nominated by him to inspect and take copies of that record in so far as it relates to the Entitlements

2.6 at all times during the Term to carry out all obligations imposed on the Tenant as occupier or producer with regard to all of the requirements of the Regulations

2.7 within fourteen days of a written request being received to furnish the Landlord (in writing if so required) with all information which the

Landlord may reasonably request concerning the Entitlements or concerning other farming activities of the Tenant which may affect or tend to affect the Entitlements whether such farming activities are carried out by the Tenant directly or indirectly and whether they relate to the Holding or to other land

2.8 prior to submitting any forms to DEFRA and/or the RPA in any year to agree with the Landlord the information required by the forms and the contents thereof and to obtain the Landlord's consent to that information and to the contents of the forms

2.9 not to dispose of transfer lend lease charge or otherwise deal with the Entitlements or any of them so as to result in the Entitlements lapsing or reducing in number or value during the Term or being transferred off the Holding or otherwise becoming unavailable for use in full on the Holding on the termination of the Tenancy

2.10 not by any direct or indirect act or omission in respect of any land or property not included in the Holding to allow the Entitlements to pass to any other person or become claimable in respect of land not comprised within the Holding

2.11 not to make or permit anyone claiming on his behalf or through him to make any claim either during or at or following termination of the Tenancy which may result in the Entitlements or any of them ceasing to belong to or be available to the Landlord and if as a result of any such claim the Entitlements or any of them are lost to the Landlord to indemnify the Landlord against all losses which the Landlord suffers thereby including (but not necessarily limited to) all costs and expenses incurred in the acquisition of replacement entitlements which are comparable in all respects to those lost

2.12 on or after quitting the Holding to take all such steps and do all such acts as may be required to procure the transfer of the Entitlements to the Landlord or such person or persons as the Landlord nominates in writing

delete as appropriate

*[at a consideration to be agreed between the parties being the open market value and in default of agreement to be determined by arbitration]

*[without payment]

*[the Landlord paying the Tenant compensation in the sum of one pound (£1)]

*[being a sum equivalent to the open market value of that part of the Entitlements that would represent the historic element of each of them at that time in the same way as if such were capable of being separately valued notwithstanding the fact that the Entitlements are not divisible AND in default of agreement to be determined by arbitration]

*[being [ ]% of the open market value of the Entitlements at the date of termination of the Tenancy]

2.13 as the provisions of 2.12 also apply where the Tenancy is terminated in respect of any part of the Holding in that case to procure the transfer pro rata to the Landlord of the number of entitlements that the Entitlements bear to the number of Eligible Hectares in respect of which the Tenancy is terminated compared with the number of Eligible Hectares comprised in the Holding and the compensation payable shall also be calculated pro rata on the same basis

2.14 during the final year of the Tenancy to obtain the Landlord's consent in respect of all matters concerning the management of the Entitlements to include cross compliance claiming the single payment and effecting the transfer of the Entitlements as the Landlord may direct

2.15 in so far as they remain to be observed and performed and are capable of taking effect after the termination of the Term howsoever determined the provisions of this Schedule shall remain in full force and effect notwithstanding the termination of the Tenancy

2.16 the Landlord covenants with the Tenant to comply with the terms of the Single Payment Scheme after the termination of the tenancy throughout the then subsisting calendar year

SIGNED by
in the presence of:-

SIGNED by
in the presence of:-

PRECEDENT 1.6 DEED OF VARIATION

[*This Deed can be used to vary the terms of an agreement granted either under AHA 1986 or ATA 1995 and most usually, of late, has been employed to incorporate into existing tenancy agreements provisions dealing with the single payment scheme where there were no such provisions in existing tenancy agreements. Clearly, this is not the only use for this particular Precedent which, suitably amended and adapted, can be used to vary any provisions in existing tenancies.*]

THIS DEED made on the [ ] day of [ ] 20[ ]

BETWEEN:-

(1) [*name*] of [*address*] ('the Landlord')

and

(2) [*name*] of [*address*] ('the Tenant')

WITNESSES as follows:-

In this Deed

1. Definitions and interpretation

1.1 'The Tenancy Agreement' means the Tenancy Agreement dated the [*date*] and made between the Landlord (1) and the Tenant (2)

1.2 'The Holding' means ALL THAT property known as [ ] situate at [ ]full particulars of which are contained in the Tenancy Agreement

1.3 Where any party to this Deed comprises two or more persons obligations expressed or implied to be made by or with that party are deemed to be made by or with the persons comprising that party jointly and severally

1.4 Words importing one gender include all other genders; words importing the singular include the plural and vice versa and words importing persons shall be construed as importing a corporate body or a partnership and vice versa

1.5 Reference to any clause, sub-clause or schedule without further designation shall be construed as a reference to the clause sub-clause or schedule in this Deed so numbered

1.6 The clause paragraph and schedule headings do not form part of this Deed and shall not be taken into account in its construction or interpretation

2. Recitals

2.1 This Deed is supplemental to the Tenancy Agreement

2.2 By the Tenancy Agreement the Landlord let to the Tenant the Holding from the [*date*] for the term as therein defined ('the Term')

2.3 In consideration of the terms hereinafter contained the Landlord and the Tenant have agreed that the terms of the Tenancy Agreement shall be varied as hereinafter appearing

3. Consideration [if any]

[*NB: as this is a deed consideration is not required although the parties may well prefer to show in this Deed what has been agreed for the avoidance of doubt at a later date.*]

[In consideration of [ ] it is mutually agreed that the provisions set out in the Schedule hereto shall be incorporated into the Tenancy Agreement and take effect from the date hereof as if they were incorporated in the Tenancy Agreement]

4. **Tenant's covenant**

The Tenant covenants with the Landlord to observe and perform the obligations on his part set out in the Schedule hereto as from the date of this Deed as if they were incorporated in the Tenancy Agreement

5. **Landlord's covenant**

The Landlord covenants with the Tenant to observe and perform the obligations on his part set out in the Schedule hereto as from the date of this Deed as if they were incorporated in the Tenancy Agreement

6. **Confirmation of the Tenancy Agreement**

The Tenancy Agreement shall continue in full force save as modified by this Deed and shall operate as if the provisions contained in the Schedule were incorporated in it. The covenants conditions and provisions of the Tenancy Agreement shall have effect as though the provisions contained in this Deed had been contained in it from the commencement of the Term

7. **Subsequent alteration**

If after the date of this Deed there are changes in the Regulations or in the interpretation of them by DEFRA or the RPA or in any other way the intention of the parties as set out in this Deed would be frustrated without amendment to this Deed either party may give notice to the other, in the absence of agreement between them on the content of any subsequent Deed to achieve their objectives as herein evidenced, to appoint an arbitrator to redraft the terms of this Deed in such a way as may fairly achieve the objectives of the parties and the costs of such appointment shall be borne equally by the parties

8. **Arbitration**

If any dispute arises under this Agreement either party may give notice to the other in writing specifying the dispute and requesting that agreement be reached on the identity of a single arbitrator to be appointed to determine the dispute in accordance with the provisions of the Arbitration Act 1996 and if no arbitrator is appointed by agreement within one month of the notice either party may make a written request to the President for the time being of the Royal Institution of Chartered Surveyors for him to appoint an arbitrator to determine the matter in dispute

IN WITNESS whereof the parties hereto have set their hands to this Deed the day and year first before written

SCHEDULE

1. In this Schedule:-

1.1 'the Regulations' means Council Regulation (EC) No 73/2009 and all relevant subsequent Commission Regulations administrative provisions statutory guidance or codes of practice implemented by the EC and/or the UK Government pursuant to the Regulations

1.2 'Entitlements' means such entitlements to the single income payment and any other entitlement to financial support of whatever kind deriving in any way from the Regulations as are held by the Tenant in his name

1.3 'DEFRA' means the Department for the Environment Food and Rural Affairs or any successor department

1.4 'RPA' means the Rural Payments Agency or any successor body carrying out similar functions

1.5 'Eligible Hectare' shall have the same meaning as in Article 34(2) of Council Regulation (EC) No 73/2009

2. The Tenant further covenants with the Landlord:-

2.1 to maintain the Entitlements upon the maximum eligible area of the Holding in accordance with the Regulations

2.2 at all times during the Term to do all acts and things which may be required to maintain the maximum number of Entitlements under the Regulations

2.3 during the Term to claim payment upon the Entitlements in accordance with the Regulations in such a way that neither the Entitlements nor any part of them are lost or charged or whereby they are removed from the Tenant whether permanently or temporarily

2.4 to use his best endeavours to obtain for the Holding any new or further entitlements or additional payment value for the Entitlements which become available under the Regulations or otherwise whether existing at or introduced after the commencement of the Tenancy and whether they become available by reference to the Entitlements, the Tenant's occupation of the Holding or otherwise

2.5 to keep and maintain records of all entitlements allocated to or transferred or leased in or out by the Tenant and their use together with any changes in their number type or payment value as may arise as a result of any official action and to permit the Landlord and any person nominated by him to inspect and take copies of that record in so far as it relates to the Entitlements

2.6 at all times during the Term to carry out all obligations imposed on the Tenant as occupier or producer with regard to all of the requirements of the Regulations

2.7 within fourteen days of a written request being received to furnish the Landlord (in writing if so required) with all information which the Landlord may reasonably request concerning the Entitlements or concerning other farming activities of the Tenant which may affect or tend to affect the Entitlements whether such farming activities are carried out by the Tenant directly or indirectly and whether they relate to the Holding or to other land

2.8 prior to submitting any forms to DEFRA and/or the RPA in any year to agree with the Landlord the information required by the forms and the contents thereof and to obtain the Landlord's consent to that information and to the contents of the forms

2.9 not to dispose of transfer lend lease charge or otherwise deal with the Entitlements or any of them so as to result in the Entitlements lapsing or reducing in number or value during the Term or being transferred off the Holding or otherwise becoming unavailable for use in full on the Holding on the termination of the Tenancy

2.10 not by any direct or indirect act or omission in respect of any land or property not included in the Holding to allow the Entitlements to pass to any other person or become claimable in respect of land not comprised within the Holding

2.11 not to make or permit anyone claiming on his behalf or through him to make any claim either during or at or following termination of the Tenancy which may result in the Entitlements or any of them ceasing to belong to or be available to the Landlord and if as a result of any such claim the Entitlements or any of them are lost to the Landlord to indemnify the Landlord against all losses which the Landlord suffers thereby including (but not necessarily limited to) all costs and expenses incurred in the acquisition of replacement entitlements which are comparable in all respects to those lost

2.12 on or after quitting the Holding to take all such steps and do all such acts as may be required to procure the transfer of the Entitlements to the Landlord or such person or persons as the Landlord nominates in writing

delete as appropriate

*[at a consideration to be agreed between the parties being the open market value and in default of agreement to be determined by arbitration]

*[without payment]

*[the Landlord paying the Tenant compensation in the sum of one pound (£1)]

*[being a sum equivalent to the open market value of that part of the Entitlements that would represent the historic element of each of them at that time in the same way as if such were capable of being separately valued notwithstanding the fact that the Entitlements are not divisible AND in default of agreement to be determined by arbitration]

*[being [ ]% of the open market value of the Entitlements at the date of termination of the Tenancy]

2.13 as the provisions of 2.12 also apply where the Tenancy is terminated in respect of any part of the Holding in that case to procure the transfer pro rata to the Landlord of the number of entitlements that the Entitlements bear to the number of Eligible Hectares in respect of which the Tenancy is terminated compared with the number of Eligible Hectares comprised in the Holding and the compensation payable shall also be calculated pro rata on the same basis

2.14 during the final year of the Tenancy to obtain the Landlord's consent in respect of all matters concerning the management of the Entitlements to include cross compliance claiming the single payment and effecting the transfer of the Entitlements as the Landlord may direct

2.15 in so far as they remain to be observed and performed and are capable of taking effect after the termination of the Term howsoever determined the provisions of this Schedule shall remain in full force and effect notwithstanding the termination of the Tenancy

2.16 the Landlord covenants with the Tenant to comply with the terms of the Single Payment Scheme after the termination of the tenancy throughout the then subsisting calendar year

SIGNED as a Deed by
in the presence of:-

SIGNED as a Deed by
in the presence of:-

PRECEDENT 1.7 SURRENDER AGREEMENT

[*To effect a surrender where a tenancy is not coming to an end immediately it is necessary for the parties to enter into a surrender agreement which sets out the terms agreed between the tenant and his landlord with the surrender agreement being accompanied by a draft of the deed of surrender which should be executed contemporaneously with the surrender agreement and then held by the landlord's solicitors, in escrow, pending the date of surrender when the tenant vacates the property, the deed can be dated and the matter completed. There is a school of thought that these agreements could be unenforceable but this is not a view shared by the authors nor, indeed, by most commentators.*

In the agreement which follows there is provision for the transfer of milk quota to one or more third parties and also reference to the possibility of entitlements being transferred from the tenant to the landlord on the date of surrender. This particular precedent, and the deed of surrender which is included at Precedent 1.8, suitably amended and adapted, can be used in most cases of a surrender by a tenant to his landlord.]

THIS AGREEMENT made the [ ] day of [ ] 20[ ]

BETWEEN:-

(1) [*name*] of [*address*] ('The Tenant')

and

(2) [*name*] of [*address*] ('the Landlord')

WHEREAS:-

1. By an agreement for tenancy ('the Tenancy Agreement') commencing on or about the [*date*] and made between the Landlord of the one part and the Tenant of the other part ALL THAT agricultural holding known as [*name of holding*] in the County of [*county*] comprising some . . . acres or thereabouts ('the Holding') was let to the Tenant on a tenancy from year to year

2. The Tenant is the Registered Producer [in the name of [*name*]] for Rural Payment Agency ('RPA') purposes in respect of . . . litres of wholesale milk quota ('the milk quota') and it has been agreed that the Tenant shall be permitted to transfer the milk quota to one or more third parties retaining the net proceeds of sale free from any claim by or on behalf of the Landlord in respect of same and further that the Landlord shall assist the Tenant to effect the said transfer or transfers by signing all necessary documentation and taking all necessary steps required by the Tenant to ensure the transfer of the milk quota as the Tenant shall direct

[3. *If there are also entitlements to be transferred from Tenant to Landlord these will need to be referred to here with appropriate definitions, and details of entitlements to be included. Reference should be made to the definitions and suggested clauses in Precedent 1.1.*]

4. It is further agreed between the parties as hereinafter appearing

NOW IT IS HEREBY AGREED as follows:-

1. In consideration of the sum of £ . . . and of the agreements on the part of the Landlord hereinafter contained and set out the Tenant with full title guarantee shall on the [*date*] ('the Surrender Date') surrender the Holding to the Landlord on the terms of a Deed of Surrender ('the Deed of Surrender')[23] in the form annexed hereto to the intent that the Tenant's interest in the same shall become merged and extinguished in the freehold reversion

2. Forthwith on the exchange of this agreement the Tenant shall execute the Deed of Surrender

2.1 As soon as practicable after the date hereof and in any event before the Surrender Date the Landlord will execute the Deed of Surrender (or a duplicate thereof)

---

[23] Do ensure that a copy of the Deed of Surrender is annexed to this Surrender Agreement and forms part of it. That particular deed, of course, does not need to be signed when the agreement itself is signed.

2.2 Until the Surrender Date the Deed of Surrender shall be held by the Landlord's solicitor/agent subject to the terms of this agreement

3. The rent payable by the Tenant to the Landlord during the period from the date hereof to the Surrender Date shall be the sum of £[ ] [per acre per annum] such rental to be paid [half yearly] in advance on the [*date*] [and [*date*] in each year] [the first] payment to be made on the date hereof in respect of the period from the date hereof to the [payment date next following]

4. The Landlord shall pay to the Tenant by way of tenant right the agreed sum of pounds (£[ ])

5. The Tenant shall pay to the Landlord in respect of dilapidations the agreed sum of pounds (£[ ])

6. The Tenant shall transfer to one or more third parties as the Tenant shall direct the milk quota the Tenant retaining the net proceeds of sale thereof free from any claim by or on behalf of the Landlord AND FURTHER the Landlord shall assist the Tenant to effect the transfer or transfers by signing all necessary documentation and at the Tenant's expense doing all necessary things to ensure the transfer of the milk quota upon being requested so to do by the Tenant and shall consent to the grant of sub tenancies provided that each of them expires on or before the Surrender Date

7. The Landlord shall not make any further representations concerning the milk quota pursuant to the Dairy Produce Quotas Regulations 2005, SI 2005/465 ('the Regulations') or otherwise nor shall he seek to register any interest in the milk quota nor make any claim with regard thereto

8. If notwithstanding the terms of this Agreement, any claims are made outside the provisions of this Agreement in respect of the milk quota consequent upon the termination of the Tenancy Agreement or otherwise resulting in an apportionment of the same in favour of the Landlord pursuant to a decision of an Arbitrator appointed under the Regulations or otherwise the Landlord will pay to the Tenant or other party or parties all lawfully recoverable losses reasonably and properly incurred by the Tenant or the other party or parties in the acquisition of milk quota of comparable butterfat base and usage to replace the milk quota together with (but without prejudice to the generality hereof) all costs incurred in such acquisition

9. The Tenant agrees that he is and will remain liable on the covenants contained in the Tenancy Agreement until the Surrender Date

10. Any provision of this agreement to which full force and effect has not been given by the Surrender Date shall continue in full force and effect after the Surrender Date

11. The provisions of this agreement shall so far as is legally possible bind the parties their successors in title assigns personal representatives and any insolvency practitioner appointed in respect of the affairs of any of them

AS WITNESS the hands of the parties hereto the day and year first before written

SIGNED by
in the presence of:-

SIGNED by
in the presence of:-

PRECEDENT 1.8 DEED OF SURRENDER

[*This deed is to be used in conjunction with the Surrender Agreement at Precedent 1.7.*]

THIS DEED OF SURRENDER made the [ ] day of [ ] 20[ ]

BETWEEN:-

(1) [*name*] of [*address*] ('the Tenant')

and

(2) [*name*] of [*address*] ('the Landlord')

WHEREAS:-

1. The Tenant is the tenant of certain premises situate and known as ('the Holding') pursuant to an agreement for tenancy ('the Tenancy Agreement') made on the [*date*] between the Tenant (1) and the Landlord (2)

2. The reversion immediately expectant upon the determination of the tenancy under the Tenancy Agreement ('the Tenancy') is vested in the Landlord

3. The Tenancy is vested in the Tenant

4. Under and by virtue of the terms of an Agreement ('the Agreement') dated the [*date*] and made between the Tenant (1) and the Landlord (2) the Tenant has agreed to the surrender of the Holding as therein referred to on the terms therein appearing

NOW THIS DEED WITNESSES as follows:-

1. In consideration of the agreements on the part of the Landlord contained in the Agreement the Tenant with full title guarantee hereby surrenders ALL THAT the Holding as the same is edged red on the plan annexed hereto to the Landlord with vacant possession to the intent that the Tenant's interest in the same shall forthwith become merged and extinguished in the freehold

2. Subject to the terms of the Agreement the parties hereby release each other from all obligations claims and liabilities arising out of or in consequence of the Tenancy Agreement

3. This Surrender shall so far as legally possible bind the parties their successors in title and assigns and any insolvency practitioner appointed in respect of the affairs of either of them

IN WITNESS whereof the parties have set their hands to this Deed the day and year first before written

SIGNED as a DEED by
in the presence of:-

SIGNED as a DEED by
in the presence of:-

PRECEDENT 1.9 AGREEMENT TO SURRENDER WHERE PART OF HOLDING HAS BEEN SUB-LET

THIS AGREEMENT is made the [ ] day of [ ] 20[ ]

BETWEEN:-

(1) [*name*] of [*address*] ('the Sub-Tenant')

(2) [*name*] of [*address*] ('the Tenant')

and

(3) [*name*] of [*address*] ('the Landlord')

WHEREAS:-

1. By an agreement for tenancy ('the Tenancy Agreement') made the [*date*] between (1) [ ] and (2) The Tenant ALL THAT land being part of [ ] Farm in the County of [ ] ('the Holding') as the same is delineated and edged red on the plan annexed hereto ('the Plan') was let to the Tenant on a tenancy from year to year ('the Tenancy')

2. On or about the [*date*] [and at the request of the Tenant] [ ] agreed that [ ] ('the Sub-Tenant') could occupy ALL THAT part of the Holding ('the Property') as the same is delineated and edged red and cross hatched green on the Plan

3. The Landlord is [now] the landlord of the Holding [in succession to]

4. The tenancy in respect of the Property is now vested in the Sub-Tenant

5. The tenancy in respect of the remainder of the Holding other than the Property ('the Remainder of the Holding') remains vested in the Tenant

[6. *If there are any entitlements to be transferred these will need to be referred to here with appropriate definitions, and details of entitlements to be included. Reference should also be made to the definitions and suggested clauses in Precedent 1.1.*]

7. The Sub-Tenant and the Tenant have agreed to surrender their respective estates, rights and interests in the Property and the Remainder of the Holding and the Landlord has agreed to accept a surrender of the same on the terms hereinafter appearing

NOW IT IS HEREBY AGREED as follows:-

1. In consideration of the agreements on the part of the Landlord hereinafter contained and set out the Sub-Tenant and the Tenant shall surrender to the Landlord with full title guarantee their respective estates rights and interests in the Property and in the Remainder of the Holding respectively on the [*date*] ('the Surrender Date') on the terms of a deed of surrender ('the Deed of Surrender') in the form annexed hereto and made between the Sub-Tenant (1) the Tenant (2) and the Landlord (3) to the intent that the estates rights and interests of the Sub-Tenant and the Tenant shall become merged and extinguished in the freehold reversion

2. Forthwith on the exchange of this Agreement the Sub-Tenant and the Tenant shall execute the Deed of Surrender

3. As soon as practicable after the date hereof and in any event before the Surrender Date the Landlord shall execute the Deed of Surrender (or a duplicate copy of it)

4. Until the Surrender Date the Deed of Surrender shall be held by the Landlord's solicitor/agent subject to the terms of this Agreement

5. In consideration of the respective agreements by the Sub-Tenant and the Tenant to surrender the Landlord will make no claims for dilapidations in respect of the Property or the Remainder of the Holding except in respect of damage (if any) caused after the date hereof wilfully or negligently by the Sub-Tenant or the Tenant or any agent employee contractor tenant or licensee of the Sub-Tenant or the Tenant

6. In consideration of the Landlord's agreements herein contained the Sub-Tenant and the Tenant respectively agree with the Landlord:-

6.1 that they will surrender the Property and the Remainder of the Holding with full vacant possession on the Surrender Date

6.2 that they will not make any claims against the Landlord for any tenant-right matters whatsoever or for improvements or otherwise under the Tenancy Agreement or the Agricultural Holdings Act 1986 or otherwise howsoever

6.3 that they will permit the Landlord or his agents at any reasonable time or times hereafter upon giving 24 hours' notice to enter upon the

Property or the Remainder of the Holding (as the case may be) to view the state of repair of the same and to ensure that no damage has been caused to it

7. The words and phrases used in this Agreement shall have the same meanings as those attributed to them in the Tenancy Agreement

8. Any provision of this Agreement to which full force and effect has not been given by the Surrender Date shall continue in full force and effect after the same

9. The provisions of this Agreement shall so far as legally possible bind the parties their successors in title assigns personal representatives and any insolvency practitioner appointed in respect of the affairs of him or any of them

10. Any dispute arising out of this Agreement shall be determined by a single arbitrator appointed in accordance with the Arbitration Act 1996 (or any legislation amending or substituted for the same)

AS WITNESS the hands of the parties hereto the day and year first before written

SIGNED by
in the presence of:-

SIGNED by
in the presence of:-

SIGNED by
in the presence of:-

PRECEDENT 1.10 DEED OF SURRENDER

[*This Deed is to be used in conjunction with the Surrender Agreement at Precedent 1.9.*]

THIS DEED OF SURRENDER made the [ ] day of [ ] 20[ ]

BETWEEN:-

(1) [*name*] of [*address*] ('the Sub-Tenant')

(2) [*name*] of [*address*] ('the Tenant')

and

(3) [*name*] of [*address*] ('the Landlord')

WHEREAS:-

1. By an agreement for tenancy ('the Tenancy Agreement') made the [*date*] between (1) [ ] and (2) [ ] ('the Tenant') ALL THAT land being part of [ ] Farm in the County of [ ] ('the Holding') as the

same is delineated and edged red on the plan annexed hereto ('the Plan') was let to the Tenant on a lease from year to year ('the Tenancy')

2. On or about the [*date*] [and at the request of the Tenant] [ ] agreed that [ ] the Sub-Tenant could occupy ALL THAT part of the Holding ('the Property') as the same is delineated and edged red and cross hatched green on the Plan

3. The Landlord is now the landlord of the Holding [in succession to]

4. The tenancy in respect of the Property is now vested in the Sub-Tenant

5. The tenancy in respect of the remainder of the Holding other than the Property ('the Remainder of the Holding') remains vested in the Tenant

6. Under and by virtue of the terms of an Agreement ('the Agreement') dated the [*date*] and made between the Sub-Tenant (1), Tenant (2) and the Landlord (3) the Sub-Tenant and the Tenant have agreed to the surrender to the Landlord with full title guarantee of all their respective estates rights and interests in the Property and the Remainder of the Holding on the terms and conditions therein referred to

NOW THIS DEED WITNESSES as follows:-

1. In consideration of the agreements on the part of the Landlord contained in the Agreement the Sub-Tenant and the Tenant with full title guarantee hereby surrender all their respective estates rights and interests in the Property and the Remainder of the Holding to the Landlord with vacant possession to the intent that the Sub-Tenant's and the Tenant's respective interests in the same shall forthwith become merged and extinguished in the freehold

2. Subject to the terms of the Agreement the parties hereby release each other from all obligations claims and liabilities arising out of or in consequence of the Tenancy Agreement

3. This Surrender shall so far as legally possible bind the parties their successors in title and assigns and any insolvency practitioner appointed in respect of the affairs of either of them

IN WITNESS whereof the parties hereto have set their hands to this Deed the day and year first before written

SIGNED as a Deed by
in the presence of :-

SIGNED as a Deed by
in the presence of:-

SIGNED as a Deed by
in the presence of:-

PRECEDENT 2 NOTICE PURSUANT TO AHA 1986, SECTION 6 BEING A REQUEST TO ENTER INTO A WRITTEN TENANCY AGREEMENT

AGRICULTURAL HOLDINGS ACT 1986
SECTION 6, SCHEDULE 1

TO: [*addressee's name and address*]

HOLDING: [*details of Holding and number*]

In accordance with section 6(1) of the Agricultural Holdings Act 1986 I/we [*Landlord's/Tenant's/Agents' name and address (as the case may be)*] [*on behalf of* [*name*] *(as the case may be)*] request you to enter into a written agreement embodying the terms of the tenancy under which I/you [*delete as applicable*] hold the Holding and containing provision for all the matters included in Schedule 1 to the said Act

Dated ....

Signed ....

Duly Authorised Signatory

on behalf of the Landlord/Landlord

Or

Signed ....

Duly Authorised Signatory

on behalf of the Tenant/Tenant

PRECEDENT 3 NOTICE UNDER LTA 1987, SECTION 48

LANDLORD AND TENANT ACT 1987 SECTION 48

To: [*name*]

Re: [*Holding address and number*]

NOTICE IS GIVEN in accordance with section 48 of the Landlord and Tenant Act 1987 that your Landlord's name and address for service of proceedings and notices in respect of your tenancy agreement [*a copy of which is annexed hereto*] is:-

....

....

....

....

....

All notices and proceedings may be served personally on the Landlord at the above address or care of his agents

Signed ....

Dated ....

PRECEDENT 4 PERMISSION TO ERECT FIXTURE UNDER AHA 1986, SECTION 10

AGRICULTURAL HOLDINGS ACT 1986
SECTION 10

Re: the Holding known as [*Holding address, details and number*]

To: [*Tenant's name and address*]

Pursuant to section 10 of the Agricultural Holdings Act 1986 I HEREBY CONSENT to the erection on the above holding of the fixture details of which are set out in the Schedule hereto upon the terms and conditions of the said Schedule

SCHEDULE

1. The Fixture(s):- [*description of fixture*]

2. Terms and conditions:-

2.1 The Fixture shall be considered a Tenant's Fixture under the provisions of the said section 10

2.2 No compensation shall be payable to the Tenant under sections 64 and 66 of the said Act at the termination of the Tenancy

2.3 The Fixture is to be sited in a position approved by me or my representative

2.4 The Tenant must at his own expense keep the Fixture in good repair and (where appropriate) decoration at all times

2.5 The Tenant will not cause any undue damage or disturbance in the erection or removal of the Fixture and must make good at his own expense any damage caused

2.6 This Consent is conditional upon the Tenant obtaining all necessary Planning and By-Law approvals in respect of the Fixture and the Tenant will comply at his own expense with any conditions attached to such approvals and will not commence work on the erection of the Fixture until all such approvals have been obtained and conditions precedent attached thereto complied with

Dated ....

Signed Signed

....

Landlord/Duly Authorised Signatory on behalf of [*Landlord's name and address*]	Tenant

PRECEDENT 5 DEMAND FOR ARBITRATION OF RENT UNDER AHA 1986, SECTION 12

AGRICULTURAL HOLDINGS ACT 1986
SECTION 12

TO: [*addressee's name and address*]

HOLDING: [*details of Holding and number*]

TAKE NOTICE THAT I DEMAND in accordance with section 12 of the Agricultural Holdings Act 1986 that the question of the rent to be paid in respect of the Holding of which you are the tenant/landlord [*delete as applicable*] as from the [*date*] being the next termination date within the meaning of sub-section (4) of the said section be referred to arbitration under the Act

THIS NOTICE is given without prejudice to any other notice or act in connection with the tenancy which has been or may after the date of this notice be given or done by me or on behalf of me or any other involved party

Dated: ....

Signed: ....

Landlord

Or

Signed: ....

Tenant

PRECEDENT 6 MEMORANDUM OF RENT REVIEW

AGRICULTURAL HOLDINGS ACT 1986

Landlord: [*Landlord's name and address*]

Tenant: [*Tenant's name and address*]

Holding: [*details of Holding and number*]

Tenancy Agreement: [*date of and parties to Tenancy Agreement*]

MEMORANDUM

IT IS MUTUALLY AGREED that the rent payable in respect of the Holding shall with effect from the [*date*] be increased to [*sum*] Pounds (£ . . .) per annum

Save in respect of the rent increase, the subject of this Memorandum, the Tenancy Agreement shall remain in full force and effect

Dated: ....

Signed: ....

Landlord

Signed: ....

Tenant

PRECEDENT 7 MEMORANDUM OF RENT INCREASE AFTER LANDLORD'S IMPROVEMENT

AGRICULTURAL HOLDINGS ACT 1986

Landlord:	[*Landlord's name and address*]
Tenant:	[*Tenant's name and address*]
Holding:	[*details of Holding and number*]
Tenancy Agreement:	[*date of and parties to Tenancy Agreement*]

MEMORANDUM

In consideration of the Landlord having carried out in agreement with the Tenant on the Holding the Improvement specified in the Schedule below it has been agreed between the parties that (pursuant to section 13 of the Agricultural Holdings Act 1986) the rent payable in respect of the Holding shall with effect from the [*date*] be increased to [*sum*] Pounds (£[ ]) per annum

Save in respect of the rent increase, the subject of this Memorandum, the Tenancy Agreement shall remain in full force and effect and for the avoidance of any doubt it is agreed that the proviso for re-entry herein contained shall apply in respect of any non-payment of the rental herein referred to or any part thereof as it does to the rental referred to in the Tenancy Agreement

SCHEDULE

Dated: ....

Signed: ....

Landlord

Signed: ....

Tenant

PRECEDENT 8 LANDLORD'S CONSENT TO TENANT'S IMPROVEMENT

AGRICULTURAL HOLDINGS ACT 1986
SECTION 67

Re: the Holding known as [*Holding address, details and number*]

To: [*Tenant's name and address*]

Pursuant to section 67 of the Agricultural Holdings Act 1986 I HEREBY CONSENT to the making on the above Holding of the Improvement details of which are set out in the Schedule hereto upon the terms and conditions of the said Schedule

SCHEDULE

1. The Improvement: [*description of improvement*]

2. Terms and conditions:-

2.1 The Improvement is to be created at the Tenant's own risk and expense

2.2 At the termination of the Tenancy the Improvement shall become the sole property of the Landlord and compensation of One Pound (£1.00) shall be payable to the Tenant in respect thereof

2.3 The Tenant must at his own expense keep the Improvement repaired and maintained at all times

2.4 The Tenant will not cause any undue damage or disturbance in the creation of the Improvement and must make good at his own expense any damage caused

2.5 This Consent is conditional upon the Tenant obtaining all necessary Planning and By-Law approvals in respect of the Improvement and the Tenant will comply at his own expense with any conditions attached to such approvals and will not commence work on the creation of the Improvement until all such approvals have been obtained and conditions precedent attached thereto complied with

Dated: ....

Signed: .... Signed: ....

Landlord Tenant

PRECEDENT 9 NOTICE OF INTENTION TO CLAIM COMPENSATION UNDER AHA 1986, SECTION 72

AGRICULTURAL HOLDINGS ACT 1986

SECTION 72

To: [*Tenant's name and address*]

Re: [*Holding address, details and number*]

Date:

We, [*name*] [*address*] as agent for [*Landlord's name and address*], give you notice of our intention on your quitting the above Holding on the termination of your tenancy to claim compensation under section 72 of the Agricultural Holdings Act 1986 for the general deterioration such as is mentioned in that section of that Act

....

Duly Authorised Signatory

on behalf of the Landlord

PRECEDENT 10 NOTICE OF INTENTION TO CLAIM COMPENSATION ON TERMINATION OF TENANCY PURSUANT TO AHA 1986, SECTION 83

AGRICULTURAL HOLDINGS ACT 1986
SECTION 83

Re: the Holding known as [*Holding address details*]

To: [*Landlord or Tenant's name and address*]

Pursuant to s 83 of the Agricultural Holdings Act 1986 I HEREBY GIVE YOU NOTICE that it is my intention to make against you the claims set out in the Schedule hereto arising out of the termination of the tenancy of the above Holding

SCHEDULE

[*Nature of claim*]	[*Statutory provision or term of agreement under which claim is made*]

Dated: ....

Signed: ....

Landlord/Tenant

PRECEDENT 11 NOTICE OF INTENTION TO RE-ENTER OR FORFEIT

NOTICE OF INTENTION TO RE-ENTER OR FORFEIT

To:	[*Tenant's name and address*]
Holding:	[*details of Holding and number*]
Landlord:	[*Landlord's name and address*]
Tenancy Agreement:	An agreement dated [*date*] and made between [*name*] (1) and [*name*] (2)

WHEREAS on [*date*] you were/[*name*] was [*delete as applicable*] adjudged bankrupt

AND WHEREAS under the provisions of clause [*forfeiture clause*] of the Tenancy Agreement it was provided that in such event it would be lawful for the Landlord after giving not less than [*figure*] weeks/months [*delete as applicable*] notice to re-enter the Holding and thereupon the tenancy thereof should automatically determine

NOW IN PURSUANCE of the provisions of the said clause [*forfeiture clause*] of the Tenancy Agreement I GIVE YOU NOTICE that by reason of your/[*name*]'s [*delete as applicable*] bankruptcy I intend to enforce the right of re-entry or forfeiture that has thereby arisen on [*date*] (being a date not less than [*as before*] weeks/months [*delete as applicable*] from the date of this Notice) and will do so by action or otherwise as I see fit AND THEREUPON your tenancy of the Holding shall absolutely determine (subject as may be set out in the said clause [*forfeiture clause*] of the Tenancy Agreement)

Dated: ....

Signed: ....

Landlord

PRECEDENT 12 NOTICE TO QUIT TO PERSONAL REPRESENTATIVES (THE CASE G NOTICE)

AGRICULTURAL HOLDINGS ACT 1986
SECTION 26 AND SCHEDULE 3, PART I, CASE G

To: The Personal Representatives of the late [*deceased Tenant's name*]

[To: The Public Trustee, The Public Trustee Office, PO Box 3010, London WC2A 1AX]

Holding: [*details of Holding*]

I GIVE YOU NOTICE to quit and deliver up the Holding which was held by the late [*deceased tenant's name*] as tenant until his/her [*delete as applicable*] death on or about the [*date*] (and which you now hold as his/her [*delete as applicable]* Personal Representatives) on [*date*] or at the expiration of the year of the tenancy which shall expire next after the end of twelve months from the service of this notice

PLEASE TAKE NOTICE that this Notice to Quit is given in pursuance of the provisions of section 26 and Schedule 3, Part 1, Case G of the Agricultural Holdings Act 1986 by reason of the death of [*deceased tenant's name*] who immediately before his/her [*delete as applicable]* death was the sole (or sole surviving) tenant under the contract of tenancy of the Holding

Dated ....

Signed ....

[Landlord]

[Duly authorised agent for and on behalf of the Landlord]

[whose address for service is ....]

PRECEDENT 13 NOTICE TO TENANT TO PAY RENT DUE IN THE PRESCRIBED FORM UNDER SI 1987/711

AGRICULTURAL HOLDINGS ACT 1986
SCHEDULE 3, PART I, CASE D
NOTICE TO TENANT TO PAY RENT DUE

Re: the holding known as [*Holding address, details and number*]

To: [*Tenant's name and address*]

IMPORTANT – FAILURE TO COMPLY WITH THIS NOTICE MAY BE RELIED ON AS REASON FOR A NOTICE TO QUIT UNDER CASE D. IF YOU WANT YOUR TENANCY TO CONTINUE YOU MUST ACT QUICKLY READ THE NOTICE AND ALL THE NOTES CAREFULLY. IF YOU ARE IN ANY DOUBT ABOUT THE ACTION YOU SHOULD TAKE, GET ADVICE IMMEDIATELY – EG FROM A SOLICITOR, SURVEYOR, OR CITIZENS ADVICE BUREAU

1. I hereby give you notice that I require you to pay within two months from the date of service of this Notice the rent due in respect of the above holding as set out below

Particulars of Rent Not Paid:-

[*Date When Due*] [*Amount Due*]

2. This Notice is given in accordance with Case D in Part I of Schedule 3 to the Agricultural Holdings Act 1986, and failure to comply with it within the period specified above may be relied on as a reason for a notice to quit under Case D

3. Your attention is drawn to the Notes following the signature to this Notice

Signed .... Date ....

(If signed by any person other than the landlord of the Holding, state in what capacity or by what authority the signature is affixed [, and if signed as duly authorised agent ensure the written authority of the landlord is annexed to the Notice. Also ensure the Notes which follow are attached to this Notice].)

Address ....
....
....
....

NOTES

1. You cannot at this stage refer to arbitration either your liability to comply with this Notice to pay rent or any other question as to the validity of the Notice. You will, however, be entitled to do so later if a

notice to quit is served on you on the ground that you have failed to comply with this Notice to pay rent. That is the only opportunity you will have to challenge this Notice

2. At that stage under article 9 of the Agriculture Holdings (Arbitration on Notices) Order 1987 (SI 1987/710) you have one month after the service of the notice to quit within which you can serve on your landlord a notice in writing requiring the question to be determined by arbitration under the Agricultural Holdings Act 1986 (c 5)

3. You will then have three months from the date of service of that notice in which to appoint an arbitrator by agreement or (in default of such an agreement) to make an application under paragraph 1 of Schedule 11 to that Act for the appointment of an arbitrator. If this is not done by you or your landlord your notice requiring arbitration ceases to be effective (see article 10 of that Order)

PRECEDENT 14 NOTICE TO TENANT TO REMEDY BREACH OF TENANCY BY DOING WORK OF REPAIR MAINTENANCE OR REPLACEMENT

AGRICULTURAL HOLDINGS ACT 1986
SCHEDULE 3, PART I, CASE D
NOTICE TO TENANT TO REMEDY BREACH OF TENANCY BY DOING WORK OF REPAIR, MAINTENANCE OR REPLACEMENT

Re: the Holding known as [*Holding details/address*]

To: [*name and address of Tenant*]

IMPORTANT – FAILURE TO COMPLY WITH THIS NOTICE MAY BE RELIED ON AS REASON FOR A NOTICE TO QUIT UNDER CASE D. IF YOU WANT YOUR TENANCY TO CONTINUE YOU MUST ACT QUICKLY. READ THE NOTICE AND ALL THE NOTES CAREFULLY. IF YOU ARE IN ANY DOUBT ABOUT THE ACTION YOU SHOULD TAKE, GET ADVICE IMMEDIATELY – EG FROM A SOLICITOR, SURVEYOR OR CITIZENS ADVICE BUREAU

1. I hereby give you notice that I require you to remedy within [ ] months from the date of service of this Notice the breaches, set out below, of the terms or conditions of your tenancy, being breaches which are capable of being remedied of terms or conditions which are not inconsistent with your responsibilities to farm the Holding in accordance with the rules of good husbandry

2. This Notice requires the doing of the work of repair, maintenance or replacement specified below

Particulars of breaches of terms or conditions of tenancy:-

[*Term or condition of tenancy*]	[*Particulars of breach and work required to remedy it*]

3. This Notice is given in accordance with Case D in Part I of Schedule 3 to the Agricultural Holdings Act 1986, and failure to comply with it within the period specified above may be relied on as a reason for a notice to quit under Case D

4. Your attention is drawn to the Notes following the signature to this Notice

Signed .... Date ....

(If signed by any person other than the landlord of the Holding, state in what capacity or by what authority the signature is affixed [, and if signed as duly authorised agent ensure the written authority of the landlord is annexed to the Notice. Also ensure the Notes which follow are attached to this Notice].)

Address ....

....

....

....

NOTES

In these Notes 'the Order' means the Agricultural Holdings (Arbitration on Notices) Order 1987 (SI 1987/710)

What to do if you wish:-

(a) to contest your liability to do the work, or any part of the work, required by this Notice to remedy (Question (a)) or

(b) to request the deletion from this Notice to remedy of any item or part of an item of work on the ground that it is unnecessary or unjustified (Question (b)) or

(c) to request the substitution in the case of any item or part of an item of work of a different method or material for the method or material which this Notice to remedy would otherwise require to be followed or used (Question (c))

1. Questions (a), (b) and (c) mentioned in the heading to these Notes can be referred to arbitration under article 3(1) of the Order. To do so you must serve a notice in writing upon your landlord within one month of the service upon you of this Notice to remedy. The notice you serve upon your landlord should specify:-

(a) if you are referring Question (a), the items for which you deny liability

(b) if you are referring Question (b), the items you wish to be deleted

(c) if you are referring Question (c), the different methods or materials you wish to be substituted and in each case should require the matter to be determined by arbitration under the Agricultural Holdings Act 1986 (c5). You will not be able to refer Question (a), (b), or (c) to arbitration later, on receipt of a notice to quit. This action does not prevent you settling the matter in writing by agreement with your landlord

Carrying out the work

2. If you refer any of these Questions (a), (b) and (c) to arbitration, you are not obliged to carry out the work which is the subject of the reference to arbitration unless and until the arbitrator decides that you are liable to do it; but you must carry out any work which you are not referring to arbitration

3. If you are referring Question (a) to arbitration you may wish to carry out any of the work which is the subject of that reference to arbitration without waiting for the arbitrator's award. If you do this and the arbitrator finds that you have carried out any such work which you were under no obligation to do, he will determine at the time he makes his award the reasonable cost of any such work which you have done and you will be entitled to recover this from your landlord (see article 8 of the Order). This provision does not apply in the case of work referred to arbitration under Question (b) or Question (c)

What to do if you wish to contest any other question arising under this Notice to remedy

4. If you wish to contest any other question arising under this Notice other than Question (a), (b) or (c), such as whether the time specified in the Notice to do work is a reasonable period in which to carry out the work, you should refer the question to arbitration in either of the following ways, according to whether or not you are also at the same time referring Question (a), (b) or (c) to arbitration:-

   (a) If you are referring Question (a), (b) or (c) to arbitration, then you must also refer to arbitration at the same time any other question relevant to this Notice which you may wish to dispute

   To do this, you should include in the Notice to your landlord referred to in Note 1 above a statement of the other questions which you require to be determined by arbitration under the Agricultural Holdings Act 1986 (see article 4(1) of the Order)

   (b) If you are not referring Question (a), (b) or (c) to arbitration, but wish to contest some other question arising under this Notice to remedy, you may refer that question to arbitration either now, on receipt of this Notice, or later, if you get a notice to quit

   To refer the question to arbitration now, you should serve on your landlord within one month after the service of this Notice to remedy a notice in writing setting out what it is you require to be determined by arbitration under the Agricultural Holdings Act 1986 (see article 4(2)(a) of the Order)

Alternatively, you have one month after the service of the notice to quit within which you can serve on your landlord a notice in writing requiring the question to be determined by arbitration under the 1986 Act (see article 9 of the Order). You will then have three months from the date of service of that notice in which to appoint an arbitrator by agreement or (in default of such an agreement) to make an application under paragraph 1 of Schedule 11 to that Act for the appointment of an arbitraror. If this is not done by you or your landlord your notice requiring arbitration ceases to be effective (see article 10 of that Order)

Warning

5. Notes 1 to 4 above outline the only opportunities you have to challenge this Notice to remedy

Extensions of time allowed for complying with this Notice to remedy

6. If you refer to arbitration now any question arising from this Notice to remedy, the time allowed for complying with the Notice will be extended until the termination of the arbitration. If the arbitrator decides that you are liable to do any of the work specified in this Notice to remedy, he will extend the time in which the work is to be done by such period as he thinks fit (see article 6(2) of the Order)

Warning as to the effect which any extension of the time allowed for complying with this Notice to remedy may have upon a subsequent notice to quit

7. If your time for doing the work is extended as mentioned in note 6 above, the arbitrator can specify a date for the termination of your tenancy should you fail to complete the work you are liable to do within the extended time. Then, if you did fail to complete that work within the extended time, your landlord could serve a notice to quit upon you expiring on the date which the arbitrator had specified, and the notice would be valid even though that date might be less than twelve months after the next term date, and might not expire on a term date. The arbitrator cannot, however, specify a termination date which is less than six months after the expiry of the extended time to do the work. Nor can he specify a date which is earlier than would have been possible if you had not required arbitration on this notice to remedy and had failed to do the work (see article 7 of the Order)

PRECEDENT 15 NOTICE TO TENANT TO REMEDY BREACH OF TENANCY (NOT BEING A NOTICE REQUIRING THE DOING OF ANY WORK FOR REPAIR, MAINTENANCE OR REPLACEMENT)

AGRICULTURAL HOLDINGS ACT 1986
SCHEDULE 3, PART I, CASE D
NOTICE TO TENANT TO REMEDY BREACH OF TENANCY (NOT BEING A NOTICE REQUIRING THE DOING OF ANY WORK OF REPAIR, MAINTENANCE OR REPLACEMENT

Re: the Holding known as [*Holding details/address*]

To: [*name and address of Tenant*]

IMPORTANT – FAILURE TO COMPLY WITH THIS NOTICE MAY BE RELIED ON AS REASON FOR A NOTICE TO QUIT UNDER CASE D. IF YOU WANT YOUR TENANCY TO CONTINUE YOU MUST ACT QUICKLY. READ THE NOTICE AND ALL THE NOTES CAREFULLY. IF YOU ARE IN ANY DOUBT ABOUT THE ACTION YOU SHOULD TAKE, GET ADVICE IMMEDIATELY – EG FROM A SOLICITOR, SURVEYOR OR CITIZENS ADVICE BUREAU

1. I hereby give you notice that I require you to remedy within . . . months from the date of service of this Notice the breaches, set out below, of the terms or conditions of your tenancy, being breaches which are capable of being remedied of terms or conditions which are not inconsistent with your responsibilities to farm the Holding in accordance with the rules of good husbandry

   Particulars of breaches of terms or conditions of tenancy:-

   [*Term or condition of tenancy*] [*Particulars of breach*]

2. This Notice is given in accordance with Case D in Part I, Schedule 3 to the Agricultural Holdings Act 1986 and failure to comply with it within the period specified may be relied on as a reason for a notice to quit under Case D

3. Your attention is drawn to the Notes following the signature to this Notice

Signed .... Date ....

PRECEDENT 16 SUGGESTED DESIGNATION FOR INCLUSION IN A WILL

'I hereby designate [*name*] as the person whom I wish to succeed me as Tenant of [*holding details*] comprised in a Tenancy Agreement made the [*date*] between [*name of landlord*] of the one part and [*name of tenant*] of the other part'

PRECEDENT 17 APPLICATION FOR DIRECTION GIVING ENTITLEMENT TO A TENANCY

AGRICULTURAL LAND TRIBUNAL

APPLICATION FOR DIRECTION GIVING ENTITLEMENT TO TENANCYOF AGRICULTURAL HOLDING

[*This is a form of application to be used when a tenant has died and a family member satisfying the close relationship test wishes to apply to succeed to the tenancy. What follows is the new form of application which was introduced with effect from 15 January 2008 by the Agricultural Land Tribunals (Rules) Order 2007, SI 2007/3105.*

The guidance notes attached to the application form have also been included.]

APPLICATION FOR DETERMINATION THAT APPLICANT BE TREATED AS AN ELIGIBLE PERSON

Please use BLOCK LETTERS and BLACK INK

* Strike out whichever is inapplicable

Strike out if inapplicable

To the Secretary of the Agricultural Land Tribunal for the .... area

1. I, ... (*full name*)

of .... (*address*)

hereby apply under Section 39(1) of the Agricultural Holdings Act 1986 for a direction entitling me to a tenancy of the holding specified in paragraph 4 below

2. The application arises on the death of ....

formerly the tenant of the holding referred to in paragraph 2, who died on (1) ....

His/her tenancy was:-

*(a) granted before 12 July 1984

*(b) obtained on or after 12 July 1984 by virtue of a direction of the Agricultural Land Tribunal under section 39 or 53 of the 1986 Act

*(c) granted on or after 12 July 1984 following a direction of the Agricultural Land Tribunal under section 39 of the 1986 Act but commenced before the relevant time for the purpose of section 45 of that Act

*(d) granted on or after 12 July 1984 by a written contract of tenancy indicating that the succession provisions in Part IV of the 1986 Act should apply

*(e) granted on or after 12 July 1984 to a person who, immediately before that date, was a tenant of the holding or of any agricultural holding which comprised the whole or a substantial part of the land comprised in the holding

*(f) granted on or after 19 October 2006 where there is a written contract of tenancy indicating that the succession provisions in Part IV of the 1986 Act should apply and the tenant previously held a 1986 Act tenancy of all or a substantial part of the land comprised in the holding

3. The landlord of the holding is .... [*name*]

Holding and existing tenancy

4. The holding in respect of which the application is made is known as:-

....

and consists of:-

(a) .... hectares of arable land (including temporary grass)

(Ordnance Survey Field Nos ....)

(b) .... hectares of permanent pasture

(Ordnance Survey Field Nos ....)

(c) .... hectares of rough grazing

(Ordnance Survey Field Nos ....)

(d) .... hectares of other land (including orchards)

(Ordnance Survey Field Nos ....)

TOTAL .... Hectares .... ANNUAL RENT: £ . . .

5. The current year of the tenancy of the holding expires on ....

6. The holding includes the following buildings (*give a general description*):-

....

#7. The holding forms part of a larger agricultural unit known as and consisting of (*give a general description*):-

....

8. I am/am not* aware that the landlord(s) has/have on ...... served a notice to quit within Case G in respect of the holding

Relationship, etc

9. I was born on ....

#10. I am the sole person validly designated by the deceased tenant of the holding in his Will as the person he wished to succeed him as tenant of the holding. A copy of the relevant part of the Will in which I am designated is attached (*attach a copy of or extract from the Will marking the relevant passage*) [(2)]

11.

#(a) I am the *wife/*husband/*civil partner/*brother/*sister/*child of the deceased tenant [(3)]

#(b) I was treated by the deceased tenant as a child of the family in relation to his/her marriage/civil partnership to .... [(4)]

on .... (*give date of marriage/civil partnership*)

Livelihood

12.

(a) *The 'livelihood test' in section 36(3)(a) (ie that in the last seven years ending with the date of the death of the deceased tenant my only or principal source of livelihood, during a period or periods together amounting to not less than five years, derived from my agricultural work (detailed below) on the holding or an agricultural unit of which the holding forms a part) is satisfied [(5)]

*(b) *The livelihood test is satisfied to a material extent and I apply, or also apply, under section 41 to the Tribunal to be treated as an eligible person for the purposes of sections 36 to 48 and contend that in all the circumstances it would be fair and reasonable for me to be able to apply under section 39 for a direction entitling me to a tenancy of the holding [(6)]

Details of the circumstances:-

....

....

(c) In relation to the livelihood test I rely on the following facts:-

Details of the agricultural unit of which the holding forms part:-

....

Details and period(s) of my agricultural work:-

....

Details and period(s) of full-time course(s) attended and the qualification(s) gained:- [(7)]

....

#(d) I rely or also rely on work (detailed below) of a description approved in writing by the landlord on the ...... (being a date on or after the 19 October 2006) comprising agricultural work carried out by me from, or other work carried out by me on or from,

*(i) the holding, or

*(ii) (where the livelihood test is satisfied) the said agricultural unit

Details and period(s) of the work (of approved description) and whether carried out on or from the holding or on or from the said agricultural unit:-

....

#(e) I rely or also rely on the work (detailed below) of the deceased tenant as my husband/civil partner

Details and period(s) of (agricultural or other) work done by husband/civil partner and whether on or from the holding or on or from the said agricultural unit:-

....

13.

*(a) During the period(s) specified above I had other source(s) of livelihood (detailed below)

Details of other sources of livelihood:-

....

*(b) During the period(s) specified above I had no other source(s) of livelihood

Joint land

#14.

(a) The following agricultural land is occupied by me, my spouse/civil partner or a company under my control, the control of my spouse/civil partner or our joint control as owner-occupier/ tenant/ licensee whether alone or jointly with others (*give particulars of any land occupied, including area* of *any land occupied joint with others*) (8):-

....

#(b) The following agricultural land is occupied by a person under a licence or such a tenancy as is mentioned in paragraph 6(1)(a) to (d) of Part I of Schedule 6 to the 1986 Act (8) granted by *me and/or *my spouse/civil partner and/or *a company controlled by me and/or my spouse/civil partner together with one or more other persons, being at the time it was granted a person or persons entitled to occupy the land otherwise than under a tenancy, or in a capacity, falling within paragraph 6(1)(a) to (f) of that part of that Schedule [9] [*give particulars of any land occupied, including area*]:-

....

#(c) I apply under paragraph 7(2) of Part I of Schedule 6 to the 1986 Act for the net annual income from the following agricultural land which is:-

*(i) jointly occupied

*(ii) deemed by virtue of paragraph 10(2) of Part I of that Schedule to be jointly occupied

by me and one or more other persons (not being only my spouse/civil partner or a company under my control, the control of my spouse/civil partner or our joint control) to be treated as limited to my appropriate share of that net annual income [*give particulars of any land in joint occupation or deemed joint occupation*] [10]:-

....

Suitability

15. I claim to be a suitable person to receive the tenancy of the holding because [11]:-

....

....

Enclosures

16. I attach the following documents which I intend to produce in support of my case:-

(a) ...... [12] copies of a map of the holding described in paragraph 4 above (and of the other land referred to in paragraph 7 and of any other land referred to in this application) at a scale of 1/10,000 (or larger) with Ordnance Survey field numbers marked on the map(s) and

(b) ...... [13] copies of:-

....

Parties

17. I confirm that the following person(s) is/are to be treated as a respondent(s) (as a person(s) against whom this application is made):-

(a) the landlord(s) named above

(b) any person or persons who make an application under section 39 in respect of the holding

(c) .... [14]

Notifications

18. I confirm that I have notified [(15)] the following respondent(s) of my application:-

(a) the landlord(s) of the holding whose name(s) and address(es) is/are:-

Name ....
Address ....

Name ....
Address ....

(b) the following other person(s) who to my knowledge has/have made or is/are eligible to make an application under section 39 of the Act to succeed to the tenancy of the holding in question [(16)]

Name ....
Address ....
Nature of interest ....

Name ....
Address ....
Nature of interest ....

19. I confirm that I have also notified [(15)] the following person(s) of my application:-

(a) The following who is/are the personal representative(s) of the deceased tenant,

or (*if there are no personal representatives*) the person(s) most eligible or likely to apply to be the personal representative of the deceased tenant or the person(s) administering his/her estate

Name ....
Address ....
Nature of interest ....

Name ....
Address ....
Nature of interest ....

(b) The following other person(s) who is/are or maybe interested in the outcome of my application:-

Name ....
Address ....
Nature of interest ....

Name ....
Address ....
Nature of interest ....

Contact details, etc

20.

(a) Contact details of my authorised representative (if any):-

.... (*name*)

.... (*address*)

.... (*profession*)

.... (*contact telephone number*)

.... (*fax number(s)*)

.... (*e-mail address*)

(b) The Tribunal should/should not* deliver notices concerning this application to the above representative instead of to me

21. I intend to attend* in person and/or be represented at any hearing

Statement of Truth

I believe (the applicant believes) that the facts stated in this document are true

I am duly authorised by the applicant to sign this statement [17]

Full Name ....

Name of applicant's Solicitor/Agents firm, if applicable ....

Date .... Signed ....

Position or office held (if signing on behalf of firm or company) ....

* Strike out whichever is inapplicable

Strike out if inapplicable

NOTES:-

(THESE NOTES PROVIDE GUIDANCE ONLY)

(1) Formal proof of the date of death will be required unless formally accepted by the respondent(s).

(2) This paragraph should be completed only if you received a specific bequest of the deceased's tenancy under his or her Will or you are specifically named in the Will as the person whom the deceased tenant wished to succeed him or her as tenant of the holding. It will be necessary for a Grant of Probate or Administration to be obtained from the Family Division of the High Court in respect of the Will before the tribunal can hear any claim to be a designated applicant. Where an applicant establishes that he or she is so designated under the deceased tenant's Will, no other application will be considered unless the Tribunal determine that the designated applicant is not eligible and suitable to become the tenant of the holding.

(3) Formal proof of the relationship to the deceased, e g by production of birth or marriage/civil partnership certificates, will be required unless formally accepted by the respondent(s). Adopted children should complete this sub-paragraph, and not paragraph 11(b) following.

(4) Paragraph 11(b) may apply where you were the stepchild or foster child of the deceased tenant or were otherwise treated by him or her as his or her child. An outline should be given of the circumstances relied on as establishing that you were treated by the deceased as his or her child in relation to the marriage/civil partnership. Production of the relevant marriage/civil partnership certificate and any relevant birth certificate will be required at the hearing unless the relevant facts are formally accepted by the respondent(s).

(5)

(i) To qualify under paragraph 12(a) you should have derived your only or principal source of livelihood from your agricultural work on the holding (or on a larger unit of which the holding forms part) during a total of five years of the seven years ending with the death of the deceased tenant. The total of five years may be made up of one continuous period, or one or more separate periods. A period of full-time education at a university, etc, may, in the circumstances set out in note (7) below, count towards the five year period of earning a livelihood from the holding, and reference should be made to paragraph 12(c) and note (6) below in deciding whether the requirements of this paragraph can be satisfied.

(ii) If you cannot satisfy the requirements of paragraph 12(a) fully but believe you can satisfy them to a material extent you should not complete paragraph 12(a) but should complete instead paragraph 12(b) together with sub paragraph (c), (d) (if relevant), (e) (if relevant) and paragraph 12. If you are in any doubt as to whether or not you can satisfy the requirement of paragraph 12(a) fully, you are advised to complete paragraphs 12(b), etc.

(iii) If you are relying on paragraph 12(d) a copy of the landlord's written approval should be enclosed with your application. By virtue of Article 5 (2) of the Regulatory Reform (Agricultural Tenancies) (England and Wales) Order 2006, SI No 2805, agricultural work on the holding includes agricultural work carried out by the potential successor from the holding or an agricultural unit of which the holding forms part, and other work carried out by him or her on or from the holding or an agricultural unit of which the holding forms part, of a description approved in writing by the landlord on or after 19 October 2006.

(iv) Paragraph 12(e) is available only to the wife/civil partner of the deceased tenant.

(6) This paragraph should be completed (together with sub-paragraphs (c), (d) (if relevant), (e) (if relevant) and paragraph 13) in any case where you, while otherwise satisfying the definition of 'eligible person' in section 36(3) of the 1986 Act cannot fully satisfy the condition as to deriving your principal or main source of livelihood from the holding contained in paragraph (a) of that sub-section. If you fully satisfy the requirements of paragraph 12(a) you need not complete this paragraph. You should state to what extent you have derived your livelihood as described in note 5 above. A summary should be given of matters relied on as establishing that it is fair and reasonable that you should be entitled to apply under section 39 of the 1986 Act for a tenancy of the holding, though not fully satisfying the condition specified in paragraph (a) of the definition of 'eligible person' in section 36(3) of that Act. The length of time you have lived on the holding, details of work done by you on the holding (apart from any details already given in paragraph 12) and any special circumstances which have prevented you from qualifying in full as an eligible person under paragraph (a) of the definition of 'eligible person' should be given. Note 5 above describes the requirements needed to qualify fully under paragraph (b) of the definition.

(7) By virtue of paragraph 2 of Part 1 of Schedule 6 to the Act any period or periods (up to an aggregate total of three years) during the seven years ending with the date of death of the deceased tenant during which you were attending a full-time course at a university, college or other establishment of further education will be treated as a period throughout which your only or principle source of livelihood was derived from your agricultural work on the holding. Any subject may have been studied. You should state the name of university, college or other establishment of further education attended and the time spent in further education.

(8) Land occupied by your spouse/civil partner or by a company controlled by your spouse/civil partner or jointly by your spouse/civil partner and yourself should not be included in paragraph 14 where either of the parties has obtained a decree of judicial separation or a decree nisi of divorce or of nullity of marriage/civil partnership and in each case that decree remains unrescinded. In addition, land should not be included in paragraph 14 if it is occupied by you, your spouse/civil partner or such controlled company:-

(a) under a tenancy approved under section 2(1) of the Agricultural Holdings Act 1986 or under a tenancy falling within section 2(3)(a) of that Act

(b) under a tenancy for more than one year but less than two years

(c) under a tenancy not falling within (a) or (b) above and not having effect as a contract of tenancy

(d) under a tenancy to which s section 3 of the 1986 Act does not apply by virtue of s section 5 of that Act

(e) as a licensee or

(f) as an executor, administrator, trustee in bankruptcy or person otherwise deriving title from another person by operation of law.

However, where you occupy land in accordance with any of (a) to (e) above under a licence or tenancy granted to you by your spouse/civil partner or by a body corporate controlled by you, that land should be included in paragraph 14.

(9) Paragraph 6(1)(a) to (f) of Part I of Schedule 6 to the Agricultural Holdings Act 1986 is described in note 8 above.

(10) If you occupy land jointly with one or more other persons (not being only your spouse/civil partner or a company under your control or your spouse/civil partner or under their joint control), or if you are deemed to occupy land jointly with one or more such persons, you may in either case complete the application set out in paragraph 13(c) of the Form for the net annual income which the land is or was capable of producing to be treated as limited to your appropriate share.

(11) All matters relied on as supporting the claim to be a suitable person to become the tenant of the holding should be summarised. These should include details of your training and practical experience of agriculture, physical health, financial standing and any educational qualifications not already listed in paragraph 12(c).

(12) Two copies of your application and maps of any land referred to in the application, any supporting written statement and any accompanying material and documents must be sent to the Secretary, and if there are more than two respondents sufficient copies of the application and supporting documents for all other respondents must be supplied and the Secretary must be informed of the name and address of each additional respondent.

(13) Mention any other document which is attached to this application. You may include with your application, or in a separate application/reply to the Tribunal, as appropriate: (a) a request for disclosure of any document or additional information about an application or reply; (b) a request for an early hearing of your application or of any question arising out of your application, and the reasons for that request; (c) consent to the matter being dealt with on written representations only; (d) a request for permission to rely on the evidence of an expert witness or expert witnesses and the name(s) and address(es) of the proposed witness or witnesses; and (e) a request for a decision on any question as a preliminary issue.

(14) Name only such other person(s), if any, as are likely to wish to make representations to the tribunal against your application.

(15) A form of notice (Form S 3) for this purpose can be obtained from the Secretary to the Tribunal.

(16) If you oppose or intend to oppose any other application under section 39 of the Act, you should include with this application your reply to the other application(s) with the following information; (a) reasons why you oppose or intend to oppose that other application; (b) a statement indicating whether you dispute another applicant's claim to be a designated applicant and, if so, why; (c) a claim to be a more suitable applicant than any other; and (d) a statement that you have agreed or might agree with one or more other applicants to request the landlord's consent to a direction entitling you and the other applicant(s) to a joint tenancy of the holding. A form (Form S 4) for this purpose is available from the Secretary to the Tribunal.

(17) If signed by any person other than the applicant him/herself, he or she should state in what capacity or by what authority he or she signs and enclose with the reply a copy of the written authority signed by the applicant himself or herself authorising that person to make the application and receive legal proceedings in connection with the application on their behalf. Any supporting written statement must be verified by a similar statement and must be signed.

PRECEDENT 18 NOTIFICATION OF THE APPLICATION TO THE LANDLORD

FORM S 3

Ref No: ....

To be inserted by the Secretary

AGRICULTURAL LAND TRIBUNAL
NOTICE OF APPLICATION FOR ENTITLEMENT TO TENANCY UNDER SECTION 39 OF THE AGRICULTURAL HOLDINGS ACT 1986

Please use BLOCK LETTERS and BLACK INK

To ....

[*full name*]

of .... [*address*]

I, ...

[*full name*]

of .... [*address*]

hereby give you notice that I applied on ...... [*date of application*] under section 39 of the above named Act for a direction entitling me to a tenancy of the agricultural holding known as ...... [*address or brief description of holding*] in succession to ...... [*name of deceased tenant of the holding*] who died on ......

Date .... Signed ....

NOTES:-

(THESE NOTES PROVIDE GUIDANCE ONLY)

This notice should be delivered to the landlord and all interested persons. These include any other person eligible to apply under section 39, the executor(s) or other personal representative(s) of the deceased tenant or person(s) most eligible or likely to apply to be his/her personal representative(s) and any person administering his/her estate

A copy of the full application will in due course be sent by the Tribunal Secretary to the landlord and to any other section 39 applicants and to any other person(s) identified in the application as a respondent(s)

Any person not yet identified as a respondent but wishing to make representations should apply through the Tribunal Secretary to the Chairman for an order that he or she be joined as a respondent in the proceedings

PRECEDENT 19 REPLY TO APPLICATION FOR DIRECTION GIVING ENTITLEMENT TO TENANCY

FORM S 1R

Ref No: ....

To be inserted by the Secretary

AGRICULTURAL LAND TRIBUNAL
REPLY TO APPLICATION FOR DIRECTION GIVING ENTITLEMENT TO TENANCY OF AGRICULTURAL HOLDING
REPLY TO APPLICATION FOR DETERMINATION THAT APPLICANT BE TREATED AS AN ELIGIBLE PERSON

Please use BLOCK LETTERS and BLACK INK

* Strike out whichever is inapplicable

Strike out if inapplicable

To the Secretary of the Agricultural Land Tribunal for the .... area

I, ... [*full name*]

of .... [*address*] landlord of .... [*name or description of holding*]

having received a copy of the application bearing the above reference number reply as follows:-

1. The facts stated in paragraphs 1 to 8 inclusive of the application are, to the best of my knowledge, information and belief, correct except# that:-

2. I dispute the claim of the applicant to be an eligible person (1) on the following grounds:-

3. I dispute the claim of the applicant to be treated as an eligible person (2) on the following grounds:-

4. I have the following comments on the suitability of the applicant to become the tenant of the above holding:-

5. I attach ...... copies of the following relevant documents (3)(4):-

6. I wish to make the following additional representations (5):-

7.

(a) Contact details of my authorised representative (if any):-

.... [*name*]

.... [*address*]

.... [*profession*]

.... [*contact telephone number*]

.... [*fax number(s)*]

.... [*e-mail address*]

(b) The Tribunal should/should not* deliver notices concerning the application to the above representative instead of to me

8. I intend to attend * in person and/or be represented at any hearing

9. The following person(s) may be interested in these proceedings (continue on a separate sheet, if necessary) (6)

Name ....
Address ....
Nature of interest ....

Name ....
Address ....
Nature of interest ....

Statement of Truth

I believe (the respondent believes) that the facts stated in this document are true

I am duly authorised by the respondent to sign this statement (7)

Full Name ....

Name of respondent's Solicitor/Agents firm, if applicable ....

Date .... Signed ....

Position or office held (if signing on behalf of firm or company) ....

* Strike out whichever is inapplicable

Strike out if inapplicable

TAKE NOTICE if you do not deliver a written reply within one month after the date on which a copy of the application was delivered to the landlord(s) by the Secretary you may not take any part in the proceedings before the tribunal on the application except: (a) to apply for an extension of time for delivering a reply; (b) to apply for a direction that the applicant provide further particulars of his or her application; (c) to apply under rule 32 for a review of the Tribunal's decision for the reason that the respondent did not receive the application or statement of reasons or was not able to deliver a reply; (d) to be called as a witness; (e) to be delivered a copy of a decision or corrected decision; (f) to make and pursue any application of your own for consent to the operation of a Case G notice to quit; and (g) to give your views on the suitability of any applicant

NOTES:-

(THESE NOTES PROVIDE GUIDANCE ONLY)

(1) The paragraphs of the application which (where completed) will be relevant to the applicant's claim to be an eligible person are paragraphs 9 to 14, inclusive

(2) The paragraphs of the application which (where completed) will be relevant to the applicant's claim to be treated as an eligible person are paragraphs 12(b)–(e) and 13

(3)

(a) Two copies of the reply and of any document which you wish to submit to the Tribunal must be sent to the Secretary, and, if there are more than two parties (e.g. if the holding or part of it is sub-let) or applicants, sufficient copies of the reply and supporting documents for all other parties and applicants must be supplied and the Secretary must be informed of the name and address of each additional party

(b) If you disagree with any map or plan attached to the application, your reply should be accompanied by two copies (or sufficient copies – see above) of a 1/10,000 (or larger) map marked with the Ordnance Survey Field Numbers showing what you consider to be the true position

(4) You may include in your reply, or in a separate application to the Tribunal, as appropriate (i) a request for disclosure of any document or additional information about an application or reply; (ii) a request

for an early hearing of the application or of any question relating to the application with the reasons for that request; (iii) consent to the matter being dealt with on written representations only; (iv) an application under rule 27(2) for permission to rely on the evidence of more than two experts; and (v) a request for a decision on any question as a preliminary issue

(5) If for example you consider that, for any reason, the applicant is not legally entitled to make his or her application you should state succinctly the grounds on which you rely

(6) You should state here the name and address of every person who appears to you to be an interested party who is not already named in the originating applications and the nature of that person's interest

(7) If signed by any person other than the respondent him/herself, he or she should state in what capacity or by what authority he or she signs and enclose with the reply a copy of the written authority signed by the respondent himself or herself authorising that person to make the reply and receive legal proceedings in connection with the reply on their behalf. Any supporting written statement must be verified by a similar statement and must be signed

PRECEDENT 20 APPLICATION FOR DIRECTION GIVING ENTITLEMENT TO TENANCY (RETIREMENT)

APPLICATION FOR DIRECTION GIVING ENTITLEMENT TO TENANCY

FORM S 5

Ref No: ....

To be inserted by the Secretary

AGRICULTURAL LAND TRIBUNAL
APPLICATION FOR DIRECTION GIVING ENTITLEMENT TO TENANCY OF AGRICULTURAL HOLDING

Please use BLOCK LETTERS and BLACK INK

* Strike out whichever is inapplicable

Strike out if inapplicable

To the Secretary of the Agricultural Land Tribunal for the ...... area

1. I, ... [*full name*]

of ....[*address*] hereby apply under section 53(1) of the Agricultural Holdings Act 1986 for a direction entitling me to a tenancy of the holding specified in paragraph 2 below

2. The application arises as a result of a retirement notice dated ...... and given by the tenant(s) to the landlord on ...... [*date*]

His/her tenancy was:-

*(a) granted before 12 July 1984

*(b) obtained on or after 12 July 1984 by virtue of a direction of the Agricultural Land Tribunal under section 39 or 53 of the 1986 Act

*(c) granted on or after 12 July 1984 following a direction of the Agricultural Land Tribunal under section 39 of the 1986 Act but commenced before the relevant time for the purpose of section 45 of that Act

*(d) granted on or after 12 July 1984 by a written contract of tenancy indicating that the succession provisions in Part IV of the 1986 Act should apply

*(e) granted on or after 12 July 1984 to a person who, immediately before that date, was a tenant of the holding or of any agricultural holding which comprised the whole or a substantial part of the land comprised in the holding

*(f) granted on or after 19 October 2006 where there is a written contract of tenancy indicating that the succession provisions in Part IV of the 1986 Act should apply and the tenant previously held a 1986 Act tenancy of all or a substantial part of the land comprised in the holding

3. The landlord(s) of the holding is/are .... [*full name(s)*]

of .... [*address*]

The tenant(s) of the holding is/are .... [*full name(s)*]

of .... [*address(es)*]

Holding and existing tenancy

4. The holding in respect of which the application is made is known as:-

....

and consists of:-

(a) .... hectares of arable land (including temporary grass)

(Ordnance Survey Field Nos ....)

(b) .... hectares of permanent pasture

(Ordnance Survey Field Nos ....)

(c) .... hectares of rough grazing

(Ordnance Survey Field Nos ....)

(d) .... hectares of other land (including orchards)

(Ordnance Survey Field Nos ....)

TOTAL .... Hectares .... ANNUAL RENT: £ . . .

5. The current year of the tenancy of the holding expires on ......

6. The holding includes the following buildings [*give a general description*]:-

....

#7. The holding forms part of a larger agricultural unit known as ...... and consisting of [*give a general description*]:-

....

Relationship, etc

8. I am the successor nominated in the retirement notice

9. I was born on ....

10.

#(a) I am the *wife/*husband/*civil partner/*brother/*sister/*child of the tenant [*where there is more than one tenant, specify which*] (1)

#(b) I am treated by the tenant [*where there is more than one tenant, specify which*] as a child of the family in relation to his or her marriage/civil partnership to ...... (2) on ...... [*give date of marriage/civil partnership*]

Livelihood

11.

(a) The 'livelihood test' in section 50(2)(a) (ie that in the last seven years ending with the date on which the retirement notice was given to the landlord(s)) my only or principal source of livelihood, during a period or periods together amounting to not less than five years, derived from my agricultural work (detailed below) on the holding or an agricultural unit of which the holding forms a part) is satisfied. (3)

(b) In relation to the livelihood test I rely on the following facts:-

Details of the agricultural unit of which the holding forms part:-

....

Details and period(s) of my agricultural work:-

....

Details and period(s) of full-time course(s) attended and the qualification(s) gained:- (4)

....

#(c) I rely or also rely on work (detailed below) of a description approved in writing by the landlord on the ...... (being a date on or after the 19 October 2006) comprising agricultural work carried out by me from, or other work carried out by me on or from,

*(i) the holding, or

*(ii) the said agricultural unit

Details and period(s) of the work (of approved description) and whether done on or from the holding or done on or from the said agricultural unit:-

....

#(d) I rely or also rely on the work (detailed below) of the retiring tenant(s) as my husband/civil partner

Details and period(s) of (agricultural or other) work done by husband/civil partner and whether on or from the holding or on or from the said agricultural unit:-

....

12.

*(a) During the period(s) specified above I had other source(s) of livelihood (detailed below)

....

*(b) During the period(s) specified above I had no other source(s) of livelihood

Joint land

#13.

(a) The following agricultural land is occupied by me, my spouse/civil partner or a company under my control, the control of my spouse/civil partner or our joint control, as owner-occupier/tenant/licensee, whether alone or jointly with others [*give particulars of any land occupied, including area and any land occupied jointly with others*] (5):-

#(b) The following agricultural land is occupied by a person under a licence or such a tenancy as is mentioned in paragraph 6(1) (a) to (d) of Part I of Schedule 6 to the 1986 Act (6) granted by *me and/or *my spouse/civil partner and/or *a company controlled by me and/or my spouse/civil partner #together with one or more other persons, being at the time it was granted a person or persons entitled to occupy the land otherwise than under a tenancy, or in a capacity, falling within paragraph 6(1) (a) to (f) of that Part of that Schedule (6) [*give particulars of any land occupied including area*]:-

....

#(c) I apply under paragraph 7(2) of Part I of Schedule 6 to the 1986 Act, as modified by paragraph 15 of Part II of that Schedule for the net annual income from the following agricultural land which is:-

*(i) jointly occupied

*(ii) deemed by virtue of paragraph 10(2) of Part I of that Schedule to be jointly occupied

by me and one or more other persons (not being only my spouse/civil partner or a company under my control, the control of my spouse/civil partner or our joint control) to be treated as limited to my appropriate share of that net annual income [*give particulars of any land in joint occupation or deemed joint occupation*] (7):-

....

Suitability

14. I claim to be a suitable person to receive the tenancy of this holding because (8):-

....

Enclosures

15. I attach the following documents which I intend to produce in support of my case:-

(a) ...... (9) copies of a 1/10,000 (or larger) map of the holding described in paragraph 2 above (and of the other land referred to in paragraph 6) with Ordnance Survey field numbers marked on the map and

(b) ...... (9) copies of the retirement notice and

(c) ...... (9) copies of (10):-

....

Parties, notifications, etc

16. I have notified the landlord(s) of this application (11)

17. I confirm that the following person(s) is/are to be treated as a respondent(s) to this application:-

    (a) the landlord(s) named above and

    (b) .... (12)

18. I confirm that I have notified the following other person(s) who may be interested in the outcome of my application:-

    Name ....
    Address ....
    Nature of interest ....

    Name ....
    Address ....
    Nature of interest ....

Contact details, etc

19.

    (a) Contact details of my authorised representative (if any):-

    .... [*name*]

    .... [*address*]

    .... [*profession*]

    .... [*contact telephone number*]

    .... [*fax number(s)*]

    .... [*e-mail address*]

    (b) The Tribunal should/should not* deliver notices concerning this application to the above representative instead of to me

20. I intend to attend * in person and/or be represented at any hearing

Statement of Truth (13)

We believe that the facts stated in this document are true

Full name of nominated successor, ie the applicant ....

Date .... Signed ....

Full name(s) of retiring tenant(s) ....

....

Signature of retiring tenant(s) .... Date ....

* Strike out whichever is inapplicable

Strike out if inapplicable

NOTES:-

(THESE NOTES PROVIDE GUIDANCE ONLY)

(1) Formal proof of your relationship to the tenant, e.g. by production of marriage/civil partnership or birth certificates, unless formally accepted by the respondent. Adopted children should complete this sub-paragraph, and not paragraph 10(b) following

(2) Paragraph 9(b) may apply where you are the stepchild or foster child of the tenant or are otherwise treated by him or her as a child. An outline should be given of the circumstances relied on as establishing that you are treated by the tenant as his or her child in relation to the marriage/civil partnership. Production of the relevant marriage/civil partnership certificate and any relevant birth certificate may be required at the hearing

(3)

(i) To qualify under paragraph 11(a) you should have derived your only or principal source of livelihood from your agricultural work on the holding (or on a larger unit of which the holding forms part) during a total of five years of the seven years ending with the date of the giving of the retirement notice. The total of five years may be made up of one continuous period, or one or more separate periods. A period of full-time education at a university, etc, may, in the circumstances set out in note (4) below, count towards the five year period of earning a livelihood from the holding, and reference should be made to paragraph 11(c) and note (4) below in deciding whether the requirements of this paragraph can be satisfied

(ii) If you are relying on paragraph 11(c) a copy of the landlord's written approval should be enclosed with your application. By virtue of article 5(2) of the Regulatory Reform (Agricultural Tenancies) (England and Wales) Order 2006, SI 2006/2805, agricultural work on the holding includes agricultural work carried out by the potential successor from the holding or an agricultural unit of which the holding forms part, and other work carried out by him or her on or from the holding or an agricultural unit of which the holding forms part, of a description approved in writing by the landlord on or after 19 October 2006

(iii) Paragraph 11(d) is available only to the wife or civil partner of a retiring tenant

(4) Any period or periods (up to an aggregate total of three years) during the seven years ending with the date on which the tenant(s) gave the retirement notice to the landlord during which you were attending a full-time course at a university, college or other establishment of further education will be treated as a period throughout which your

only or principal source of livelihood was derived from your agricultural work on the holding. Any subject may have been studied

(5) Land occupied by your spouse/civil partner or by a company controlled by that person or jointly by that person and yourself should not be included in paragraph 13 where either of the parties has obtained a decree of judicial separation or a decree nisi of divorce or of nullity of marriage/civil partnership and in each case that decree remains unrescinded. In addition, land should not be included in paragraph 13 if it is occupied by you, your spouse/civil partner or such a controlled company:-

(a) under a tenancy approved under section 2(1) of the Agricultural Holdings Act 1986 or under a tenancy falling within section 2(3)(a) of that Act

(b) under a tenancy for more than one year but less than two years

(c) under a tenancy not falling within (a) or (b) above and not having effect as a contract of tenancy

(d) under a tenancy to which section 3 of the 1986 Act does not apply by virtue of section 5 of that Act

(e) as a licensee or

(f) as an executor, administrator, trustee in bankruptcy or person otherwise deriving title from another person by operation of law

However, where you occupy land in accordance with any of (a) to (e) above under a licence or tenancy granted to you by your spouse/civil partner or by a body corporate controlled by you, that land should be included in paragraph 13

(6) Paragraph 6(1)(a)–(f) of Part I of Schedule 6 to the Agricultural Holdings Act 1986 is described in note 5 above

(7) If you occupy land jointly with one or more other persons (not being only your spouse/civil partner or a company under your control or under the control of your spouse/civil partner or under your joint control), or if you are deemed to occupy land jointly with one or more such persons, you may in either case complete the application set out in paragraph 13(c) of the Form for the net annual income which the land is or was capable of producing to be treated as limited to your appropriate share

(8) All matters relied on as supporting your claim to be a suitable person to become the tenant of the holding should be summarised. These should include details of your training and practical experience of agriculture, physical health, financial standing and any educational qualifications not already listed in paragraph 11(b)

(9) Two copies of the application and of any map, any supporting written statement and any accompanying material and documents must be sent to the Secretary, and if there are more than two respondents (eg if the holding or part of it is sub-let), sufficient copies of the application and supporting documents for all other respondents .must be supplied and the Secretary must be informed of the name and address of each additional respondent

(10) Mention any other document which is attached to your application. You may include in your application, or in a separate application to the Tribunal, as appropriate:-

(a) a request for disclosure of any document or additional information about an application or reply

(b) a request for an early hearing of the application or of any question arising out of the application, and the reasons for that request

(c) consent to the matter being dealt with on written representations only

(d) a request for permission to rely on the evidence of an expert witness or expert witnesses and the name(s) and address(es) of the proposed witness or witnesses and

(e) a request for a decision on any question as a preliminary issue

You are required to send to the landlord(s) of the holding notice of this application. A form (Form S 6) for this purpose can be obtained from the Secretary to the Tribunal

(11) Name only such other person(s), if any, as are likely to wish to make representations to the tribunal against your application

(12) By virtue of section 53(3) of the 1986 Act this application must be signed by both the nominated successor (ie the applicant) and the retiring tenant or, where the retirement notice was given by joint tenants, by each of the retiring tenants. Any supporting written statement must be verified by a similar statement and must be signed

PRECEDENT 21 NOTICE TO LANDLORD

FORM S 6

Ref No: ....

To be inserted by the Secretary

AGRICULTURAL LAND TRIBUNAL
NOTICE OF APPLICATION FOR ENTITLEMENT TO TENANCY UNDER SECTION 53 OF THE AGRICULTURAL HOLDINGS ACT 1986

Please use BLOCK LETTERS and BLACK INK

To .... [*full name*]

of .... [a*ddress*]

I .... [*full name*]

of .... [a*ddress*]

hereby give you notice that I applied on ...... [*date of application*] under section 53 of the above named Act for a direction entitling me to a tenancy of the agricultural holding known as .... [*address or brief description of holding*] in succession to ...... [*name of present tenant(s) of the holding*] who served on you his/her/their retirement notice on ......

Date .... Signed ....

NOTES:-

(THESE NOTES PROVIDE GUIDANCE ONLY)

This notice should be delivered to the landlord(s) and the retiring tenant(s) of the holding and any other interested person(s). A copy of the full application will in due course be sent by the secretary of the tribunal to the landlord(s) and any other person identified in the application as a respondent. Any other person not identified as a respondent but wishing to make representations to the tribunal should apply to the Tribunal Secretary for an order that he or she be joined as a respondent in the proceedings

Chapter 4

TENANCIES UNDER THE AGRICULTURAL TENANCIES ACT 1995

INTRODUCTION

PRECEDENTS

1 Landlord's notice complying with notice condition
2 Tenant's notice complying with notice condition
3 Farm business tenancy (long form)
4 Farm business tenancy (short form)
5 Farm business tenancy specifying entitlements compensation for tenant at valuation
6 Farm business tenancy letting with entitlements
7 Farm business tenancy letting with transfer of entitlements and transfer back
8 Letting without entitlements
9 Letting specifying entitlements
10 Notice to terminate fixed term tenancy of over 2 years
11 Landlord's notice to terminate periodic tenancy
12 Tenant's notice to terminate periodic tenancy
13 Notice exercising option to terminate tenancy or part of tenancy over 2 years
14 Tenant's notice demanding statutory review of rent
15 Landlord's notice demanding statutory review of rent
16 Notice of proposal to appoint an arbitrator
17 Notice of reference of dispute to a third party

INTRODUCTION

Farm Business tenancies are governed by the Agricultural Tenancies Act 1995 (the 1995 Act), as amended by the Regulatory Reform (Agricultural Tenancies) (England and Wales) Order 2006[1] (RRO 2006). The RRO amendments relate to notice to terminate the tenancy, rent and compensation.

---

1 SI 2006/2805.

All tenancies created after 1 September 1995 will be farm business tenancies unless they fall within one of the exceptions set out in the 1995 Act.[2] The Act covers both oral and written tenancies. The underlying principle of the Act is freedom of contract with maximum flexibility to enable the parties to enter into agreements which support their own individual needs. No security of tenure is given to the tenant other than those provided by the contract itself. There are very few mandatory provisions. Those there are relate to:

(1) definition of a farm tenancy;

(2) termination of a farm tenancy;

(3) rent;

(4) fixtures and compensation; or

(5) resolution of disputes.

Apart from these matters there are no underlying rules and regulations. For instance, liability for repairs is a matter for negotiation between the parties. Accordingly, it is important when drafting an agreement to make sure that everything the parties need has been included. Farm business tenancies need to be very much bespoke.

Most short-term tenancies are dealt with by land agents on standard forms, but a short form of agreement as Precedent 4 is included for information. However, some tenancies exist for a much longer period, for example, 20 or 30 years. These arise either from sale and lease-back arrangements or where, for some reason, the parties want to substitute something rather like an Agricultural Holdings Act 1986 tenancy but are forced to do so by means of a farm business tenancy. Accordingly, a long form of agreement is also included as Precedent 3. Clearly, the variety of possible terms is extremely great.

Precedents 5–9 are short tenancies intended for use with single fields for single crop lettings with a variety of entitlement arrangements.

Notices

There are no prescribed notices. However various notices are required by the 1995 Act and precedent notices are therefore given in this section.

In order to create a farm business tenancy two conditions must be fulfilled. These are the business condition and either the agriculture condition or the notice condition. The advantage of the notice condition is that provided the character of the tenancy was primarily or wholly agricultural at the beginning of the tenancy the tenancy will continue to be governed by the 1995 Act even if

---

2 Section 4. See Introduction to Chapter 2.

there is subsequent diversification. Both the landlord and the tenant must give each other written notices, identifying the land to be comprised in the tenancy and stating that the tenancy is to be and remain a farm business tenancy.

Other precedent notices relate to the termination of the tenancy, rent reviews and the appointment of an arbitrator or third party to settle disputes.

Termination

The 1995 Act as amended provides that a tenancy for a term of more than 2 years shall continue as a tenancy from year to year unless, at least 12 months before the term date, a written notice has been given by either party to the other of his intention to terminate the tenancy.[3] The Act originally provided that a notice to quit the holding must be given at least 12 months but less than 24 months before the date on which it was to take effect. The upper limit has now been removed.[4]

A tenancy from year to year will be terminable only by notice of at least 12 months, taking effect at the end of a year of the tenancy. Similarly, the notice to operate the break in a long lease containing a break clause must be at least 12 months in length.

Tenancies for 2 years or less will expire without any notice. Periodic tenancies shorter than yearly tenancies will not be covered by the rules and will therefore be subject to the usual common law notice period.

If a break clause is to be included, it is important to consider in what circumstances either party may wish to break the tenancy. For example, if the tenant dies, his executors might wish to bring the tenancy to an end. Unless there is an option to terminate a fixed term in the event of the death of the tenant, his death will not ipso facto lead to the termination of the tenancy. The landlord may therefore wish to have a right to break in these circumstances. Every case must be negotiated.

Rent

The rent clause will need careful consideration. Generally speaking, if the tenancy lasts for more than 3 years, the statutory review procedure contained in the 1995 Act, ss 9–14 will apply. If this is not intended, it is necessary to contract out. The parties are free to agree that the rent should not be reviewed at all.[5] If this is what they intend, the agreement must expressly state that there will be no rent review.

3 Agricultural Tenancies Act 1995, s 5.
4 Article 13 of RRO 2006 amends ss 5(1), 6(1)(c) and 7(1) of the 1995 Act.
5 Agricultural Tenancies Act 1995, s 9(a).

It is possible for the parties to agree what date should be the rent review date for purposes of statutory reviews.[6]

The RRO 2006[7] provides for greater freedom of contract with regard to rents. It enables the parties to opt out of the rent review provisions in Part 11 of the 1995 Act by an express provision to that effect in the tenancy agreement or by a provision for rent reviews to be referred to an independent expert whose decision is final, provided in either case the agreement does not preclude a reduction in rent. Where the tenancy agreement provides that rent reviews are to be referred to an independent expert, arbitration on rent reviews is excluded. Care must be taken with the drafting as all aspects, for example, review notice periods, need to be covered.

Section 13(2) of the 1995 Act provided that on a statutory rent review, the arbitrator must ascertain the open market rent taking into account the terms of the tenancy, but not those 'relating to the criteria by reference to which any new rent is to be determined'. The RRO 2006[8] substitutes for these words but not those 'which (apart from this section) preclude a reduction in the rent during the tenancy'.

Compensation

Section 20(1) of the 1995 Act provides that the amount of end of tenancy compensation payable in respect of a tenant's improvement is the increase in value of the holding attributable to the improvement. Because the compensation provisions as originally enacted were mandatory and could not be altered by agreement nor the value written down, landlords were reluctant to give consent to improvements. If they gave consent they would be uncertain how much compensation they would be liable to pay when the tenancy came to an end. The RRO 2006[9] has provided that where consent is given after the Order came into force the liability for compensation can be capped. The amount of compensation will be either the amount by which the rental value of the holding has been increased by the improvement or the compensation limit, whichever is less. The compensation limit is the amount agreed in writing by the parties or, where the parties are unable to agree an amount, an amount equal to the cost of making the improvement. The parties can therefore agree at the time the consent is being sought on the maximum amount of compensation.

Where there is a resumption of part of the holding for which the tenant received compensation, that amount is taken into consideration when assessing any further compensation at the end of the tenancy so that the compensation does not exceed the agreed limit.[10]

---

6 Ibid, s 10(4) and (5).
7 RRO 2006, art 14.
8 Ibid, art 15.
9 Ibid, art 16.
10 Ibid, art 17.

PRECEDENTS

PRECEDENT 1 LANDLORD'S NOTICE COMPLYING WITH NOTICE CONDITION

AGRICULTURAL TENANCIES ACT 1995
SECTION 1(1)(a), (4)

HOLDING [*name and identification of holding by address, ordnance survey number(s) or plan*]

TO TENANT

[*Name and address*]

I, [*name and address of Landlord*], give you notice that I intend the proposed tenancy of the above holding to be and remain a farm business tenancy and that at its beginning the character of the tenancy will be primarily or wholly agricultural

Signed: ....

Dated: ....

PRECEDENT 2 TENANT'S NOTICE COMPLYING WITH NOTICE CONDITION

AGRICULTURAL TENANCIES ACT 1995
SECTION 1(4)(a)

HOLDING [name and identification of holding by address, ordnance survey number(s) or plan]

TO LANDLORD

[*Name and address*]

I, [*name and address of Tenant*], give you notice that I intend the proposed tenancy of the above Holding to be and remain a farm business tenancy and that at its beginning the character of the tenancy will be primarily or wholly agricultural

Signed: ....

Dated: ....

PRECEDENT 3 FARM BUSINESS TENANCY (LONG FORM)

[*NOTE: If the Agreement to be entered into is for a term of more than seven years the Land Registry prescribed clauses must also be included – Land Registration Act 2002.*]

FARM BUSINESS TENANCY AGREEMENT (LONG FORM)

PARTICULARS

Date of Agreement	:	....
The Landlord	:	.... of ....
The Tenant	:	....
[The Surety	:	.... of ....]
Title Guarantee	:	Full/Limited/Limited except that the covenant implied by section 3(3) of the Law of Property (Miscellaneous Provisions) Act 1994 shall apply only to a charge incumbrance or third party right created by the Landlord personally (delete as appropriate)
Commencement Date	:	....
Term Date	:	[*date*]
Term	:	The period of [ ] years from the Commencement Date/The period from the Commencement Date to the day preceding the next following Term Date and thereafter from year to year until determined in accordance with clause 2.2 (delete as appropriate)
Break Date	:	[*date*]/the anniversary of the first Term Date during the Term/the [ ] and [ ] anniversaries of the first Term Date during the Term (delete as appropriate) [*Note: Delete clause 2.2 if appropriate*]
Initial Holding Rent	:	[£ . . .] per year
Rent Days	:	[*date*] and [*date*]
Rent Suspension Period	:	[of] years
Resolution of disputes by	:	Arbitrator/expert (delete as appropriate)
Grain Holdover Date	:	[*date*] next following the termination of the Tenancy
Sugar Beet Holdover Date	:	[*date*] next following the termination of the Tenancy
Valuation Association	:	Central Association of Agricultural Valuers

Landlord's Address	:	....
for service of notices	:	....
(Landlord and Tenant		....
Act 1987, section 48(1))		....
***Landlord's Milk Quota**		
– litres	:	[of]
– butterfat	:	[of]
		and any milk quota subsequently transferred to the Tenant without payment but as to the whole subject to any addition or reduction due to any general adjustment (delete if inapplicable)
*** Tenant's Milk Quota**		
– litres	:	[of]
– butterfat	:	[of]
		and any milk quota subsequently transferred to the Holding by the Tenant at his expense in accordance with the provisions hereof but as to the whole subject to any addition or reduction due to any general adjustment (delete if inapplicable)
Single Payment Scheme	:	the single payment scheme governed by Council Regulation (EC) No 73/2009 and all the relevant subsequent Commission and other regulations administrative provisions statutory guidance or codes of practice implemented by the EC and/or the UK Government ('the Regulations')
Entitlements	:	such entitlements to the single income payment and any other entitlement to financial support of whatever kind deriving in any way from the Regulations and made available by the Landlord to the Tenant by reference to the area of the Holding subject to this Tenancy Agreement and no other holding and comprising the following entitlements held by the Landlord in his name pursuant to the Regulations namely

Type(s)	Number	Value

and reference to Type(s) of entitlements has the same meaning(s) as that contained in the Regulations

THIS FARM BUSINESS TENANCY AGREEMENT is made on the Date of Agreement

BETWEEN:-

(1) [*name*] of [*address*] ('the Landlord')

(2) [*name*] of [*address*] ('the Tenant')

and

[(3) [*name*] of [*address*] ('the Surety')]

1. Particulars definitions and interpretation

1.1 The attached Particulars are incorporated in this Agreement and unless the context otherwise requires the expressions in the left-hand column of the Particulars have the respective meanings set against them in the Particulars

1.2 In this Agreement unless the context otherwise requires the following phrases shall have the respective meanings set under them:-

1.2.1 'the Department'

the Department for the Environment Food and Rural Affairs or (as the case may be) the Welsh Office Agricultural Department or any substitute Department for either and shall include where the context so admits the authority responsible for administering IACS or any alternative scheme as aforesaid introduced in addition thereto or in substitution therefor

1.2.2 'the Fixed Equipment'

the buildings structures fixtures and fittings (if any) affixed to and any works constructed on in over or under the Holding from time to time except any which the outgoing tenant is entitled to remove and any which are to be regarded as belonging to the Tenant

1.2.3 'the Holding'

the premises described in the Holding Description Schedule including the Fixed Equipment TOGETHER WITH the rights specified in the Schedule of Rights Granted but EXCEPT AND RESERVING the rights specified in the Schedule of Reservations

1.2.4 'the Holding Rent'

the yearly rent payable in accordance with paragraph 1 of the Rents Schedule

1.2.5 'the Holdover Buildings'

the grain drying and storage buildings and bins on the Holding at the termination of the Tenancy

1.2.6 'the Holdover Plant'

all drying plant conveyors and other machines and machinery in the Holdover Buildings

1.2.7 'IACS'

the Integrated Administration and Control System or any alternative scheme for the support of agricultural incomes or the control of agricultural production or the restriction of land use which may be introduced in addition thereto or in substitution therefor and administered by the Department or the European Commission or any other public or statutory body whether jointly or individually

1.2.8 'the Insured Risks'

fire storm lightning tempest flood aircraft in peacetime and articles dropped therefrom impact explosion earthquake subsidence riot malicious damage and civil commotion (or such of them as the Landlord may from time to time specify in writing in relation to any particular insurance) and such other risks (including terrorism) as the Landlord in his discretion from time to time specifies by notice in writing to the Tenant

1.2.9 'the Interest Rate'

four per centum per annum above the base lending rate from time to time of [*name*] Bank plc or in the event of that base rate ceasing to be published such other rate as the Landlord shall reasonably specify by notice in writing to the Tenant, and the Planning and Compensation Act 1991

1.2.10 'the Land'

the land comprised in the Holding but excluding the Fixed Equipment

1.2.11 'the Landlord's Surveyor'

any qualified person or corporation (including an employee of the Landlord) appointed by or acting for the Landlord to perform the function of a surveyor or manager for any purpose of this Agreement

1.2.12 'the Planning Acts'

the Town and Country Planning Act 1990, the Planning (Listed Buildings and Conservation Areas) Act 1990, the Planning (Hazardous Substances) Act 1990 and the Planning (Consequential Provisions) Act 1990

1.2.13 'Redundant Items'

those items which are redundant to the farming of the Holding as specified or subsequently deemed to be specified in the Schedule of Redundant Items

1.2.14 'the Referee'

the person appointed as an arbitrator or independent expert (as the case may be) in accordance with the clause of this Agreement headed 'Resolution of Disputes'

1.2.15 'Relevant Quotas'

such (if any) quotas (but excluding the Tenant's Milk Quota) contracts premium quota rights or similar arrangements (as permitted by the context) as are from time to time attaching to or available in respect of the Holding or any produce thereof or to which the Tenant or any other person is entitled in respect of such produce whether existing at the commencement of the Tenancy or introduced thereafter and 'Relevant Quota' shall be construed accordingly

1.2.16 'Restricted Use'

use for agricultural purposes/as a dairy farm/as an arable farm/in accordance with a system of farming approved in writing by the Landlord (delete as appropriate)

1.2.17 'Restricted Use Period'

the first [*no*] months/years (delete as appropriate) of the Term

1.2.18 'the Tenancy'

the tenancy hereby created

1.2.19 'Tenant's Fixtures'

those specified in the Schedule of Tenant's Fixtures and any other tenant's fixtures affixed to the Holding during the Tenancy

1.2.20 'the 1995 Act'

the Agricultural Tenancies Act 1995

1.3 In this Agreement unless the context otherwise requires:-

1.3.1 words importing one gender only include the other genders; words importing the singular include the plural number and vice versa; words importing persons include corporations and vice versa

1.3.2 the expression 'the Landlord' includes the person for the time being entitled to the reversion immediately expectant on the Tenancy

1.3.3 the expression 'the Tenant' includes the person for the time being entitled to the Tenancy

[1.3.4 the expression 'the Surety' includes the personal representatives of the named Surety]

1.3.5 where there are two or more persons included in the expression 'the Landlord' 'the Tenant' or 'the Surety' the obligations of such persons under this Agreement shall be joint and several

1.3.6 any reference to 'the Holding' includes any part or parts thereof

1.3.7 any matter 'referred for decision under this Agreement' shall be determined in accordance with the clause of this Agreement headed 'Resolution of Disputes'

1.3.8 any reference to any tax includes that tax as from time to time amended or extended and any tax by which it is replaced

1.3.9 any provision not to do an act or thing imports an obligation not to cause or permit such act or thing to be done

1.3.10 any reference to a statutory provision regulation or code of practice includes any later provision substituted for or amending the same and any subsidiary provision made under powers thereby conferred

1.3.11 the clause and paragraph headings are for convenient reference only and shall not affect its construction

2. Letting

2.1 The Landlord lets to the Tenant with the Title Guarantee ALL THAT the Holding TO HOLD the same unto the Tenant for the Term (subject to any provision for earlier determination herein contained) on the terms of this Agreement YIELDING AND PAYING therefor the rents specified in the Rents Schedule

2.2 (*Optional*) [Either the Landlord or the Tenant may determine the Tenancy by giving to the other notice in writing of not less than twelve months expiring on the day preceding the second or any subsequent Term Date both the Commencement Date and the expiry date being included in the Term]

[2.3 *See Precedent 4, clause 3 in connection with the leasing of Entitlements (if applicable)*]

3. Matters affecting the holding

The Holding is let subject to:-

3.1 such public or private rights liabilities provisions agreements and other matters as affect the Holding

3.2 the rights of the outgoing tenant (if any)

3.3 any existing tenancies or occupancies of any dwellings on the Holding

4. Tenant's covenants

The Tenant hereby covenants with the Landlord to the intent that the obligations shall continue throughout the Tenancy as follows:-

4.1 *Pay rent*

to pay the rents specified in the Rents Schedule at the times and in the manner therein provided and if the Landlord so requires to make any payment or payments of such rents by banker's standing order or direct debit to such account as the Landlord from time to time notifies to the Tenant in writing

4.2 *Pay VAT*

to pay within fourteen days after demand (against a proper Value Added Tax invoice where appropriate) all Value Added Tax which may be payable on any taxable supplies made to the Tenant under or in connection with this Agreement (including Value Added Tax or replacement tax on any rent payable hereunder)

4.3 *Pay outgoings*

4.3.1 to pay all existing and future rates taxes assessments and outgoings of every kind (whether or not of a novel nature) which are now or may at any time hereafter during the Tenancy be charged or imposed upon the Holding or upon the owner or occupier thereof except only such as the Landlord is by law bound to pay notwithstanding any contract to the contrary and any taxation payable either by reason of the Landlord's ownership of the Holding or in respect of any dealing with the Landlord's reversion in the Holding

4.3.2 if any payment falling within sub-clause 1 of this clause is payable in respect of or calculated by reference to both the Holding and other property the Tenant shall pay a proper and reasonable proportion thereof

4.4 *Pay interest and costs of recovery*

4.4.1 in relation to any rent or other sum due to the Landlord but unpaid for seven days to pay interest at the Interest Rate from the time when such rent or other sum became due (whether formally demanded or not) until the time when it is actually paid (as well after as before any judgment) and the aggregate of such sum and interest shall at the option of the Landlord be recoverable by action or as rent in arrear

4.4.2 to pay costs and disbursements incurred by the Landlord in seeking to recover any rent or other payment in arrear by any appropriate steps including instructing a bailiff to levy distress for that rent (whether or not any distress in the event be levied)

4.5 *Pay incoming valuation*

to pay to the Landlord:-

4.5.1 the amount payable by the Landlord to the outgoing tenant in respect of the Holding (except only any compensation for disturbance and game damage and any compensation in respect of milk quota unless otherwise expressly agreed)

4.5.2 (where there is no outgoing tenant) such sum as would have been payable under sub-clause 1 of this clause had there been an outgoing tenant such payment to be made within fourteen days of the Tenant being notified by the Landlord in writing of the amount payable under this clause

4.6 *Notice of charges*

to give notice in writing to the Landlord forthwith upon giving any charge over his farming stock under the Agricultural Credits Act 1928 or any other similar enactment and on request to supply the Landlord with copies of all charges given

4.7 *Death of Tenant*

to take all such steps as are necessary to ensure that:-

4.7.1 if the Tenant or any of them dies during the currency of the Tenancy notice in writing of such death and the date thereof is given to the Landlord within one month of the death

4.7.2 an office or duly certified copy of any probate of a will letters of administration order of court or other instrument affecting the Holding or the Tenancy is served on the Landlord within one month from the date thereof

4.8 *Landlord's proceedings*

to permit the Landlord to take proceedings or do any other act in the name of the Tenant which the Landlord considers appropriate to preserve or protect his interest in the Holding and to co-operate with the Landlord in relation thereto the Landlord indemnifying the Tenant against any costs charges and expenses the Tenant may incur in connection with any such matter

4.9 *Insuring dealing and other obligations in schedules*

to comply at all times with the obligations of the Tenant:-

4.9.1 as to insurance contained in the Insurance Schedule

4.9.2 as to dealings with the Holding contained in the Dealings Schedule

4.9.3 as to holdover to the Tenant contained in the Holdover Schedule

4.9.4 as to the Single Payment Scheme contained in the Single Payment Scheme Schedule

4.9.5 as to any other matter not otherwise specifically provided for or contained in any other schedule to this Agreement

4.10 *Waste*

4.10.1 not to commit any wilful or voluntary waste spoil or destruction on the Holding

4.10.2 to prevent any encroachment easement or footpaths from being made in or acquired over the Holding

4.10.3 not to give any acknowledgement as to the entitlement to any right or easement over or in respect of the Holding

4.10.4 where damage results from the laying underground of any pipeline sewer or other apparatus or from the erection of any pole pylon or other apparatus above the ground to procure that without delay such damage is fully made good and the Holding is fully restored

4.10.5 not to do or omit any act or thing in relation to the Holding (including the making of any application or the giving of any notice of any kind) which may:-

4.10.5.1 render the Landlord liable for the payment of any tax chargeable by reference to the realisation of development value

4.10.5.2 entitle any authority to acquire the Holding or take any other step adverse to the interest of the Landlord under any then existing statute or subsidiary legislation

EITHER

4.11 *Repairs*

4.11.1 at all times during the Tenancy to keep in good and substantial repair order and condition (which where appropriate includes keeping the same clean and properly decorated and/or protected) all items from time to time on the Holding or on other property but serving the Holding which are stated in the Repairs Schedule to be the responsibility of the Tenant using the best and most suitable materials for the purpose

4.11.2 the provisions of sub-clause 1 of this clause shall not apply to:-

4.11.2.1 any Redundant Item

4.11.2.2 any repair or replacement which is required by reason of loss or damage by any risk against which the Landlord has or should have insured in accordance with the provisions of this Agreement save to the extent that any such insurance has been vitiated by some act or default of the Tenant or any person for whose acts or defaults the Tenant may be responsible

OR

4.11 *Repairs*

4.11.1 at all times during the Tenancy to keep in good and substantial repair order and condition (which where appropriate includes keeping the same clean and properly decorated and/or protected) all items (other than any Redundant Items) from time to time on the Holding or on other property but serving the Holding using the best and most suitable materials for the purpose

4.12 *Common items*

to pay to the Landlord on demand an appropriate proportion of the cost from time to time incurred by the Landlord in the maintenance and repair of all items used in common with adjoining or neighbouring property

4.13 *Alterations*

4.13.1 not to carry out any alteration or make any addition to the Holding and specifically but without prejudice to the generality hereof the farmhouse buildings or cottages (if any) nor erect any new buildings or make any other improvement on the Holding

4.13.2 before making any permitted Tenant's Improvement to use his best endeavours to obtain any grant subsidy or other payment which may be available in respect thereof under any local national or European scheme

4.13.3 to indemnify the Landlord against any liability to any adjoining owner or occupier or the local authority or any member of the public which may be incurred by reason of the execution by or on behalf of the Tenant of any works which the Tenant is permitted or required to carry out on the Holding

4.13.4 to remove (and make good all damage occasioned by such removal) any erection addition or alteration made by the Tenant without the consent required by this Agreement or in respect of which the permission of the planning authority is refused withdrawn or lapses within one month after being required in writing by the Landlord to do so

4.14 *Planning*

4.14.1 to comply in all respects with the provisions and requirements of the Planning Acts and all licences consents permissions and conditions (if any) granted or imposed thereunder so far as the same relate to or affect the Holding or any operations works acts or things already or hereafter to be carried out executed done or omitted thereon or the use thereof for any purpose and at all times hereafter to keep the Landlord fully indemnified against all liabilities actions costs claims losses and expenses in respect of any act matter or thing contravening any of those provisions and requirements

4.14.2 as often as occasion requires at the expense in all respects of the Tenant to obtain all such licences consents and permissions and serve all such notices as may be required for the carrying out by the Tenant of any operation on the Holding or the institution or continuance by the Tenant thereon of any act or use which may constitute development within the meaning of the Planning Acts

4.14.3 not to make any application for planning permission without the previous consent in writing of the Landlord and to deliver to the Landlord within five days after the service or giving of the same a copy of any notice or application served or given by the Tenant on or to any planning authority under the Planning Acts

4.14.4 to pay and satisfy any charge that may hereafter be imposed under the Planning Acts in respect of the carrying out or maintenance by the Tenant of any such operations or the institution or continuance by the Tenant of any such use as aforesaid

4.14.5 notwithstanding any consent which may be granted by the Landlord under this Agreement not to carry out or make any alteration or addition to the Holding or any change of use thereof before any requisite planning permission has been produced to the Landlord and acknowledged by the Landlord in writing as satisfactory to the Landlord PROVIDED THAT the Landlord may refuse so to express satisfaction with any such planning permission on the ground that the period thereof or anything contained therein or omitted therefrom in the reasonable opinion of the Landlord or the Landlord's Surveyor would be or be likely to be prejudicial to the Landlord's interest in the Holding whether during or following the expiration of the Tenancy

4.14.6 unless the Landlord otherwise directs to carry out before the termination of the Tenancy any works stipulated to be carried out on the Holding by a date subsequent to that termination as a condition of any planning permission wholly or partly implemented during the Tenancy

4.14.7 if and when called upon so to do to produce to the Landlord or the Landlord's Surveyor all such plans documents and other evidence as the Landlord may reasonably require in order to satisfy the Landlord that the provisions of this covenant have been complied with in all respects

4.14.8 not to do or omit any act matter or thing in on or respecting the Holding which is required to be done or omitted (as the case may be) by the Planning Acts or to do or suffer to be done any act or thing which contravenes any provision of those Acts and at all times to keep the Landlord fully indemnified against all liabilities actions costs claims losses and expenses in respect of any such matter or thing contravening the provisions of the Planning Acts

4.15 *Legislation*

to observe and comply fully with the provisions of any statute order regulation rule direction code of practice or other thing which extends to or affects the Holding or anything done or omitted or to be done or omitted thereon by the Landlord or the Tenant in relation to the Holding

4.16 *Notices*

in relation to any direction notice proposal or order served on him under or by virtue of any statutory provision or otherwise:-

4.16.1 forthwith to give the Landlord notice in writing thereof and if so required to produce to the Landlord such direction notice proposal or order and allow the Landlord to make copies thereof

4.16.2 to comply immediately with and give sufficient effect to every such direction notice or order and keep the Landlord fully indemnified against all liabilities actions costs claims losses and expenses arising out of or in connection with the same

4.17 *Fire precautions*

4.17.1 to maintain adequate fire fighting equipment throughout the Holding and to comply in all respects with any requirements or recommendations made by the insurers or a fire officer or other authority

4.17.2 to keep the Holding supplied and equipped with proper and suitable fire fighting and extinguishing apparatus and appliances appropriate to the use of and the business carried on on the Holding

4.18 *Environmental practice*

4.18.1 to use the Tenant's best endeavours to prevent pollution of whatever kind on the Holding or any neighbouring land

4.18.2 when using spray chemicals to take all reasonable care to ensure that hedges trees and crops on the Holding or any adjacent property are not adversely affected and to make good or pay full compensation for any damage done

4.18.3 in performing his obligations hereunder and his obligation to comply with the rules of good husbandry so far as possible to select chemicals and to adopt farming practices which cause the least harm to game birds wildlife (other than vermin and pests) and the environment

4.19 *Ancient monuments*

not in any way to interfere with or injure any tumulus stone circle or other ancient remains structure or monument (including known buried archaeological remains) or any actual or proposed site of special scientific interest (whether statutorily protected or not) and not to permit any person to enter on the Holding (with or without metal detectors tools hammers or other equipment) for the purpose of searching for or removing any object or specimen of antiquarian archaeological or scientific interest (save under the rights hereby reserved to the Landlord) but if it comes to the knowledge of the Tenant that any such object or specimen has been found or observed on the Holding forthwith to report the same to the Landlord in writing and to give such assistance as the Tenant reasonably can to enable possession thereof to be given to the Landlord or as he shall direct

4.20 *Refuse*

4.20.1 not to deposit any refuse waste redundant material or redundant machinery of any kind on the Holding

4.20.2 within seven days of receipt thereof to comply with any notice issued by the Landlord requiring removal from the Holding of any such material or item as is referred to in this clause in accordance with all statutory provisions relating to such removal then in force

4.21 *Recordings*

4.21.1 not without the written approval of the Landlord to permit or authorise any person to take for hire or reward any film or other recording (in whatever form and whether of sight or sound) of or on the Holding or any natural or man-made feature thereon or anything done or occurring thereon but so that this provision shall not restrict the liberty of the Tenant to take such film or other recording as he thinks fit for the personal and domestic use and enjoyment of any of himself his family and friends or in direct connection with his agricultural business

4.21.2 forthwith to pay to the Landlord any money received by the Tenant in consequence of any breach of this provision such money to be deemed for the purpose of calculating interest thereon to have become due on the date of the breach

4.22 *Non-permitted users*

4.22.1 not to use the Holding for any illegal or immoral purpose

4.22.2 not to carry on on the Holding any offensive noisome noxious or dangerous occupation or activity

4.22.3 not to do on the Holding any act matter or thing whatsoever which may be or become a nuisance or annoyance or cause damage disturbance or inconvenience to the Landlord or the owner or occupier of any adjoining or neighbouring property or the users of any road abutting the Holding

4.22.4 not to hold on the Holding any sale by auction or public meeting

4.22.5 not to use the Holding for any trade other than the Restricted Use or for advertising purposes or for the holding of any show race match game or contest whether of animals birds vehicles or persons

4.22.6 not to place any tent or caravan upon the Holding nor to take in lodgers in the farmhouse or other dwellings on the Holding

4.22.7 not to burn any crop residues on land from which a crop has been harvested

4.23 *Personal farming and residence*

personally to farm the Holding and reside in the farmhouse and make the same his usual place of residence but if the Tenant comprises more than one person residency by one of them shall satisfy this residence requirement

4.24 *User*

during the Restricted Use Period to use the Holding only for the Restricted Use and:-

4.24.1 not at any time to change the use of the Holding without the Landlord's consent in writing but any such consent shall not be deemed to be a consent for the purposes of this Agreement or of sections 17 and 18 of the 1995 Act to any Tenant's Improvement

4.24.2 in the last year of the Tenancy to farm and cultivate the Holding in accordance with such reasonable requirements as are notified to him in writing by the Landlord but subject thereto:-

4.24.2.1 to leave the arable land in a course normally practised in the district

4.24.2.2 at the termination of the Tenancy to leave upon the Holding all the unconsumed hay straw roots or green crops (other than peas beans potatoes sugar beet carrots or other vegetable crops) produced on the Holding during the last year of the Tenancy

4.25 *Cultivation*

4.25.1 to cultivate and manage the agricultural land on the Holding according to the best standards and system of cultivation used in the district and so as to comply with the rules of good husbandry as defined by section 11 of the Agriculture Act 1947 and so as not to impoverish or deteriorate the agricultural land and to leave the same clean and in good condition

4.25.2 to take all reasonable precautions to prevent the introduction of any pest or disease on the Holding and in particular not to grow on the same land during any period of four consecutive years more than one crop of potatoes and one crop of sugar beet (including sugar beet grown for seed)

4.25.3 not to make or keep rabbit warrens on the Holding and to destroy any that may exist

4.25.4 when so required by and at the expense of the Landlord to have taken a proper analysis of the soil of the Holding (or such part of it as may be specified by the Landlord) by the Ministry or such other body as the Landlord may specify

4.26 *Grass land*

4.26.1 to manage the grass land in a proper manner and also to destroy and prevent from seeding and spreading all rushes thistles nettles thorns gorse ragwort and other weeds and to keep down moles and rats and other vermin and spread all molehills and ant hills

4.26.2 not to remove from the Holding any turf or topsoil without the consent in writing of the Landlord and forthwith to pay to the Landlord any money received by the Tenant in consequence of a breach of this provision such money to be deemed for the purpose of calculating interest thereon to have become due on the date of the breach

4.26.3 not to break up or convert into tillage any land which is considered permanent pasture and for this purpose (and for the purposes of paragraph 2 of the Second Schedule to the Agriculture Act 1947) there shall be so considered the whole of the land described in the Cultivation column of the Holding Description Schedule as pasture meadow grass or ley and also any land subsequently laid to grass for which the Landlord has prescribed seeds

4.26.4 AND shall be at liberty to break up and convert into tillage any of the remainder of the grass land on the Holding PROVIDED THAT:-

4.26.4.1 he shall at least one year before the termination of the Tenancy sow down to temporary pasture in a husbandlike manner with a seed mixture approved in writing by the Landlord the whole of the land so broken up so that such pasture is fully established before the termination of the Tenancy

4.26.4.2 with the consent in writing of the Landlord the Tenant may sow down to temporary pasture in substitution for any grass land ploughed up an equal area of arable land elsewhere on the Holding upon such terms as the Landlord may direct in writing but such consent shall not imply any obligation on the part of the Landlord to provide fencing or a water supply for such substituted land nor shall it imply the Landlord's consent to the provision of fencing or water supplied by the Tenant as an improvement under the terms of: section 17 of the 1995 Act

4.26.4.3 no compensation shall be payable to him in respect of any temporary pasture laid down by him in accordance with this clause

4.26.5 after giving or receiving notice to quit not to plough up temporary grass land sown more than five years prior to the end of the Tenancy

4.26.6 if the Tenancy terminates between 1 January and 30 June in any year not to stock the new seeds after 1 October immediately preceding such termination

4.26.7 not to sell off or remove but to consume on the Holding all grasses and clover and forage crops (whether harvested or not) and all fodder straw and roots grown on the Holding in the last year of the Tenancy or remaining unconsumed from previous years and to leave for the use of the Landlord or incoming tenant all such grasses clover and forage crops fodder straw and roots remaining unconsumed on the Tenant quitting the Holding

4.26.8 to return to the Holding the full equivalent manurial value of any grasses clover or forage plants (whether green or conserved) or any fodder straw roots or other produce of the Holding sold or removed from it at any time prior to the last year of the Tenancy

4.26.9 to spread annually on the Holding all farmyard manure and compost produced and made on the Holding and at the termination of the Tenancy to leave carefully stored all such farmyard manure and compost not already spread

4.27 *Trees*

to use his best endeavours to protect from damage any trees coppice wood or underwood growing on the Holding and in particular:-

4.27.1 not to cut down cut top or lop or drive nails into or otherwise injure trees growing on the Holding

4.27.2 not to cut any tellars whether growing from stools or otherwise without the consent in writing of the Landlord

4.27.3 to protect all trees from damage by livestock machinery or otherwise howsoever

4.27.4 not without the Landlord's consent in writing to plant with trees any area of the Holding not so planted at the commencement of the Tenancy

4.28 *Orchards*

to keep all orchards and gardens in a proper state of cultivation well manured and in good heart to protect from damage by livestock or otherwise howsoever and to prune at proper seasons all fruit trees and bushes and at his own expense to replace dead or worn out fruit trees with other fruit trees suitable to the district PROVIDED THAT the Tenant may at his option in lieu of replacing individual trees grub up and replace the whole or a substantial part of an orchard in accordance with a scheme approved in writing by the Landlord

4.29 *Game*

4.29.1 to use his best endeavours to protect from damage all animals fish and protected birds (with their nests and eggs) reserved by the relevant paragraph of the Schedule of Reservations and not to interfere with measures taken by the Landlord or persons authorised by him to rear and preserve game

4.29.2 to warn off all poachers and trespassers and allow his name to be used in any proceedings which the Landlord may take against any poachers or trespassers (the Landlord indemnifying the Tenant against any costs charges and expenses he may thereby incur) and so far as possible to stop all encroachments

4.30 *Boundaries*

save in accordance with the rules of good husbandry and with the Landlord's prior written consent not to alter or destroy any hedge fence or boundary on the Holding nor do anything whereby the size or shape of any field is rendered different from its present size or shape

4.31 *Records*

4.31.1 to keep on the Holding and maintain a field book showing how every field or parcel thereof has been stocked cropped and cultivated in each year of the Tenancy and to make the same available to the Landlord at all reasonable times

4.31.2 to supply a copy to the Landlord of any return or other form which the Tenant makes or completes pursuant to any local national or European Union requirement or directive and if so required by the Landlord to consult with the Landlord in regard to the completion of such return or form to the extent required to preserve and safeguard in all respects the interest of the Landlord

4.31.3 to keep true accounts of all crops or produce sold off and of the equivalent manurial value returned to the Holding and to produce such accounts to the Landlord forthwith upon receiving at any time a request in writing to do so

4.32 *Disease*

4.32.1 to give the Landlord immediate notification (to be confirmed in writing forthwith) of any outbreak or suspected outbreak of any notifiable disease of crops or livestock on the Holding

4.32.2 to comply at his own expense with any recommendation from time to time made by any appropriate authority or qualified person (including the Landlord's veterinary surgeon) for the prevention or treatment of any such disease

4.32.3 not to bring any diseased plant material or animal on to the Holding and to compensate the Landlord for all losses incurred as a result of any diseased plant material or animal being on the Holding

4.33 *Water*

4.33.1 on entering the Holding to give all necessary notices and make all necessary applications and payments to preserve any existing licence granted in respect of the Holding under the Water Resources Act 1991 or any similar enactment and subsequently to continue the licence in force during the Tenancy and not at any time by any act or omission to jeopardise the licence nor to surrender it without the consent in writing of the Landlord

4.33.2 on the termination of the Tenancy to make such application for the continuance or renewal of the licence as may be requisite for the time being under regulations then in force to preserve the licence for the benefit of the Landlord or his nominee without payment

4.33.3 not to enter into any agreement or licence for the supply of water from any spring or well pond stream watercourse or underground water supply on the Holding to any adjoining or neighbouring premises

4.33.4 to reimburse to the Landlord on demand all charges costs and expenses reasonably incurred by the Landlord in connection with the provision or the improved provision of water to the Holding

4.34 *New schemes*

except as expressly required to do so by or under the provisions hereof not without the Landlord's consent in writing to enter into any voluntary scheme or arrangement for setting aside land or imposing limitations on cropping stocking or production or requirements for public access

4.35 *IACS*

4.35.1 promptly to give to the Landlord a copy of any document submitted to the Department under IACS relating to the Holding and a copy of any communication received from the Department concerning any such document

4.35.2 to confirm in writing to the Landlord no later than one month before the closing date in any year for applications under IACS whether or not he has submitted or will be submitting an area aid application under IACS for that year

4.35.3 to submit prior to the deadline date in each year properly completed applications under IACS and to take such steps as are required to ensure that all payments due under IACS are received and that the requirements relevant to the Holding are met

4.35.4 to allow the Landlord and any person authorised by the Landlord or the Department to enter on any part of the Holding at all reasonable times for any reasonable purpose in connection with IACS

4.35.5 forthwith on the Landlord's request to instruct the Department to give such information to the Landlord as the Landlord may require in respect of the Holding any return made for the purpose of IACS and any representations made by or to the Department in respect thereof AND the Tenant hereby appoints the Landlord as his agent for the purpose of obtaining any such information as is referred to in this sub-clause

4.36 *Genetically modified organisms*

4.36.1 not [without the Landlord's previous consent in writing] to grow or permit to be grown on the Holding any plant which has been 'genetically modified' within the meaning of Part VI of the Environmental Protection Act 1990

[4.36.2 In respect of any such plants grown on the Holding:-

4.36.2.1 to observe all statutory provisions Regulations Codes of Practice and Guidelines

4.36.2.2 to indemnify the Landlord against all claims proceedings and demands of whatsoever nature and howsoever arising from the presence of such plants on the Holding or the escape from the Holding of any 'organism' (within the meaning of Part VI of the Environmental Protection Act 1990) being part of or derived from such plants]

[4.36.3 to use all reasonable endeavours to remove from the Holding on the determination of the tenancy all such plants and every part thereof]

4.37 *Groundwater Regulations*

to comply with the provisions of the Groundwater Regulations 1998, SI 1998/2746 ('the Groundwater Regulations') and in particular:-

4.37.1 to supply the Landlord with details of disposal and tipping on the Holding in each year to 31 March within one month of the expiration thereof

4.37.2 to maintain an authorisation in respect of the Holding under the Groundwater Regulations

4.37.3 to take all necessary steps to transfer such authorisation to the Landlord or the incoming occupier on the termination of the tenancy and to give notice of such transfer to the Environmental Agency within twenty one days thereof

4.37.4 to indemnify the Landlord against all costs claims proceedings and demands of whatsoever nature and howsoever arising from any breach of the Groundwater Regulations

4.38 *Nitrate Vulnerable Zones*

4.38 if the Holding or any part of it is situated within a Nitrate Vulnerable Zone:-

4.38.1 to retain all records required by the Nitrate Pollution Prevention Regulations 2008, SI 2008/2549 and on the termination of the tenancy to hand to the Landlord or the incoming occupier such records for the period of five years preceding the end of the tenancy

4.38.2 to comply with the provisions of the said Regulations and to indemnify the Landlord against all claims proceedings and demands of whatever nature and howsoever arising from any failure to comply therewith

4.39 *Farm assurance schemes*

4.39.1 at all times at his own expense to ensure that the requirements of the Farm Assurance Schemes specified in Schedule 7 or any schemes replacing the same are met in respect of the Holding and the crops produced thereon

4.39.2 if he wishes to register the Holding or for the purposes of any Farm Assurance Scheme to comply at his own expense with the requirements thereof including any works required in respect of any fixed equipment on the Holding

4.39.3 at his own expense to ensure that the requirements of any Farm Assurance Scheme available in respect of crops grown on the Holding are met

4.39.4 on the expiration of the Tenancy to hand to the Landlord all records required as evidence that he has complied with the requirements of all such Schemes

4.40 *Sewage sludge*

4.40.1 not to use or permit to be used sludge as defined in the Sludge (Use in Agriculture) Regulations 1989 on the Holding or any part thereof

OR

4.40.2 not to use or permit to be used sludge as defined in the Sludge (Use in Agriculture) Regulations 1989 on the Holding or any part thereof otherwise than:-

4.40.2.1 in accordance with the said Regulations

4.40.2.2 in accordance with any code of practice or guidelines relating thereto from time to time published by the Department or any other department or agency

4.40.2.3 after obtaining the Landlord's consent in writing and in compliance with any conditions specified in writing by the Landlord

4.41 *Health and safety*

4.41.1 to comply in respect of the Holding and the Tenant's activities thereon with the Regulations relating to Health and Safety at Work and all codes of practice in respect thereof for the time being in force

4.41.2 forthwith to notify the Landlord of any defects under the Health and Safety Regulations in respect of the Holding or any area used by the Tenant in common with other persons

4.41.3 to indemnify the Landlord in respect of any liability arising in respect of the Holding under the Health and Safety Regulations and in respect of any areas used in common to reimburse the Landlord a fair payment according to user of the costs of any works required to be undertaken by the Landlord to comply with such Regulations

4.42 *Relevant Quota*

4.42.1 at all times during the Tenancy to do all acts and things (including producing a sufficient quantity of any relevant product and signing and lodging within any applicable time-limits such documents as are required) which may be required to maintain all Relevant Quota

4.42.2 not to do or omit to do anything during the Tenancy which results or may result in all or any part of a Relevant Quota being reduced lost or charged or whereby it is removed from the Holding whether permanently or temporarily

4.42.3 to notify the Landlord in writing of any change in the level of any Relevant Quota within seven days of the Tenant becoming aware of that change

4.42.4 to use his best endeavours to obtain for the Holding any new or further Relevant Quota which becomes available under any local national or European scheme whether existing at or introduced after the commencement of the Tenancy and which he believes is suitable for the Holding

4.42.5 at all times during the Tenancy to carry out all obligations imposed on the Tenant as occupier or producer or quota or contract owner in regard to notices compliance with directions payments of levies maintenance of records or otherwise in connection with every Relevant Quota

4.42.6 forthwith to furnish to the Landlord (in writing if required) all information which the Landlord may reasonably request concerning any Relevant Quota or concerning other farming activities of the Tenant which may affect or tend to affect any Relevant Quota whether such farming activities are carried out by the Tenant directly or indirectly and whether they relate to the Holding or to other land

4.42.7 if the Landlord at any time or times wishes to agree or record any figures facts or calculations concerning any Relevant Quota and the relative rights of the parties thereto under any local national or European scheme legislation or directive from time to time in force then to endeavour to agree the same with the Landlord and in default of such agreement the matter shall (unless referred to arbitration in accordance with the arbitration provisions of such scheme legislation or directives) be referred for decision under this Agreement

4.42.8 not so to dispose of transfer lend lease or otherwise deal with the whole or any part of a Relevant Quota as to result in the Relevant Quota concerned lapsing or reducing during the Tenancy or the Relevant Quota concerned being transferred off the Holding or otherwise becoming unavailable for use in full on the Holding on the termination of the Tenancy

4.42.9 not by any direct or indirect act or omission in respect of any land or property not included in the Holding to allow any Relevant Quota or the benefit of any Relevant Quota to pass to any other person whatsoever or become attached to land not comprised in the Holding

4.42.10 not to make or permit anyone claiming on his behalf or through him to make any claim either during or at or following termination of the Tenancy which may result in any Relevant Quota or any part thereof ceasing to belong to or be available for the Landlord or the Holding and if as a result of any such claim a Relevant Quota or any part thereof is lost to the Landlord or the Holding to indemnify the Landlord against all losses which the Landlord suffers thereby including (but not necessarily limited to) all costs incurred in the acquisition of replacement Relevant Quota which is comparable in all respect to that lost

4.42.11 on quitting the Holding to take all such steps as may be required to procure the transfer of each of the Relevant Quotas to the Landlord or such person as the Landlord nominates in writing as his successor in each case [without charge or payment] [*payment provision to be included? – market value?*]

4.43 *Landlord's Milk Quota*

in relation to the Landlord's Milk Quota (if any):-

4.43.1 to fulfil and maintain such quota with no less litres (subject only to generalised cuts) and the same butterfat base by milk produced on the Holding

4.43.2 not to use any land not comprised in the Holding for the production of milk or ancillary purposes

4.43.3 not to sign any contract for the supply of milk from the Holding without ensuring appropriate provisions for the continuation of the acquisition of the same quantity of milk from the Holding in the event of the Tenancy being terminated and on the termination of the

Tenancy to use his best endeavours to transfer any contract for the supply of milk to the Landlord or his nominee if the Landlord so requires

4.43.4 in relation to the period which falls both prior to the termination of the Tenancy and within the milk quota year current at that termination not to produce a quantity of milk which is greater than the due proportion for that period as calculated by apportioning the litreage for that quota year on a daily basis subject to any appropriate adjustment for seasonal variations

4.43.5 if the Tenant produces a quantity of milk in excess of the proportion calculated in accordance with sub-clause 4 of this clause to indemnify the Landlord and any other occupier of the Holding against any levy payable in respect of such excess

4.43.6 forthwith on demand to give to the Landlord copies of any documentation relating to the Landlord's Milk Quota

4.44 *Indemnities*

4.44.1 to indemnify the Landlord against all solicitors' surveyors' architects' and agents' costs and fees and all other expenses properly incurred by the Landlord in connection with:-

4.44.1.1 any application by the Tenant for any consent or approval under or pursuant to the terms of this Agreement whether or not such consent is given

4.44.1.2 the preparation and service of all notices and schedules relating to wants of repair on the Holding whether the same be served during or after the termination of the Tenancy and in particular of all notices under section 146 or 147 of the Law of Property Act 1925 notwithstanding that forfeiture is avoided otherwise than by relief granted by the court

4.44.2 to indemnify the Landlord against all liabilities actions costs claims losses and expenses arising out of or in connection with a breach of any obligation of the Tenant herein contained or implied and such indemnity shall extend to and cover all costs and expenses incurred by the Landlord in connection with any steps which the Landlord may (but without being in any way obliged so to do) take to remedy any such breach and shall be without prejudice to any other right or remedy of the Landlord in respect of such breach

4.44.3 to indemnify the Landlord against all liabilities actions costs claims losses and expenses instituted or arising out of or in connection with the use or occupation of the Holding the existence of any article in or about the Holding or the execution or omission of any works upon the Holding except and insofar as the same may be due to the Landlord's own act or default or the act or default of any employee or agent of the Landlord

4.44.4 to indemnify the Landlord against the Landlord's Solicitors' costs of and incidental to the preparation and completion of this Agreement and a counterpart hereof and the stamp duties thereon

4.45 *Re-letting*

to permit the Landlord during the six months immediately preceding the termination of the Tenancy to affix and retain without interference upon any part of the Holding a notice for re-letting the same and at any time during the Tenancy a notice for the sale of the Landlord's reversion in the Holding and to permit all persons with the authority in writing of the Landlord or the Landlord's agents at all reasonable times of the day to view the Holding

4.46 *Yielding up*

4.46.1 in the last year of the Tenancy to permit (to the extent required by the Landlord) the Landlord or his ingoing tenant to enter on directly after harvest and cultivate any part of the land from time to time cleared of crops before the termination of the Tenancy without any reduction or abatement of the Holding Rent by reference to such entry

4.46.2 to take every available measure (including the giving of any necessary notice) to enable vacant possession of the Holding (including any cottage or house comprised therein) to be given to the Landlord upon the termination of the Tenancy whether by notice forfeiture or otherwise

4.46.3 on the termination of the Tenancy (whether by notice forfeiture or otherwise) to return to the Landlord the Holding as to the Land in good heart and condition and as to the Fixed Equipment in good repair and (where appropriate) decorated in accordance with the provisions of the Repairs Schedule

5. Landlord's agreements

The Landlord hereby agrees with the Tenant as follows:-

5.1 *Compensation*

to pay to the Tenant on quitting the compensation specified in the Compensation Schedule save that the Landlord may set off:-

5.1.1 any sum due to the Landlord for rent breaches of agreement or otherwise

5.1.2 if the year of quitting does not coincide with a year during which the external painting or internal painting and decoration must be executed then a due proportion of the cost of such painting or decoration the deduction for external painting being one-fifth and that for internal painting papering and decorating one-seventh of the estimated total cost in respect of each year that has elapsed since the last repainting papering or decorating or since the commencement of the Tenancy (whichever is the later)

5.1.3 the estimated cost of such repairs and replacement to the Fixed Equipment as may in the opinion of the Landlord be required at the time of quitting

5.2 *Insuring repairing and other obligations in schedules*

to comply at all times with the obligations of the Landlord:-

5.2.1 as to insurance contained in the Insurance Schedule

5.2.2 as to repairs contained in the Repairs Schedule (*only if first alternative for 4.11 is used*)

5.2.3 as to the Single Payment Scheme contained in the Single Payment Schedule

5.2.4 as to allowing the Tenant to holdover after the termination of the Tenancy contained in the Holdover Schedule

5.2.5 as to any other matter not otherwise specifically provided for or contained in any other schedule to this Agreement

5.3 *Make good damage*

save to the extent otherwise expressly or impliedly provided in the reservation concerned to cause as little damage as practicable in the exercise of the reservations herein contained and to pay reasonable compensation to the Tenant for any damage or interruption to the Tenant's business so caused

5.4 *Quiet enjoyment*

to permit the Tenant on paying the rents hereby reserved and performing and observing the several covenants by the Tenant herein contained peaceably to hold the Holding without any lawful interruption by the Landlord or any person deriving title under him

6. Provisos and conditions

Provided always and it is hereby agreed and declared as follows:-

6.1 *Re-entry*

at any time after the happening of any of the events specified in the following paragraphs of this clause it shall be lawful for the Landlord after giving two months' notice to the Tenant of his intention to do so to re-enter the Holding or any part thereof in the name of the whole and thereupon the Tenancy shall absolutely determine but without prejudice to any right of action or remedy of the Landlord in respect of any antecedent breach by the Tenant of any of his obligations in relation to the Tenancy namely:-

6.1 if any rent hereby reserved is unpaid for fourteen days after becoming due (whether legally demanded or not)

6.2 if the Tenant at any time fails or neglects to perform or observe any of his obligations in relation to the Tenancy

6.3 if the Tenant has any distress or process of execution levied on his goods

EITHER (Company Tenant)

6.4 if the Tenant:-

6.4.1 enters into liquidation whether compulsorily or voluntarily (not being a voluntary liquidation for the sole purpose of reconstruction or amalgamation of a solvent company)

6.4.2 suffers an administrative receiver to be appointed

6.4.3 enters into a voluntary arrangement as defined in section 1 of the Insolvency Act 1986

6.4.4 does anything which would entitle a receiver to be appointed to take possession of any of its assets

6.4.5 does anything which would entitle any person to present a petition for winding up the Tenant or apply for an administration order in relation to the Tenant

6.4.6 ceases to exist or is dissolved

OR (Individual Tenant)

6.5 if the Tenant:-

6.5.1 is unable to pay a debt or has no reasonable prospect of being able to pay a debt within the meaning of section 268 of the Insolvency Act 1986

6.5.2 makes an application to the court for an interim order under section 253 of the Insolvency Act 1986

6.5.3 makes any assignment for the benefit of or enters into any arrangement with his creditors either by composition or otherwise

6.5.4 does anything which would entitle a petition for a bankruptcy order to be made against him

6.2 *Non-waiver*

the demand for or acceptance of rent by the Landlord or anyone acting on his behalf shall not constitute or be construed as a waiver of any breach of the Tenant's obligations under this Agreement or of the Landlord's remedies in respect thereof

6.3 *Distress*

upon seizure by the Landlord under distress for rent the Landlord shall not be obliged to sell any hay straw silage or crops (except any crops which the Tenant is permitted to sell under the terms hereof without returning an equivalent manurial value to the Holding) upon the terms that the same may be removed from the Holding but may exercise the power of sale on distrained goods conferred by statute by selling the same subject to the condition that such produce shall be consumed on the Holding or subject to some other condition which secures that the manurial value of such hay straw or crops is returned to the Holding AND upon any such seizure and sale it shall be lawful for the Landlord to grant to any purchaser of such hay straw or crops and for such period or periods as the Landlord may think fit the use of such part or parts of the Holding as the Landlord may think necessary for the purpose of harvesting storing consuming or otherwise dealing with such hay straw or crops and without making any compensation to the Tenant in respect thereof

6.4 *Exclusion of implied rights*

nothing herein contained shall by implication of law or otherwise operate to confer on the Tenant any easement right or privilege whatsoever (except those hereby expressly granted) over or against any adjoining or other property belonging to the Landlord which might restrict or prejudicially affect such adjoining or other property for any present or future use or purpose nor shall the Tenant be entitled to compensation for any damage or disturbance caused by or suffered through any present or future use of or operation on such adjoining or other property

6.5 *Provisions in schedules*

the provisions of the several schedules to this Agreement (in addition to the respective obligations of the parties hereto under those schedules) shall have effect and be enforceable between the parties hereto

6.6 *Tenant's fixtures*

6.6.1 the fixtures and fittings (if any) specified in the Schedule of Tenant's Fixtures are fixtures belonging to the Tenant which the Tenant is entitled under section 8 of the 1995 Act to remove subject to the provisions of that section

6.6.2 any Tenant's Fixtures which the Tenant does not on or before the termination of the Tenancy remove from the Holding shall immediately and without more become the property of the Landlord unless the Landlord otherwise elects in which case:-

6.6.2.1 the Landlord shall notify the Tenant in writing of such election

6.6.2.2 the Tenant shall forthwith remove such fixtures and fittings from the Holding and make good any damage occasioned as a result for which purposes the Landlord shall afford reasonable access and

6.6.2.3 if the Tenant fails so to remove such fixtures and fittings within one month from the date of the Landlord having so notified him the Landlord shall be entitled to arrange the removal of such fixtures and fittings and the cost of so doing shall be recoverable by the Landlord from the Tenant by action or as if it were rent in arrears

6.7 *Redundant items*

6.7.1 the items specified in the Schedule of Redundant Items are agreed as obsolete or redundant to the farming of the Holding

6.7.2 if at any time or times during the Tenancy either party wishes to treat any building or other item of fixed equipment on the Holding (other than those included in the Schedule of Redundant Items) as obsolete or redundant to the proper requirements of the Holding then in default of agreement such party shall be entitled on giving two months' written notice to the other to have the question whether such buildings or other items of fixed equipment or any of them are obsolete or redundant referred for decision under this Agreement and if it is agreed or awarded that such building or other item of fixed equipment is obsolete or redundant then as from the date of such agreement or award (as the case may be) the building or other item of fixed equipment so agreed or awarded to be obsolete or redundant shall be deemed to be included in the Schedule of Redundant Items

6.7.3 in relation to the items specified or deemed to be included in the Schedule of Redundant Items both parties shall be relieved from all liability to repair or insure the same save that the Tenant shall keep them in a safe condition

6.7.4 the Tenant may use any redundant building or other item of fixed equipment at his own risk and without any obligation (save as aforesaid) to carry out any repairs to such building or item but he must keep it in a clean and tidy condition

6.7.5 the Landlord may at any time or times enter and remove at his own expense any obsolete or redundant building or item of fixed equipment whether or not the Tenant is using it in accordance with the immediately preceding sub-clause

6.8 *Record of condition*

if any party requires any record of condition of the Holding or of anything on the Holding to be made that party shall bear the cost of making the record

6.9 *Break clause – Landlord*

the Landlord may by notice in writing of twelve months (expiring at any time except as otherwise provided) terminate this Agreement to the extent and subject to the provisions of this clause:-

6.9.1 as to the entire Holding by notice expiring on the Break Date or any of the Break Dates if there is more than one

6.9.2 as to the entire Holding by notice given within three months after the death of the Tenant or the last surviving Tenant or if later within two months after receiving notice of that death in accordance with the provisions of this Agreement

6.9.3 as to the entire Holding by notice given at any time after the Tenant (or any of them) has become insolvent

6.9.4 as to any dwelling comprised in the Tenancy which has at any time been unoccupied for a continuous period of not less than six months (other than by reason of the temporary absence of an employee or former employee of the Tenant who can demonstrate an intention to return) or occupied in breach of the provisions hereof by notice given at any time before the dwelling has become again occupied in accordance with the provisions hereof. In respect of any such property the Landlord may require the Tenant (at the Tenant's expense) to apply to the court for the purpose of obtaining such orders as may be necessary to secure vacant possession thereof and relief ancillary thereto

6.9.5 as to the whole or any part of the Holding for any purpose not being the use of the Holding or such part for agriculture and whether the use thereof for such purpose is to be made by the Landlord or by a purchaser lessee assignee or other person deriving title from or through the Landlord

6.9.6 as to any part of the Holding which the Tenant is at the time the notice is given using for any purpose other than agriculture

6.10 *Break clause – Tenant*

the Tenant may terminate this Agreement by written notice of twelve months given to expire on the Break Date or any of the Break Dates if there is more than one

6.11 *Determination as to part*

if there is a termination of the Tenancy as to any part of the Holding (whether under the foregoing provisions hereof or by surrender or otherwise) then subject to any contrary agreement the Tenant shall be entitled to a reduction of the Holding Rent proportionate to that part of the Holding and in respect of any depreciation of the value to him of the residue of the Holding caused by the severance or by the use to be made of the part severed and in the event of the parties hereto failing to agree as to the amount of such reduction of the Holding Rent the same shall be referred for decision under this Agreement

6.12 *Set-off*

6.12.1 the Tenant shall not be entitled to exercise any right of set-off against any rents or other monies payable by the Tenant pursuant to this Agreement

6.12.2 the Landlord shall be entitled to set off against any sum payable by the Landlord to the Tenant in respect of the Holding any moneys payable to the Landlord by the Tenant in respect of the Holding in each case whether of a liquidated character or not

6.13 *Custom*

the rights of the parties under this Agreement or otherwise in respect of the Tenancy shall not depend on or be affected by any custom of the country

6.14 *Tenant's Milk Quota*

6.14.1 the Tenant's Milk Quota shall belong to the Tenant free from any interest of the Landlord or any third party or parties claiming through him or on behalf of the Landlord

6.14.2 the Tenant shall not transfer any Tenant's Milk Quota to the Holding whether at the commencement of the Tenancy or subsequently without obtaining the Landlord's prior consent in writing which shall neither be unreasonably delayed nor withheld

6.14.3 if at any time during the Tenancy the Tenant wishes to dispose of all or any part of the Tenant's Milk Quota by way of permanent transfer he shall first give the Landlord notice in writing of his intention so to do and the Landlord may within fourteen days thereafter notify the Tenant in writing of his wish to acquire such quota or part thereof in which event such quota or part shall be purchased by the Landlord at the then current market value thereof AND the Landlord and the Tenant shall sign all necessary forms and do all necessary things to ensure the transfer of the same to the Landlord or as he directs with all due despatch

6.14.4 the Tenant shall be free to sell any such quota which the Landlord does not duly notify the Tenant he wishes to purchase to one or more third parties and the Landlord and the Tenant shall sign all necessary forms and do all necessary things to ensure the transfer of the same with all due despatch

6.14.5 during the last fifteen months of the Tenancy the Tenant may transfer the Tenant's Milk Quota or any part of it to another holding or holdings and the Landlord hereby consents to the Tenant transferring such quota to such alternative holding or holdings free from any claim by the Landlord or anyone claiming on his behalf or through him and confirms that he will sign all necessary documents and do all necessary things to ensure the transfer to such alternative holding or holdings including (notwithstanding the other terms of this Agreement) the creation of any interests in land which may be required for this purpose so long as they do not attract any security of tenure in any form they have been created by documentation which has first been approved in writing by the Landlord and they terminate prior to the end of the Tenancy

6.14.6 if at the determination of the Tenancy any part of the Tenant's Milk Quota remains available to the owner or occupier of the land formerly comprised in the Tenancy the Landlord shall pay to the Tenant the market value of such quota valued as at the date of the termination of the Tenancy

6.14.7 unless agreed by the parties the market value of any quota required to be ascertained for the purposes of this clause shall be determined by an independent expert to be appointed failing agreement between the parties as to his identity by the President of the Royal Institution of Chartered Surveyors

6.14.8 the Landlord confirms that except in accordance with the provisions of this Agreement neither he nor his successors in title shall make any claim in respect of the Tenant's Milk Quota either during or at the termination of the Tenancy howsoever determined

6.15 *Sale*

in the event of the sale of the Holding the responsibility for the Landlord's obligations hereunder shall pass to the purchaser and no action shall lie against the Landlord (vendor) for their non-observance after the date on which the rent accrues to the purchaser

6.16 *Notices and agency*

6.16.1 where any notice is required under this Agreement to be given by or to the Landlord such notice may be given by or to his duly authorised agent; any notice (including notices to quit) to be given by either party under this Agreement may be served in the manner provided by section 36 of the 1995 Act, or as regards any notice to the Tenant by being sent to him by registered post or recorded delivery to the farmhouse or by handing it to the Tenant's agent or by affixing the notice to some conspicuous part of the Holding and any notice to the Tenant shall be sufficient though the same is not addressed to the Tenant by name but only addressed to him by the description of the Tenant of the Holding

6.16.2 any right or power under this Agreement granted to the Landlord shall be exercisable by the Landlord or the Landlord's duly authorised agent or servant

6.17 *Resolution of disputes*

6.17.1 any question valuation or claim which is in dispute between the Landlord and the Tenant arising out of this Agreement and not compulsorily referred to arbitration by any statutory provision shall (except as hereby otherwise expressly provided) be decided by an arbitrator or expert as stated in the Particulars

6.17.2 any dispute to be decided by an arbitrator shall be referred to arbitration under the provisions of the Arbitration Acts 1950 to 1979 by a single arbitrator and any matter to be decided by an expert shall be referred to an independent expert

6.17.3 either party may give notice in writing to the other of his intention to refer the matter to the Referee pursuant to the provisions of this clause

6.17.4 the parties may agree on the identity of the Referee in which event they shall jointly appoint him but in the absence of such agreement within one month of the serving of the notice of intention either party may apply to the President for the time being of the Royal Institution of Chartered Surveyors for the appointment of a suitably qualified person to act as the Referee

6.17.5 the Referee's decision shall be final and binding on the parties

6.17.6 the fees and expenses of the Referee and (as the case may be) of the President of the Royal Institution of Chartered Surveyors shall be borne equally or as the Referee otherwise decides in his discretion

6.17.7 if the Referee dies or becomes unable or unwilling to act or fails to deliver a decision within four months (or such other period as the parties agree in writing and notify to the Referee) of his appointment the procedure set out in this clause shall be repeated and may be repeated as often as is necessary to achieve the resolution of the dispute

7. Farm business tenancy

The parties hereto acknowledge that prior to the execution of this Agreement notices were served under section 1(4) of the 1995 Act and that it is intended that the Tenancy shall be and remain at all times a Farm Business Tenancy within the meaning of the 1995 Act

[8. Surety's covenants

The Surety in consideration of the letting of the Holding having been made at the request of the Surety hereby covenants with the Landlord as follows:-

8.1 the Tenant will observe and perform all the Tenant's obligations contained or implied in this Agreement as well after as before any disclaimer of this Agreement by a liquidator or trustee in bankruptcy and during any extension or continuance of the Term and that in case of default in such observance and performance of any of those obligations as aforesaid the Surety will pay and make good to the Landlord on demand all losses damages costs and expenses thereby arising or incurred by the Landlord PROVIDED ALWAYS and the parties hereby agree that any neglect or forbearance of the Landlord in endeavouring to obtain payment of any rent when the same becomes payable or to enforce performance of any of the Tenant's obligations contained or implied in this Agreement and any time which may be given to the Tenant by the Landlord or any variation in the terms of this Agreement which may be agreed between the Landlord and the Tenant shall not release or exonerate or in any way affect the liability of the Surety under this covenant

8.2 if during the Tenancy the Tenant (being a corporation) enters into liquidation or (being an individual) becomes bankrupt and the liquidator or the trustee in such bankruptcy (as the case may be) disclaims this Agreement or if this Agreement is determined by forfeiture or by the exercise of the Landlord's right of re-entry hereunder and if the Landlord within the period of six months after such disclaimer forfeiture or determination serves upon the Surety a notice in writing so to do the Surety will accept from the Landlord a lease of the Holding for a term equal in duration to the residue remaining unexpired of the Term at the time of such disclaimer forfeiture or determination such new lease to take effect from the date of such disclaimer forfeiture or determination and to reserve the same rents and to contain the like Landlord's and Tenant's covenants respectively and the like provisos agreements and conditions in all respects (including the proviso for re-entry) as are reserved and contained in this Agreement]

9. Certificate

It is hereby certified that there is no Agreement for Lease to which this Lease gives effect

IN WITNESS whereof this Deed has been executed by the parties hereto the day and year first before written

THE HOLDING DESCRIPTION SCHEDULE

ALL THAT Holding shortly described as [*description*] and more particularly described below by reference to the Ordnance Survey plan (County Series Edition)/(Metric Edition sheets [*nos*]) (delete as appropriate) of the district

	Ordnance Survey			Area		
Parish	Sheet No	Enclosure No	Field Name	Cultiva-tion	Hec-tares	Acres

THE SCHEDULE OF RIGHTS GRANTED

THE SCHEDULE OF RESERVATIONS

There are excepted and reserved out of the Holding:-

1. Game

subject to the provisions of the Ground Game Act 1880 (as amended by section 12 of the Wildlife and Countryside Act 1981) and the Ground Game (Amendment) Act 1906 all deer foxes game wildfowl woodcock snipe fish hares and rabbits and all birds listed in Part 1 of the Second Schedule to the Wildlife and Countryside Act 1981 and the nests and eggs of all birds referred to in this paragraph with the right for the purposes of shooting fishing hawking sporting taking preserving and rearing such wildlife and destroying pigeons and other pests to enter onto the Holding with such equipment and persons as the Landlord authorises and also the right to hunt or permit hunting on foot or with horses and hounds

2. Water

all springs of water or other wells ponds streams and water courses (including any underground water) with the right for the Landlord and all persons authorised by the Landlord to have access thereto and to take water therefrom by means of pumps and other constructions and equipment and pipes laid or to be laid on across or under the Holding or otherwise (subject to sufficient water being left for the Tenant for domestic and agricultural purposes other than irrigation)

3. Timber

all timber timber-like trees fruit trees tellers pollards saplings and underwood with the right for the Landlord and all persons duly authorised to make fell cut and remove the same and to plant upon any coppice or woodland such quantity of young trees as the Landlord from time to time thinks fit including the right to enter the Holding for such purposes with or without motor vehicles plant and equipment

4. Minerals

all mines minerals quarries stones sand brick-earth clay gravel petroleum and its relative hydrocarbons and all other gases and substances in upon or under the Holding including the air space void created from time to time by the winning working getting and carrying away of any of such substances whether opened or unopened within upon or under the Holding together with the right for the Landlord and all persons authorised by the Landlord and his lessees and agents and all persons authorised by them or any of them with workmen and others from time to time and at all times thereafter whether by entry on the surface or by underground working to win work get and carry away the said substances and any like substances within upon or under any adjacent or other lands with full powers for those purposes:-

4.1 to withdraw vertical and lateral support from the surface of the Holding and from any buildings or works now erected or hereafter to be erected thereon or from any crops standing thereon

4.2 to sink pits or shafts open quarries and erect buildings and construct place and maintain any plant machinery equipment apparatus or materials in on or under the Holding

4.3 to construct any requisite access tracks roads or railways fences pipes drains sewers wires cables conduits or other apparatus in under or over the Holding

4.4 to use the surface of the Holding for the temporary storage or deposit of any of the said substances or any waste

4.5 to do all other things necessary or convenient for such purposes or any of them notwithstanding any substance or other injury or damage which may thereby be occasioned to the Holding or any building or works or crops as aforesaid or any other injury or damage or loss whatsoever arising whether directly or indirectly from any such working or operation as aforesaid which may be sustained by the Tenant save that the Landlord shall pay to the Tenant reasonable compensation for damage thereby done to the surface buildings or crops and shall allow an abatement of rent in respect of the surface land of the Holding of which the Tenant is deprived by reason of the exercise of any such rights and liberties as aforesaid

5. Roads

the right at all times and for all purposes for the Landlord and the occupiers and tenants of any adjoining or neighbouring land of the Landlord and all other persons authorised by the Landlord or by such occupiers or tenants to use the roads tracks and footpaths within the Holding together with all other ways hitherto used or enjoyed across the Holding and in the case of each such road track or way so used (other than footpaths) either with or without animals vehicles implements machinery or plant

6. **Services**

6.1 the right for the Landlord and the tenants or other occupiers of any adjoining or neighbouring land of the Landlord to the free passage of drainage water gas oil telecommunications and electricity from and to such other land through the watercourses channels drains sewers pipes cables and wires belonging to or running through the Holding

6.2 the right for the Landlord and all persons authorised by the Landlord to erect lay maintain inspect repair renew and use within upon under or over the land comprised in the Holding telecommunications and electricity lines and cables and gas oil water drainage and sewage pipes and any necessary apparatus in connection therewith respectively with full power for the Landlord and such other persons as aforesaid and the agents and workmen of the Landlord and such other persons to enter onto the Holding for the purpose of inspecting repairing replacing and renewing such lines cables pipes and apparatus respectively

7. **Repairs**

the right to enter and execute repairs for which the Tenant is liable under any provision hereof and which he has failed to start within two months or complete within three months of being required so to do by notice in writing given by the Landlord and the whole or the proportionate part for which the Tenant is liable (as the case may be) of the expense thereof shall be payable by the Tenant forthwith after such repairs have been executed and shall at the option of the Landlord be recoverable by action or as rent in arrears PROVIDED THAT nothing in this paragraph shall relieve the Tenant from his liability under any provision hereof to execute repairs

8. **Cultivation**

the right after three days' written notice to enter and execute acts of husbandry or cultivation which the Tenant after giving or receiving notice to quit is in the opinion of the Landlord or his agent neglecting or performing in a dilatory or otherwise unsatisfactory manner and to take such manure as the Landlord shall require for such purpose and the Tenant shall at the Tenant's entire expense provide accommodation for the Landlord's workmen and free of cost accommodation for any tractors machinery fertiliser seed or other items brought onto the Holding by the Landlord in exercise of this right

9. **Entry**

9.1 the right for the Landlord and all persons authorised by the Landlord (where appropriate with or without workmen animals carts motor vehicles plant machinery implements guns and other articles):-

9.1.1 for the due and proper exercise and enjoyment of his rights hereunder

9.1.2 for the due and proper performance of his obligations hereunder

9.1.3 to make good any breach by the Tenant of any of his obligations under this Agreement

9.1.4 for any other reasonable purpose

9.2 the right to permit any party involved in any arbitration or other legal proceedings for which an inspection of the Holding is or may be relevant to inspect the Holding whether or not the Landlord is one of the parties to the arbitration or proceedings

10. Wayleaves

10.1 the right to grant any wayleave contract easement or licence to any public or local authority or public utility company or other company or person with the right to authorise agents and servants of the grantee with or without motor vehicles plant and equipment to enter upon the Holding and carry out all appropriate works and in particular but without prejudice to the generality of the foregoing the right to run pipes drains conduits cables wires or other works (either already existing or any new ones) for the benefit of any adjoining or neighbouring land of the Landlord and the right to carry out works for the benefit of such adjoining or neighbouring land

10.2 the benefit of all such contracts easements or licences existing at the date hereof or granted or made under this paragraph and all rents and other payments reserved thereby (except annual payments made for disturbance to the occupier) and the benefit of any other similar agreements or contracts now or at any time affecting the Holding

11. Holdover

if the Commencement Date falls between 1 July and 31 December the right for the Landlord and all persons authorised by the Landlord with or without motor vehicles plant and equipment to enter at any time or times upon the Holding for the purpose of storing any of the following crops produced on the Holding in appropriate buildings or on appropriate sites on the Holding nominated by the Landlord until the next following respective dates set opposite them:-

Grain	:	the Grain Holdover Date
Sugar Beet	:	the Sugar Beet Holdover Date

on the same respective terms (mutatis mutandis) as those set out in the Holdover Schedule (delete if inapplicable)

12. Antiquities

the right to all antiquities on the Holding together with the right for persons authorised by the Landlord to carry out scientific and research studies of whatsoever nature on the Holding

13. Recordings

the right for persons authorised by the Landlord to enter on the Holding with or without motor vehicles plant equipment cameras and visual and sound recording equipment and to use the Holding (other than the interiors of dwellings) for the purpose of making sound or visual recordings of whatever nature

THE RENTS SCHEDULE
PART 1
The Holding Rent

EITHER

1. Amount

The Holding Rent shall be the Holding Rent specified in the Particulars which shall be payable throughout the Term and there shall be no review

OR

1. Amount

1.1 The Holding Rent for the first [ ] years of the Term shall be the Initial Holding Rent specified in the Particulars

1.2 The Holding Rent payable for the next following [ ] year[s] of the Term shall be £ . . . (repeat this sub-clause as necessary)

1.3 If at any time at which a variation of the Holding Rent is due to take place under the provisions of this paragraph legislation restricts the right to vary the Holding Rent then the Holding Rent shall be varied on the day following the removal of the legislative restriction to the same amount as that to which it would have been at that time had the legislative restriction not been imposed

1.4 Except as provided in this clause the Holding Rent shall not be capable of variation

OR

1. Amount

1.1 In this paragraph:-

1.1.1 'Review Date' means (subject to sub-paragraph 1.5) [*date*] and every [ ] anniversary thereof during the Term

1.1.2 'Review Period' means successively:-

1.1.2.1 each period of the Term beginning on a Review Date and ending on the day before the next Review Date

1.1.2.2 the period of the Term commencing on the last Review Date and ending on the expiry of the Term

1.1.3 'Base Figure' means the figure of the Index for [*date*]

1.1.4 'Index' means the 'All Items' Index of Retail Prices published by the Department of Employment or any successor Ministry or Department

1.2 The Holding Rent payable for the first years of the Term shall be the Initial Rent specified in [*clause*]

1.3 During each successive Rent Period the Holding Rent shall be the particular amount which bears the same direct proportion to the Initial Rent as the figure of the Index for the month preceding the relevant Review Date bears to the Base Figure

1.4 If the reference base used to compile the Index changes after the date hereof the figure taken to be shown in the Index after the change shall be the figure which would have been shown in the Index if the reference base current at the date hereof had been retained

1.5 If on any Rent Day the Holding Rent payable remains unquantified by reason of non-publication of the relevant data or is otherwise undetermined the Tenant shall pay rent at the rate contractually payable immediately prior to the relevant Review Date and within fourteen days of publication of the relevant data or determination the Tenant shall pay to the Landlord the amount of the shortfall and/or the Landlord shall repay to the Tenant the amount of any arrears (delete as appropriate)

1.6 If at the relevant Review Date legislation restricts the right of either party to require a review of the rent payable hereunder then the day following the removal of the legislative restriction shall be treated for all purposes as the relevant Review Date

1.7 After each occasion on which the Reviewed Rent is determined pursuant to this paragraph a record or memorandum thereof shall be made and signed by or on behalf of the Landlord and the Tenant in such manner and form as the Landlord may require the parties paying their own costs of this procedure

1.8 Except as provided in this paragraph the rent hereunder shall not be capable of variation

OR

1. **Amount**

The Holding Rent shall be the Initial Holding Rent specified in the [*place specified*] or any rent substituted therefor in accordance with the provisions of the 1995 Act

2. **Payment**

The Holding Rent shall be paid without any deductions whatsoever by equal payments in advance on the Rent Days the first payment (being a proportionate amount in respect of the period from the Commencement Date to the next Rent Day) to be made on the date of this Agreement

3. **Suspension**

If the Landlord has or should have effected insurance against loss of rent and if any part of the Holding is damaged or destroyed by any of the Insured Risks so as to be unfit for occupation and use then (save to the extent that the relevant insurance has been vitiated by some act or default of the Tenant or any person for whose acts or defaults the Tenant may be responsible) a fair proportion of the Holding Rent (assessed according to the nature and extent of the damage sustained) shall be suspended until the part of the Holding concerned is again rendered fit for occupation and use or (if earlier) the expiration of the Rent Suspension Period calculated from the date of such damage or destruction and any dispute concerning this paragraph shall be referred for decision under this Agreement

OR

1. The Holding Rent payable during the Tenancy shall be the rent specified in the Particulars until such time as a different rent is substituted in accordance with the following provisions of this Schedule

2. Part II of the Agricultural Tenancies Act 1995 shall not apply

3. Nothing contained in this Schedule shall preclude a reduction in the Holding Rent from time to time payable

4. In this Schedule 'Review Date' shall mean:-

4.1 [ ] day of [ ] 20[ ]; and thereafter

4.2 [ ] in any year

Provided in each case that either the Landlord or the Tenant shall first have given not less than twelve months' notice in writing requiring that the Holding Rent payable shall be reviewed and further provided in the cases of a notice pursuant to paragraph 4.2 that the date so specified shall not be earlier than the end of three years from the date from which there took effect a previous increase or reduction of rent

5. The Holding Rent payable from a Review Date shall be such amount as may from time to time be agreed in writing by the parties or by their respective authorised agents or, in the absence of agreement, as may be determined in accordance with paragraph 6 of this Schedule

6. If no arbitrator has been appointed under an agreement to determine the question of the rent on a basis agreed by the parties either party may at any time during the period of six months ending with the Review Date apply to the President of the Royal Institution of Chartered Surveyors for the appointment of an arbitrator by him

7. The Holding Rent to be agreed or determined as above shall be such amount as would be the Rent for the Holding with effect from the Review Date assessed in accordance with paragraphs 1, 2 and 3 of Schedule 2 to the Agricultural Holdings Act 1986 for a yearly tenancy subject to that Act and with the same assumptions and disregards as in those paragraphs, modified so far as necessary in order to apply to this Agreement

8. If the Holding Rent payable from a Review Date is not agreed or determined as above by that date the Holding Rent shall provisionally continue at the rate in force immediately prior to that date. Within fourteen days after the date on which the Holding Rent payable is so agreed or determined as aforesaid the Tenant shall pay an appropriate amount in respect of any increase in the rate of Holding Rent or (as the case may require) the Landlord shall refund an appropriate amount in respect of any decrease in the rate of Holding Rent

PART 2
The Insurance Rent

1. Amount

The Insurance Rent is a sum equal to the gross premium or premiums which the Landlord from time to time pays to effect or maintain the insurance which he is required to effect and maintain under this Agreement

2. Payment

The Insured Rent shall be paid without deduction within fourteen days after demand in writing

THE INSURANCE SCHEDULE

[*To be adjusted to meet the case.*]

PART 1
(Landlord's Obligations)

The obligations of the Landlord in respect of insurance are:-

1. **Property**

to insure and keep insured the undermentioned items (except any buildings and fixtures belonging to the Tenant and any Redundant Items) against destruction or damage by any of the Insured Risks and in the event of destruction or damage by an Insured Risk (except to the extent that such insurance has been vitiated by some act or default of the Tenant or any person for whose acts or defaults the Tenant may be responsible) to reinstate or replace the same (and to make good any shortfall from his own resources) if its reinstatement or replacement is required for the fulfilment of his responsibilities to manage the Holding in accordance with the rules of good estate management:-

dwellings

other buildings

fixed equipment

plant and machinery

2. **Rent**

if the Landlord in his discretion so decides or if the Tenant so requires by notice in writing to the Landlord to insure and keep insured the Holding against loss of the Holding Rent for the Rent Suspension Period by reason of damage or destruction due to an Insured Risk

PART 2
(Tenant's Obligations)

The obligations of the Tenant in respect of insurance are:-

1. **The Premises**

to insure and keep insured the undermentioned items (except any buildings and fixtures belonging to the Tenant and any Redundant Items) against destruction or damage by any of the Insured Risks and in the event of destruction or damage by an Insured Risk to reinstate or replace the same:-

dwellings

other buildings

fixed equipment

plant and machinery

2. **Crops and stock**

2.1 to insure and keep insured against loss destruction or damage by any of the Insured Risks (including consequential loss relating to such loss destruction or damage) all live and dead farming stock and crops for the time being on the Holding to the full market value thereof or the cost of replacement (if greater) and so that every policy of such insurance shall:-

2.1.1 not be avoidable on account of the storing handling or keeping of petrol or any other inflammable substance on the Holding or for any other reason connected with the use of the Holding

2.1.2 provide cover for the return to the land of the full equivalent manurial value of any crops destroyed or damaged insofar as the return thereof is required for the fulfilment by the Tenant of his obligations under this Agreement

3. Liabilities

3.1 to maintain adequate insurance in respect of all employer's risks for which the Tenant may be liable

3.2 to maintain insurance in respect of all third party liability risks in respect of which liability may fall on the Landlord in relation to the Holding the policy or policies of such insurance to provide cover in respect of each and every claim of not less than two million pounds or such higher sum as the Landlord may from time to time direct in writing

4. Landlord's insurance

in relation to each insurance in respect of the Holding and/or any other property effected by the Landlord:-

4.1 not to effect any insurance cover against any risks covered by the policy effected by the Landlord

4.2 not to do on the Holding any act or thing which shall or may invalidate such insurance or cause such insurance to be offered on less favourable terms or cause the premium thereon to be increased or increase the risk of any insured peril to any part of the Holding and/or any other property but to take full and proper precautions to protect the Holding and/or any other property against risk from any Insured Risk

4.3 save as permitted by any relevant insurance not to store or bring on the Holding any articles of a specifically combustible inflammable radio-active or dangerous nature and in particular to take all responsible precautions when installing petrol oil gas or electric engines and when storing petrol paraffin oil or similar fuel or lubricants which shall be kept in proper containers and where reasonably practicable in a detached building and in such cases to observe and perform every requirement of any statutory or other relevant provision

4.4 to indemnify the Landlord against any excess payable in relation to any claim made

5. **Tenant's insurance**

in relation to each insurance effected by the Tenant:-

5.1 to ensure that it is effected in the joint names of the Landlord and the Tenant with an insurer approved in writing by the Landlord

5.2 to produce to the Landlord on demand the policy or policies of insurance and the receipt for the premium in respect of the current period of insurance

5.3 to apply all money received through any such policy for the purpose in respect of which the same has been paid and to make good any shortfall from his own resources

6. **Repay premiums**

if any insurance which the Tenant is obliged to maintain hereunder is not effected or maintained or if the policy or premium receipt relating to it is not produced in accordance with the terms hereof and the Landlord effects such insurance cover as the Landlord may determine to be appropriate to make good such default (which he is hereby authorised to do) to repay to the Landlord on demand all premiums paid for such purpose and all other expenses incurred by the Landlord in connection therewith

THE REPAIRS SCHEDULE

[*If first alternative for 4.11 is used.*

To be adjusted to meet the case.]

PART 1
(Landlord's Repairing Obligations)

The Landlord shall be responsible for the maintenance repair and replacement of the following:-

1. **Dwellings**

Roofs (including chimneys)

External walls

Ceilings and internal plastering

Ceilings and floor joists

Floors

Staircases

Doors

Windows and skylights

Gutters and downpipes

Baths, toilets and other sanitary fittings

Electrical installations including fittings

Water pipes

Foul drainage systems

Boilers and heating apparatus

Internal decorations and treatments

External decorations and treatments

Fire detection and security systems

2. **Other buildings and fixed equipment**

Roofs (including chimneys)

Structural frames and walls

Cladding

Floors

Doors and gates

Windows

Staircases and fixed ladders

Gutters and downpipes

Electrical installation including fittings

Water supplies and fittings

Foul drainage system

Fixtures and fittings

External decorations and treatments

Internal decorations and treatments

Timber and other infestations

3. **External work and services**

Rainwater drainage systems

Foul drainage systems

Sewage disposal

Slurry systems

Water supply systems

Electrical supply systems

Gas supply systems

Garden walls and fences

Yard walls, fences and gates

Roads and yards

Cattle grids

Field gates and posts

Bridges and culverts

Field drains, ditches and associated works

Field boundaries

Watercourses, ponds, lakes and associated systems

Signs and notices

PART 2
(Tenant's Repairing Obligations)

The Tenant shall be responsible for the repair and maintenance of the following:-

1. Dwellings

Roofs (including chimneys)

External walls

Internal walls

Ceilings and internal plastering

Ceiling and floor joists

Floors

Staircases

Doors

Windows and skylights

Gutters and downpipes

Baths, toilets and other sanitary fittings

Electrical installations including fittings

Water pipes

Foul drainage systems

Boilers and heating apparatus

Internal decorations and treatments

External decorations and the treatments

Fire detection and security systems

2. Other buildings and fixed equipment

Roofs (including chimneys)

Structural frames and walls

Cladding

Floors

Doors and gates

Windows

Staircases and fixed ladders

Gutters and downpipes

Electrical installation including fittings

Water supplies and fittings

Foul drainage system

Fixtures and fittings

External decorations and treatments

Internal decorations and treatments

Timber and other infestations

3. External works and services

Rainwater drainage systems

Foul drainage systems

Sewage disposal

Slurry systems

Water supply systems

Electrical supply systems

Gas supply systems

Garden walls and fences

Yard walls, fences and gates

Roads and yards

Cattle grids

Field gates and posts

Bridges and culverts

Field drains, ditches and associated works

Field boundaries

Water courses, ponds, lakes and associated systems

Signs and notices

THE DEALINGS SCHEDULE

The Tenant's obligations in this Schedule are:-

1. if any dwelling on the Holding (other than any farmhouse) is not required by the Tenant for occupation by a person employed on the Holding but the Tenant wishes the dwelling to remain in the Holding so as to be available in the future for a person employed on the Holding then on giving the Landlord written notice thereof the Tenant may let the dwelling on an assured shorthold tenancy under the provisions of the Housing Act 1996 or any statutory modification thereof for a fixed term not exceeding twelve months at a full market rent

2. not to make any contract for any producer to grow a specialist crop on the Holding except on condition that such contract:-

2.1 is in writing and signed by both parties before the producer enters upon the Holding

2.2 does not confer on the producer any interest greater than a tenancy for a period of less than two years

2.3 does not confer on the producer any security of tenure

2.4 does not (or if more than one contract is granted all the contracts together do not) relate to a greater area in any one year than one-fifth of the total arable area of the Holding

2.5 does not relate to any part of the last year of the Tenancy except with the consent in writing of the Landlord

3. save as permitted by the preceding paragraphs of this Schedule:-

3.1 not to assign charge or underlet any part or parts (as opposed to the whole) of the Holding

3.2 not to part with the possession of or share the occupation of the Holding or any part or parts thereof

4. not to assign the Holding as a whole without the previous consent in writing of the Landlord but such consent shall not be unreasonably withheld

[5. upon every assignment to procure:-

5.1 that the assignee enters into a direct covenant with the Landlord to pay the rents and observe and perform the covenants and conditions on the part of the Tenant herein contained

5.2 if the assignee is a limited liability company and if the Landlord reasonably so requires to procure that a guarantor or guarantors of adequate standing enter into direct covenants with the Landlord in a form similar to those in the clause headed 'Surety's Covenants']

[6. not to underlet the whole of the Holding without the previous consent in writing of the Landlord which shall not be unreasonably withheld PROVIDED THAT:-

6.1 the rent for the Holding to be reserved by the underlease shall not be less than the Holding Rent and shall be the best rent obtainable without taking a fine or premium

6.2 the underlease shall contain such provisions as are requisite to secure that it is in all respects consistent with the provisions hereof including in particular provision for the review of the rent thereby reserved at the same time as the review of the Holding Rent hereunder provision that the Landlord shall approve the amount of any reviewed rent and an absolute prohibition against any further underletting]

[7. to use his best endeavours to enforce the performance and observance of all the covenants and conditions on the part of the undertenant contained in such underlease]

THE SINGLE PAYMENT SCHEME SCHEDULE

1. In this Schedule:-

1.1 'DEFRA' means the Department for the Environment Food and Rural Affairs or any successor department

1.2 'RPA' means the Rural Payments Agency or any successor body carrying out similar functions

1.3 'Eligible Hectare' shall have the same meaning as in Article 34(2) of Council Regulation (EC) No 73/2009

2. The Tenant further covenants with the Landlord:-

2.1 to maintain the Entitlements upon the maximum eligible area of the Holding in accordance with the Regulations

2.2 at all times during the Term to do all acts and things which may be required to maintain the maximum number of Entitlements under the Regulations

2.3 during the Term to claim payment upon the Entitlements in accordance with the Regulations in such a way that neither the Entitlements nor any part of them are lost or charged or whereby they are removed from the Tenant whether permanently or temporarily

2.4 to use his best endeavours to obtain for the Holding any new or further entitlements or additional payment value for the Entitlements which become available under the Regulations or otherwise whether existing at or introduced after the commencement of the Tenancy and whether they become available by reference to the Entitlements, the Tenant's occupation of the Holding or otherwise

2.5 to keep and maintain records of all entitlements allocated to or transferred or leased in or out by the Tenant and their use together with any changes in their number type or payment value as may arise as a result of any official action and to permit the Landlord and any person nominated by him to inspect and take copies of that record in so far as it relates to the Entitlements

2.6 at all times during the Term to carry out all obligations imposed on the Tenant as occupier or producer with regard to all of the requirements of the Regulations

2.7 within 14 days of a written request being received to furnish the Landlord (in writing if so required) with all information which the Landlord may reasonably request concerning the Entitlements or concerning other farming activities of the Tenant which may affect or tend to affect the Entitlements whether such farming activities are carried out by the Tenant directly or indirectly and whether they relate to the Holding or to other land

2.8 prior to submitting any forms to DEFRA and/or the RPA in any year to agree with the Landlord the information required by the forms and the contents thereof and to obtain the Landlord's consent to that information and to the contents of the forms

2.9 not to dispose of transfer lend lease charge or otherwise deal with the Entitlements or any of them so as to result in the Entitlements lapsing or reducing in number or value during the Term or being transferred off the Holding or otherwise becoming unavailable for use in full on the Holding on the termination of the Tenancy

2.10 not by any direct or indirect act or omission in respect of any land or property not included in the Holding to allow the Entitlements to pass to any other person or become claimable in respect of land not comprised within the Holding

[2.11 not to make or permit anyone claiming on his behalf or through him to make any claim either during or at or following termination of the Tenancy which may result in the Entitlements or any of them ceasing to belong to or be available to the Landlord and if as a result of any such claim the Entitlements or any of them are lost to the Landlord to indemnify the Landlord against all losses which the Landlord suffers thereby including (but not necessarily limited to) all costs and expenses incurred in the acquisition of replacement entitlements which are comparable in all respects to those lost]

[2.12 on or after quitting the Holding to take all such steps and do all such acts as may be required to procure the transfer of the Entitlements to the Landlord or such person or persons as the Landlord nominates in writing]

delete as appropriate

*[at a consideration to be agreed between the parties being the open market value and in default of agreement to be determined by arbitration]

*[without payment]

*[the Landlord paying the Tenant compensation in the sum of one pound (£1)]

*[being a sum equivalent to the open market value of that part of the Entitlements that would represent the historic element of each of them at that time in the same way as if such were capable of being separately valued notwithstanding the fact that the Entitlements are not divisible AND in default of agreement to be determined by arbitration]

*[being [ ]% of the open market value of the Entitlements at the date of termination of the Tenancy]

[2.13 as the provisions of clause 2.12 also apply where the Tenancy is terminated in respect of any part of the Holding in that case to procure the transfer pro rata to the Landlord of the number of entitlements that the Entitlements bear to the number of Eligible Hectares in respect of which the Tenancy is terminated compared with the number of Eligible Hectares comprised in the Holding and the compensation payable shall also be calculated pro rata on the same basis]

2.14 During the final year of the Tenancy to obtain the Landlord's consent in respect of all matters concerning the management of the Entitlements to include cross compliance claiming the single payment and effecting the transfer of the Entitlements as the Landlord may direct

2.15 In so far as they remain to be observed and performed and are capable of taking effect after the termination of the Term howsoever determined the provisions of this Schedule shall remain in full force and effect notwithstanding the termination of the Tenancy

THE SCHEDULE OF TENANT'S FIXTURES
(Tenant's Fixtures and Fittings)

THE SCHEDULE OF TENANT'S IMPROVEMENTS

THE SCHEDULE OF REDUNDANT BUILDINGS

THE HOLDOVER SCHEDULE

PART 1

Grain

1. Where at the termination of the Tenancy grain produced on the Holding is being stored in the Holdover Buildings the Landlord shall permit the Tenant to continue to use the Holdover Buildings and Holdover Plant for the purpose of drying and storing that grain until

the Grain Holdover Date or (if earlier) until all his grain has been removed from the Holding and to have appropriate access with or without vehicles to the Holdover Buildings

2. The Tenant shall reimburse the Landlord for the actual cost of fuel or electricity consumed and the cost of any men employed by the Landlord (and approved by the Tenant) who operate the Holdover Plant and/or load the grain

3. During the period of holdover under this Part of this Schedule the Landlord shall be responsible for the insurance of the Holdover Buildings and the Holdover Plant but not for the grain therein

4. The Landlord shall pay compensation for any damage arising from the negligence of the Landlord or any employee of his but subject thereto the Tenant shall be liable for the cost of any repair required to remedy any breakdown in the Holdover Plant

5. The Landlord shall be entitled to defer until the end of the period of holdover under this Part of this Schedule the payment to the Tenant of all compensation to which he is entitled for any of the Holdover Buildings or for any of the Holdover Plant which the Landlord has elected to acquire at valuation and any valuation of such items shall take into account the use of the Holdover Buildings and the Holdover Plant by the Tenant under this Part of this Schedule and any expenditure on repairs during the period of holdover under this Part of this Schedule

6. The Landlord shall be entitled to have all capital or other tax allowances apportioned from the date of termination of the Tenancy

PART 2

Sugar Beet

7. After the termination of the Tenancy the Landlord shall permit the Tenant to enter upon the Holding to raise the remainder of the his crop of sugar beet and cart the same to heaps on convenient sites at the road or drove-side in a good and husbandlike manner

8. The work of lifting and carting off shall be completed by the Sugar Beet Holdover Date next following the termination after which date the Landlord shall have the right to harvest and cart off any unharvested beet receiving payment for such work in accordance with the current prevailing agricultural contractors' charges in the district

9. The Tenant shall be responsible for arranging the loading and delivery of his crop to the factory and for the removal of the beet from the heaps

PART 3

General

10. No rent or other charge shall be payable by the Tenant for any right of holdover under this Schedule

11. At the end of the relevant holdover period the Tenant shall leave all parts of the Holding used in connection with the holdover in a clean and tidy condition

12. In this Schedule references to 'the Landlord' shall where the context requires or permits include any incoming tenant

THE COMPENSATION SCHEDULE
(Tenant's Compensation on Termination)

1. On the termination of this Agreement the Tenant shall be entitled to receive compensation in respect of the matters and on the terms and conditions set out in and subject to the provisions of Part III of the 1995 Act

2. Such compensation shall be calculated having regard to the methods of practice and prices of the Valuation Association

[3. Where the Tenant has submitted an IACS form and any set-aside payable in consequence thereof has not been paid:-

3.1 the Landlord may retain from the compensation due to the Tenant a sum equal to the estimated amount of such aid pending payment

3.2 for the purposes of calculating the compensation payable to the Tenant under this paragraph the Landlord agrees with the Tenant that:-

3.2.1 the value which may be attributed to the growing crops or severed or harvested crops and produce shall include a sum to reflect a fair proportion of the payments payable under IACS in respect of these crops

3.2.2 the compensation shall include a sum to reflect a fair proportion of the payments payable under IACS which relate to the Holding (including set-aside) but which do not relate to a specific crop]

4. Where sheep belonging to the Tenant have become acclimatised hefted or settled on the Holding the Tenant shall be entitled to receive compensation in the amount of a sum equal to the increase attributable to such acclimatisation hefting or settlement in the value of the Holding at the termination of this Agreement as land comprised in a tenancy

5. Where the Landlord requires by notice in writing to the Tenant given at any time prior to the termination of this Agreement that the Tenant leaves on the Holding severed crops unconsumed hay straw silage and farmyard manure the Tenant shall be entitled on leaving such items on the Holding to compensation equal to the market value of the items left subject to paragraph 3.1 above

SIGNED as a DEED by
in the presence of:-

SIGNED as a DEED by
in the presence of:-

PRECEDENT 4 FARM BUSINESS TENANCY (SHORT FORM)

PARTICULARS

Date of Agreement	:	....
The Landlord	:	....
		of ....
		....
The Tenant	:	....
		of ....
		....
Commencement Date	:	....
Term Date	:	[*date*]
Term	:	The period of [*no*] years from and including the Commencement Date/The period from and including the Commencement Date to and including the next following Term Date and thereafter from year to year until determined in accordance with clause [*no*] (*delete as appropriate*)
Rent	:	[£ . . .] per year (or such other yearly amount as may be substituted from time to time) (*delete if inapplicable*)
Rent Instalments	:	Equal instalments (except in respect of a broken period) payable in arrear/advance (*delete as appropriate*) on the Rent Days
Rent Days	:	[*date*] and [*date*] in each [*period*]
Permitted Use	:	Use in the trade or business of agriculture
Review Date	:	Any day which is a day next following a Term Date and which complies with the conditions set out in paragraph (b) of section 10(6) of the 1995 Act.

Single Payment Scheme	:	the single payment scheme governed by Council Regulation (EC) No 73/2009 and all the relevant subsequent Commission and other regulations, administrative provisions, statutory guidance or codes of practice implemented by the EC and/or the UK Government ('the Regulations')
Entitlements	:	such entitlements to the single income payment and any other entitlement to financial support of whatever kind deriving in any way from the Regulations and made available by the Landlord to the Tenant by reference to the area of the Holding subject to this Tenancy Agreement and no other holding and comprising the following entitlements held by the [Landlord/Tenant] in his name pursuant to the Regulations namely Type(s) Number Value and reference to Type(s) of entitlements has the same meaning(s) as that contained in the Regulations

THIS FARM BUSINESS TENANCY AGREEMENT is made on the Date of Agreement

BETWEEN:-

(1) [*name*] of [*address*] ('the Landlord')

and

(2) [*name*] of [*address*] ('the Tenant')

1. Particulars and definitions

1.1 The attached Particulars are incorporated in this Agreement and to the extent allowed by the context the expressions in the left-hand column of the Particulars have the respective meanings shown opposite them

1.2 In this Agreement to the extent allowed by the context the following phrases have the respective meanings set under them:-

1.2.1 'the Fixed Equipment'

the buildings structures fixtures and fittings (if any) affixed to and any works constructed on in or over the Holding from time to time except any which are to be regarded as belonging to the Tenant

1.2.2 'the Holding'

the property described in the Holding Description Schedule

1.2.3 'Insured Risks'

such risks in relation to the insurance concerned as the Landlord may from time to time reasonably require

1.2.4 'the Tenancy'

the tenancy hereby created

1.2.5 'the 1995 Act'

the Agricultural Tenancies Act 1995

2. Interpretation

In this Agreement to the extent allowed by the context:-

2.1 'the Holding' includes any part of the Holding

2.2 'the Landlord' includes whoever for the time being owns the interest in the Holding which gives the right to possession of it at the end of the Tenancy

2.3 'the Tenant' includes the Tenant's successors in title

2.4 'Value Added Tax' includes any substituted or additional tax of a similar nature

2.5 unless stated otherwise any reference to a sum of money is to that sum exclusive of Value Added Tax and any Value Added Tax charged on it is payable in addition

2.6 a right given to the Landlord to enter the Holding extends to anyone the Landlord authorises to enter and includes the right to bring workmen and appliances onto the Holding for the stated purposes

2.7 authority given to a person to enter the Holding after giving notice extends if the circumstances justify it to entry after giving less notice than specified or without giving any notice

2.8 any obligation not to do any act or thing includes an obligation not to cause or allow anyone else to do that act or thing

2.9 whenever any party in relation to the Tenancy comprises two or more persons the obligations of that party may be enforced against any one of those persons alone or against all or some of them jointly

2.10 the act or default of any person on the Holding with the express or implied consent of the Tenant shall be deemed the act or default of the Tenant

2.11 words importing the singular number include the plural and vice versa; words importing persons include corporations and vice versa; words importing only one gender include the other genders

2.12 any reference to this Agreement includes its schedules

2.13 any reference to a page clause or schedule is to that page clause or schedule of this Agreement

2.14 the clause and paragraph headings are for convenient reference only and shall not affect its construction

2.15 any reference to a statute regulation order or code of practice includes any later provision substituted for or amending the same and any subsidiary provision made under powers thereby conferred

3. Letting

EITHER

3.1 The Landlord lets to the Tenant ALL THAT the Holding TO HOLD the same unto the Tenant for the Term (subject to any provision for earlier determination herein contained) on the terms of this Agreement YIELDING AND PAYING therefor the Rent in accordance with the provisions of this Agreement

[3.2 The Landlord shall also lease the Entitlements to the Tenant for the Term together with the Holding on the terms hereinafter appearing at an additional rental of pounds per annum (£) plus VAT @ 15% (if such is applicable) such additional rental to be paid in the same way and at the same times as the Rent and subject to the same provisions for enforcement of payment as if they were separately set out below

[*NOTE: A landlord may let a holding and entitlements to a tenant for a global sum and not separately identify the two elements as here. In such a case the definitions of Holding and/or Rent can be amended to make reference to the fact that the entitlements are being let together with the Holding.*]]

OR

3.1 The Landlord lets to the Tenant ALL THAT the Holding TO HOLD the same unto the Tenant for the Term (subject to any provision for earlier determination herein contained) on the terms of this Agreement YIELDING AND PAYING therefor the Rent in accordance with the provisions of this Agreement

[3.3 Either the Landlord or the Tenant may determine the Tenancy by giving to the other notice in writing of not less than twelve months expiring on the second or any subsequent Term Date]

4. Tenant's covenants

The Tenant covenants with the Landlord:-

4.1 *Rent*

to pay the Rent by the Rent Instalments on the Rent Days

4.2 *Outgoings*

to pay all existing and future rates taxes assessments and outgoings of any kind charged or imposed upon the Holding or upon the owner or occupier thereof

4.3 *Value added tax*

to pay all Value Added Tax which may be payable on any taxable supply made to the Tenant in relation to the Tenancy or on any payment by the Landlord which the Tenant is required to reimburse to the Landlord

4.4 *Insurance*

4.4.1 to insure the Fixed Equipment against damage or destruction by the Insured Risks to the full cost of rebuilding and reinstatement (including the cost of demolition and fees) and to have the interest of the Landlord duly noted on the policy

4.4.2 if so required by the Landlord to produce reasonably satisfactory evidence that the insurance required by this clause is in force

4.5 *Repair*

4.5.1 at all times to maintain and keep in good repair and condition and (where appropriate) neat and tidy the Holding and all fixtures and fittings (whether belonging to the Landlord or the Tenant) on it or on any other property but serving the Holding

4.5.2 to keep clear and free from obstruction all land drains and ditches on the Holding

4.5.3 to keep all boundary features on the Holding in such state as is adequate for their purpose and (in the case of hedges) reasonably cut back

4.5.4 to pay to the Landlord on demand an appropriate proportion of the costs from time to time incurred by the Landlord in the maintenance and repair of any items used in common with any adjoining or neighbouring property

4.6 *Good husbandry*

to farm in accordance with the requirements of good husbandry so as to keep the land at all times in good heart and condition

4.7 *Permitted use*

to use the Holding for the Permitted Use only and in particular not to use any part of the Holding as a market garden

4.8 *Alteration*

the Tenant shall not on the Holding:-

4.8.1 commit waste of any kind

4.8.2 plant any trees coppice saplings pollards or underwood

4.8.3 make any alteration any addition or any tenant's improvement falling within the obligations on his part contained in Part III of the 1995 Act other than any routine improvement as defined in section 19(10) of the 1995 Act

4.9 *Statutory obligations*

at the Tenant's own expense to do everything in relation to the Holding or its use which is required to comply with any statutory or other obligation (whether such obligation is imposed on the owner or the occupier) and in particular those relating to:-

4.9.1 the health safety and welfare of the persons using or employed on or about or visiting the Holding

4.9.2 waste effluent manure or other matter or thing of a noxious dangerous or offensive nature

4.10 *Not to obstruct roadway*

not to obstruct any roadway or track by which access to and from the Holding or any neighbouring property is gained

4.11 *No nuisance*

not to do or allow to remain upon the Holding anything which may be or become or cause a nuisance disturbance inconvenience injury or damage to the Landlord or the owner or occupier of any neighbouring property

4.12 *Adverse rights*

to take all reasonable steps to prevent any right over the Holding being acquired and to prevent trespass on the Holding

4.13 *Indemnity – loss*

4.13.1 not to do or omit to do on or near the Holding any act or thing which may render the Landlord liable to pay any penalty damages compensation costs charges or expenses or to carry out any work of whatever nature

4.13.2 at all times to keep the Landlord fully indemnified against all losses claims actions and liabilities arising (whether directly or indirectly) as a result of any act or default of the Tenant

4.14 *Dilapidations notice*

to pay to the Landlord on an indemnity basis all expenses incurred by the Landlord in the preparation and service of a notice under section 146 of the Law of Property Act 1925 or in relation to proceedings under section 146 or 147 of that Act notwithstanding that forfeiture is avoided otherwise than by relief granted by the Court

4.15 *Alienation*

not without the consent of the Landlord to assign underlet charge or part with possession of the whole or any part of the Holding or permit anyone else to occupy the whole or any part of the Holding

4.16 *Notices*

to give full particulars to the Landlord of any notice direction order or proposal relating to the Holding made given or issued to the Tenant within seven days of receipt and to take all necessary steps to comply with it

4.17 *Yield up*

on the determination of the Tenancy to yield up the Holding in such repair and so cultivated as to be in accordance with the terms of this Agreement and with vacant possession

4.18 *Single Payment Scheme*

to comply with the provisions of the Single Payment Scheme Schedule

[4.19 *Costs of agreement*

on the signing hereof to reimburse the Landlord the legal costs and expenses incurred by the Landlord of and incidental to the preparation of this Agreement and a counterpart thereof]

5. Landlord's agreement

The Landlord agrees with the Tenant:-

5.1 to permit the Tenant on paying the Rent and performing and observing the several obligations of the Tenant in relation to the Tenancy peaceably and quietly to hold and enjoy the Holding during the Tenancy without any interruption or disturbance from or by the Landlord or any person claiming under or in trust for the Landlord

5.2 to comply with the obligations on his part contained in the Single Payment Scheme schedule

6. Re-entry

At any time after the happening of any of the events specified in the following paragraphs of this clause it shall be lawful for the Landlord after giving two months' notice to the Tenant of his intention to do so to re-enter the Holding or any part thereof in the name of the whole and thereupon the Tenancy shall absolutely determine but without prejudice to any right of action or remedy of the Landlord in respect of any antecedent breach by the Tenant of any of his obligations in relation to the Tenancy namely:-

6.1 if any rent hereby reserved is unpaid for fourteen days after becoming due (whether legally demanded or not)

6.2 if the Tenant at any time fails or neglects to perform or observe any of his obligations in relation to the Tenancy

6.3 if the Tenant has any distress or process of execution levied on his goods

EITHER (Company Tenant)

6.4 if the Tenant:-

6.4.1 enters into liquidation whether compulsorily or voluntarily (not being a voluntary liquidation for the sole purpose of reconstruction or amalgamation of a solvent company)

6.4.2 suffers an administrative receiver to be appointed

6.4.3 enters into a voluntary arrangement as defined in section 1 of the Insolvency Act 1986

6.4.4 does anything which would entitle a receiver to be appointed to take possession of any of its assets

6.4.5 does anything which would entitle any person to present a petition for winding up the Tenant or apply for an administration order in relation to the Tenant

6.4.6 ceases to exist or is dissolved

OR (Individual Tenant)

6.5 if the Tenant:-

6.5.1 is unable to pay a debt or has no reasonable prospect of being able to pay a debt within the meaning of section 268 of the Insolvency Act 1986

6.5.2 makes an application to the court for an interim order under section 253 of the Insolvency Act 1986

6.5.3 makes any assignment for the benefit of or enters into any arrangement with his creditors either by composition or otherwise

6.5.4 does anything which would entitle a petition for a bankruptcy order to be made against him

EITHER

7. Rent review

For the purposes of Part II of the 1995 Act the Review Date from and including which the Rent may be varied is the Review Date

OR

8. No rent review

The Rent specified in the Particulars shall be payable throughout the Term and there shall be no review

[9. ***Break clause***

A periodic break provision is referred to above at clause 3.3. This may be all the parties require. Alternatively, there could be a provision permitting a break on the happening of a particular event, eg death of the Tenant, the Landlord obtaining planning permission for development of all or part of the Holding. In such a case clause 3.3 can be omitted and that provision can be included here. The notice can relate to all or part of the Holding, but only if the Term is for two years or less can the notice period be less than twelve months.]

10. Custom

The rights of the parties under this Agreement or otherwise in respect of the Tenancy shall not depend upon or be affected by any custom of the country

11. Certificate

It is hereby certified that there is no agreement for lease to which this document gives effect

AS WITNESS the hands of the parties hereto the day and year first before written

THE HOLDING DESCRIPTION SCHEDULE

ALL THAT farm shortly described as [*address*] and more particularly described below by reference to the Ordnance Survey Plan ([ ] Edition) of the district

TOGETHER with the benefit of any rights which are appurtenant to the Holding or which are reasonably needed for its proper enjoyment by the Tenant and capable of grant by the Landlord

SUBJECT TO any rights which are expressly excepted or reserved out of the Holding and any rights or quasi-rights which have been enjoyed until now over the Holding by any occupier of any neighbouring property or the Landlord

SUBJECT ALSO TO any subsisting obligations under any covenants affecting the Holding but with any subsisting benefit under any such covenants

OS No	Area (hectares)

THE SINGLE PAYMENT SCHEME SCHEDULE

1. In this Schedule:-

1.1 'DEFRA' means the Department for the Environment Food and Rural Affairs or any successor department

1.2 'RPA' means the Rural Payments Agency or any successor body carrying out similar functions

1.3 'Eligible Hectare' shall have the same meaning as in Article 34(2) of Council Regulation (EC) No 73/2009

2. The Tenant further covenants with the Landlord:-

2.1 to maintain the Entitlements upon the maximum eligible area of the Holding in accordance with the Regulations

2.2 at all times during the Term to do all acts and things which may be required to maintain the maximum number of Entitlements under the Regulations

2.3 during the Term to claim payment upon the Entitlements in accordance with the Regulations in such a way that neither the Entitlements nor any part of them are lost or charged or whereby they are removed from the Tenant whether permanently or temporarily

2.4 to use his best endeavours to obtain for the Holding any new or further entitlements or additional payment value for the Entitlements which become available under the Regulations or otherwise whether existing at or introduced after the commencement of the Tenancy and whether they become available by reference to the Entitlements, the Tenant's occupation of the Holding or otherwise

2.5 to keep and maintain records of all entitlements allocated to or transferred or leased in or out by the Tenant and their use together with any changes in their number type or payment value as may arise as a result of any official action and to permit the Landlord and any person nominated by him to inspect and take copies of that record in so far as it relates to the Entitlements

2.6 at all times during the Term to carry out all obligations imposed on the Tenant as occupier or producer with regard to all of the requirements of the Regulations

2.7 within fourteen days of a written request being received to furnish the Landlord (in writing if so required) with all information which the Landlord may reasonably request concerning the Entitlements or concerning other farming activities of the Tenant which may affect or tend to affect the Entitlements whether such farming activities are carried out by the Tenant directly or indirectly and whether they relate to the Holding or to other land

2.8 prior to submitting any forms to DEFRA and/or the RPA in any year to agree with the Landlord the information required by the forms and the contents thereof and to obtain the Landlord's consent to that information and to the contents of the forms

2.9 not to dispose of transfer lend lease charge or otherwise deal with the Entitlements or any of them so as to result in the Entitlements lapsing or reducing in number or value during the Term or being transferred off the Holding or otherwise becoming unavailable for use in full on the Holding on the termination of the Tenancy

2.10 not by any direct or indirect act or omission in respect of any land or property not included in the Holding to allow the Entitlements to pass to any other person or become claimable in respect of land not comprised within the Holding

[2.11 not to make or permit anyone claiming on his behalf or through him to make any claim either during or at or following termination of the Tenancy which may result in the Entitlements or any of them ceasing to belong to or be available to the Landlord and if as a result of any such claim the Entitlements or any of them are lost to the Landlord to indemnify the Landlord against all losses which the Landlord suffers thereby including (but not necessarily limited to) all costs and expenses incurred in the acquisition of replacement entitlements which are comparable in all respects to those lost]

[2.12 on or after quitting the Holding to take all such steps and do all such acts as may be required to procure the transfer of the Entitlements to the Landlord or such person or persons as the Landlord nominates in writing]

delete as appropriate

*[at a consideration to be agreed between the parties being the open market value and in default of agreement to be determined by arbitration]

*[without payment]

*[the Landlord paying the Tenant compensation in the sum of one pound (£1)]

*[being a sum equivalent to the open market value of that part of the Entitlements that would represent the historic element of each of them at that time in the same way as if such were capable of being separately

valued notwithstanding the fact that the Entitlements are not divisible AND in default of agreement to be determined by arbitration]

*[being [ ]% of the open market value of the Entitlements at the date of termination of the Tenancy]

[2.13 as the provisions of clause 2.12 also apply where the Tenancy is terminated in respect of any part of the Holding in that case to procure the transfer pro rata to the Landlord of the number of entitlements that the Entitlements bear to the number of Eligible Hectares in respect of which the Tenancy is terminated compared with the number of Eligible Hectares comprised in the Holding and the compensation payable shall also be calculated pro rata on the same basis]

2.14 during the final year of the Tenancy to obtain the Landlord's consent in respect of all matters concerning the management of the Entitlements to include cross compliance claiming the single payment and effecting the transfer of the Entitlements as the Landlord may direct

2.15 in so far as they remain to be observed and performed and are capable of taking effect after the termination of the Term howsoever determined the provisions of this Schedule shall remain in full force and effect notwithstanding the termination of the Tenancy

3. the Landlord covenants with the Tenant to comply with the requirements of the Single Payment Scheme after the expiration of the tenancy during the then subsisting calender year

SIGNED by
in the presence of:-

SIGNED by
in the presence of:-

PRECEDENT 5 FARM BUSINESS TENANCY SPECIFYING ENTITLEMENTS COMPENSATION FOR TENANT AT VALUATION

FARM BUSINESS TENANCY

PARTICULARS

Date of Agreement	:	....
The Landlord	:	....
The Tenant	:	....
Commencement Date	:	....
Term	:	from and including the Commencement Date

Rent	:	£ . . .
Rent Instalments	:	Equal instalments payable on the Rent days
Rent Days	:	and
Permitted Use	:	Use in the trade or business of agriculture
Holdover Period	:	[ ] months following the end of the Term
The Relevant Entitlements		• Normal Entitlements having a value in [ ] of [ ] each • Former set-aside Entitlements

THIS FARM BUSINESS TENANCY AGREEMENT is made on the Date of Agreement

BETWEEN:-

(1) [*name*] of [*address*] ('the Landlord')

and

(2) [*name*] of [*address*] ('the Tenant')

1. Particulars and definitions

1.1 The attached Particulars are incorporated in this Agreement and to the extent allowed by the context the expressions in the left-hand column of the Particulars have the respective meanings shown opposite them

1.2 In this Agreement to the extent allowed by the context the following phrases have the respective meanings set under them:-

1.2.1 'Environmental Stewardship Schemes' means national or European schemes for the conservation or enhancement of the environment under which the Landlord has assumed obligations or is subject to restrictions in respect of the Holding

1.2.2 'the Holding' means the property described in Schedule 1

1.2.3 'the Landlord' includes whoever for the time being owns the interest in the Holding which gives right to possession of it at the end of the Tenancy

1.2.4 'the Tenancy' means the tenancy hereby created

1.2.5 'the Tenant' includes the Tenant's successors in title

1.2.6 'the 1995 Act' means the Agricultural Tenancies Act 1995

2. Interpretation

In this Agreement to the extent allowed by the context:-

2.1 'the Holding' includes any part of the Holding

2.2 'the Single Payment Scheme' means the Scheme governed by Council Regulation (EC) No 73/2009 and any Commission Regulations supplementary thereto and any related European or UK regulations statutory guidance or codes of practice

2.3 'Value Added Tax' includes any substituted or additional tax of a similar nature

2.4 unless stated otherwise any reference to a sum of money is to that sum exclusive of Value Added Tax and any Value Added Tax charged on it is payable in addition

2.5 any obligation not to do any act or thing includes an obligation not to cause or allow anyone else to do that act or thing

2.6 any reference to this Agreement includes its schedules

2.7 any reference to a page clause or schedule is to that page clause or schedule in this Agreement

2.8 the clause and paragraph headings are for convenient reference only and shall not affect its construction

2.9 any reference to a statute regulation order or code of practice includes any later provision substituted for or amending the same and any subsidiary provision made under powers thereby conferred

3. Letting

3.1 The Landlord lets the Holding to the Tenant for the Term

4. Tenant's covenants

The Tenant covenants with the Landlord:-

4.1 *Rent*

to pay the Rent by the Rent Instalments on the Rent Days

4.2 *Value added tax*

to pay all Value Added Tax which may be payable on any taxable supply made to the Tenant in relation to the Tenancy or on any payment by the Landlord which the Tenant is required to reimburse to the Landlord

4.3 *Repair*

4.3.1 at all times to maintain and keep in good condition and (where appropriate) neat and tidy the Holding

4.3.2 to keep clear and free from obstruction all land drains and ditches on the Holding

4.3.3 to keep all boundary features on the Holding in such state as is adequate for their purpose and (in the case of hedges) reasonably cut back

4.4 *Good agricultural and environmental practice*

to keep the Holding in good agricultural and environmental condition as defined under the Single Payment Scheme

4.5 *Environmental Stewardship Schemes*

to ensure that its activities on the Holding do not breach the requirements of any Environmental Stewardship Schemes written details of which have been provided to the Tenant by the Landlord

4.6 *Permitted use*

to use the Holding for the Permitted Use only

4.7 *Alteration*

not on the Holding to:-

4.7.1 commit waste of any kind

4.7.2 plant any trees, coppice, saplings, pollards or underwood

4.7.3 make any alteration, any addition or any tenant's improvement falling within the provision of Part III of the 1995 Act other than by any routine improvements as defined in section 19(10) of the 1995 Act

4.8 *Statutory obligations*

at the Tenant's own expense to do everything in relation to the Holding or its use which is required to comply with any statutory or other obligation relating to the occupation of the Holding and in particular those relating to:-

4.8.1 the health safety and welfare of the persons using or employed on or about or visiting the Holding

4.8.2 waste effluent manure or other matter or thing of a noxious, dangerous or offensive nature

4.9 *Not to obstruct roadway*

not to obstruct any roadway or track by which access to and from the Holding or any neighbouring property is gained

4.10 *No nuisance*

not to do or allow to remain upon the Holding anything which may be or become or cause a nuisance, disturbance, inconvenience, injury or damage to the Landlord or the owner or occupier of any neighbouring property

4.11 *Adverse rights*

to take all reasonable steps to prevent any right over the Holding being acquired and to prevent trespassers on the Holding

4.12 *Indemnity – loss*

4.12.1 not to do or omit to do on or near the Holding any act or thing which may render the Landlord liable to pay any penalty, damages, compensation, costs, charges or expenses or to carry out any work of whatever nature

4.12.2 at all times to keep the Landlord fully indemnified against all losses, claims, actions and liabilities arising (whether directly or indirectly) as a result of any act or default of the Tenant

4.13 *Dilapidations notice*

to pay to the Landlord on an indemnity basis all expenses incurred by the Landlord in the preparation and service of a notice under section 146 of the Law of Property Act 1925 or in relation to proceedings under section 146 or 147 of the Act notwithstanding that forfeiture is avoided otherwise than by relief granted by the Court

4.14 *Alienation*

not without the consent of the Landlord to assign underlet charge or part with possession of the whole or any part of the Holding or permit anyone else to occupy the whole or any part of the Holding

4.15 *Notices*

to give full particulars to the Landlord of any notice, direction, order or proposal relating to the Holding made given or issued to the Tenant within seven days of receipt

4.16 *Single Payment Scheme*

to comply with the provisions of Schedule 2

4.17 *Yield up*

on the determination of the Tenancy to yield up the Holding in such repair and so cultivated as to be in accordance with the terms of this Agreement and with vacant possession

5. Landlord's agreement

5.1 The Landlord agrees with the Tenant to permit the Tenant on paying the Rent and performing and observing the several obligations of the Tenant in relation to the Tenancy peaceably and quietly to hold and enjoy the Holding during the Tenancy without any interruption or disturbance from or by the Landlord or any person claiming under or in trust for the Landlord

5.2 The Landlord agrees to comply with the provisions of Schedule 2

6. Landlord's reservations

The Landlord reserves the right to go onto the Holding to do anything reasonably necessary to secure compliance with any Environmental Stewardship Schemes (including carrying out works and accompanying officials from any competent body responsible for such schemes)

7. Re-entry

At any time after the happening of any of the events specified in the following paragraphs of this clause it shall be lawful for the Landlord after giving two months' notice to the Tenant of his intention to do so to re-enter the Holding or any part thereof in the name of the whole and thereupon the Tenancy shall absolutely determine but without prejudice to any right of action or remedy of the Landlord in respect of any antecedent breach by the Tenant of any of his obligations in relation to the Tenancy

7.1 If the Tenant:-

7.1.1 is unable to pay a debt or has no reasonable prospect of being able to pay a debt within the meaning of section 268 of the Insolvency Act 1986

7.1.2 makes an application to the court for an interim order under section 253 of the Insolvency Act 1986

7.1.3 makes any assignment for the benefit of or enters into any arrangement with his creditors either by composition or otherwise

7.1.4 does anything which would entitle a petition for a bankruptcy order to be made against him

8. Custom

The rights of the parties under this agreement or otherwise in respect of the Tenancy shall not depend upon or be affected by any custom of the country

9. Early entry

9.1 Following clearance of the Tenant's crop from the holding and provided that the Landlord complies with relevant rules or regulations

of the Single Payment Scheme (including cross-compliance) the Landlord will be permitted access to the Holding before the termination of the Tenancy for the purpose of carrying out cultivations and subject to any reasonable limitations or restrictions specified by the Tenant

9.2 For the avoidance of doubt the Tenant retains possession of the Holding and remains responsible under paragraph 8 of Schedule 2 for the requirements of the Single Payment Scheme and the Landlord enters as licensee only and agrees to comply with the Tenant's reasonable requirements

10. Holdover

At the end of the Term the Landlord shall permit the Tenant free of charge to enter on the Holding in a good and husbandlike manner during the Holdover Period to harvest and remove any crop then in the ground

AS WITNESS the hands of the parties hereto the day and year first before written

SCHEDULE 1
HOLDING DESCRIPTION

ALL THAT land described below TOGETHER with the benefit of any rights which are appurtenant to the Holding or which are reasonably needed for its proper enjoyment by the Tenant and capable of grant by the Landlord SUBJECT TO any rights which are expressly excepted or reserved out of the Holding and any rights or quasi-rights which have been enjoyed until now over the Holding by any occupier of any neighbouring property or the Landlord SUBJECT ALSO TO any subsisting obligations under any covenants affecting the Holding but with any subsisting obligations under any covenants

OS Ref	NG No	Total field area	Eligible Area

SCHEDULE 2
THE SINGLE PAYMENT SCHEME

1. In this Schedule:-

1.1 'the Council Regulation' means Council Regulation (EC) No 73/2009

1.2 'Normal Entitlements' shall have the same meaning as in the Council Regulation

1.3 'Former Set-Aside Entitlements' shall have the same meaning as in the Council Regulation

1.4 'the Entitlements' means Normal Entitlements and/or former Set-Aside Entitlements (as the case may be)

1.5 'the Transferred Entitlements' means the Entitlements to be transferred to the Landlord or his nominee in accordance with the terms hereof

2. On quitting the Holding the Tenant will take all such steps and do all such acts as are required to procure the transfer to the Landlord or his nominees of the Transferred Entitlements

3. The Tenant shall be entitled to compensation in respect of the Transferred Entitlements of the market value thereof at the end of the term agreed between the parties and in default of agreement determined by arbitration to be paid forthwith on ascertainment

4. At the end of the tenancy the Tenant shall retain the payments in respect of the Transferred Entitlements for the then current harvest year or as otherwise agreed

5. In so far as they remain to be observed and performed and are capable of taking effect after the termination of the tenancy created by this Agreement the provisions of this clause shall remain in full force and effect notwithstanding such termination

6. The Tenant so long as the tenancy subsists created by this Agreement and thereafter the Landlord will ensure compliance with the rules and regulations of the Single Payment Scheme in respect of the Holding

SIGNED by
in the presence of:-

SIGNED by
in the presence of:-

PRECEDENT 6 FARM BUSINESS TENANCY LETTING WITH ENTITLEMENTS

FARM BUSINESS TENANCY

PARTICULARS

Date of Agreement	:	....
The Landlord	:	....
The Tenant	:	....
Commencement Date	:	....
Term	:	from and including the Commencement Date
Rent	:	£[ ]

Rent Instalments	:	Equal instalments payable on the Rent days
Rent Days	:	and
Permitted Use	:	Use in the trade or business of agriculture
Entitlements	:	As described in Schedule 3
Entitlement Charge	:	A sum equal to the amount payable to the Tenant under the Single Payment Scheme in respect of the Entitlements on the assumption that the rules of the Single Payment Scheme have been fully complied with by the Tenant, payable on the due date for receipt from the Rural Payments Agency ('the Due Date')
Holdover Period	:	[ ] months following the end of the Term

THIS FARM BUSINESS TENANCY AGREEMENT is made on the Date of Agreement BETWEEN:-

(1) [*name*] of [*address*] ('the Landlord')

and

(2) [*name*] of [*address*] ('the Tenant')

1. Particulars and definitions

1.1 The attached Particulars are incorporated in this Agreement and to the extent allowed by the context the expressions in the left-hand column of the Particulars have the respective meanings shown opposite them

1.2 In this Agreement to the extent allowed by the context the following phrases have the respective meanings set under them:-

1.2.1 'Environmental Stewardship Schemes' means national or European schemes for the conservation or enhancement of the environment under which the Landlord has assumed obligations or is subject to restrictions in respect of the Holding

1.2.2 'the Holding' means the property described in Schedule 1

1.2.3 'the Landlord' includes whoever for the time being owns the interest in the Holding which gives right to possession of it at the end of the Tenancy

1.2.4 'the Tenancy' means the tenancy hereby created

1.2.5 'the Tenant' includes the Tenant's successors in title

1.2.6 'the 1995 Act' means the Agricultural Tenancies Act 1995

2. Interpretation

In this Agreement to the extent allowed by the context:-

2.1 'the Holding' includes any part of the Holding

2.2 'the Single Payment Scheme' means the single payment scheme governed by Council Regulation (EC) No 73/2009 and all the relevant subsequent Commission and other regulations, administrative provisions, statutory guidance or codes of practice implemented by the EC and/or the UK Government ('the Regulations')

2.3 'Value Added Tax' includes any substituted or additional tax of a similar nature

2.4 unless stated otherwise any reference to a sum of money is to that sum exclusive of Value Added Tax and any Value Added Tax charged on it is payable in addition

2.5 any obligation not to do any act or thing includes an obligation not to cause or allow anyone else to do that act or thing

2.6 any reference to this Agreement includes its schedules

2.7 any reference to a page clause or schedule is to that page clause or schedule in this Agreement

2.8 the clause and paragraph headings are for convenient reference only and shall not affect its construction

2.9 any reference to a statute regulation order or code of practice includes any later provision substituted for or amending the same and any subsidiary provision made under powers thereby conferred

3. Letting

3.1 The Landlord lets the Holding to the Tenant for the Term

4. Tenant's covenants

The Tenant covenants with the Landlord:-

4.1 *Rent*

to pay the Rent by the Rent Instalments on the Rent Days

4.2 *Value Added Tax*

to pay all Value Added Tax which may be payable on any taxable supply made to the Tenant in relation to the Tenancy or on any payment by the Landlord which the Tenant is required to reimburse to the Landlord

4.3 *Repair*

4.3.1 at all times to maintain and keep in good condition and (where appropriate) neat and tidy the Holding

4.3.2 to keep clear and free from obstruction all land drains and ditches on the Holding

4.3.3 to keep all boundary features on the Holding in such state as is adequate for their purpose and (in the case of hedges) reasonably cut back

4.4 *Good agricultural and environmental practice*

to keep the Holding in good agricultural and environmental condition as defined under the Single Payment Scheme

4.5 *Environmental Stewardship Schemes*

to ensure that its activities on the Holding do not breach the requirements of any Environmental Stewardship Schemes written details of which have been provided to the Tenant by the Landlord

4.6 *Permitted use*

to use the Holding for the Permitted Use only

4.7 *Alteration*

not on the Holding to:-

4.7.1 commit waste of any kind

4.7.2 plant any trees, coppice, saplings, pollards or underwood

4.7.3 make any alteration, any addition or any tenant's improvement falling within the provision of Part III of the 1995 Act other than by any routine improvements as defined in section 19(10) of the 1995 Act

4.8 *Statutory obligations*

at the Tenant's own expense to do everything in relation to the Holding or its use which is required to comply with any statutory or other obligation relating to the occupation of the Holding and in particular those relating to:-

4.8.1 the health safety and welfare of the persons using or employed on or about or visiting the Holding

4.8.2 waste effluent manure or other matter or thing of a noxious, dangerous or offensive nature

4.9 *Not to obstruct roadway*

not to obstruct any roadway or track by which access to and from the Holding or any neighbouring property is gained

4.10 *No nuisance*

not to do or allow to remain upon the Holding anything which may be or become or cause a nuisance, disturbance, inconvenience, injury or damage to the Landlord or the owner or occupier of any neighbouring property

4.11 *Adverse rights*

to take all reasonable steps to prevent any right over the Holding being acquired and to prevent trespassers on the Holding

4.12 *Indemnity – loss*

4.12.1 not to do or omit to do on or near the Holding any act or thing which may render the Landlord liable to pay any penalty, damages, compensation, costs, charges or expenses or to carry out any work of whatever nature

4.12.2 at all times to keep the Landlord fully indemnified against all losses, claims, actions and liabilities arising (whether directly or indirectly) as a result of any act or default of the Tenant

4.13 *Dilapidations notice*

to pay to the Landlord on an indemnity basis all expenses incurred by the Landlord in the preparation and service of a notice under section 146 of the Law of Property Act 1925 or in relation to proceedings under section 146 or 147 of the Act notwithstanding that forfeiture is avoided otherwise than by relief granted by the Court

4.14 *Alienation*

not without the consent of the Landlord to assign underlet charge or part with possession of the whole or any part of the Holding or permit anyone else to occupy the whole or any part of the Holding

4.15 *Notices*

to give full particulars to the Landlord of any notice, direction, order or proposal relating to the Holding made given or issued to the Tenant within seven days of receipt

4.16 *Single Payment Scheme*

to comply with the provisions of Schedule 2

4.17 *Yield up*

on the determination of the Tenancy to yield up the Holding in such repair and so cultivated as to be in accordance with the terms of this Agreement and with vacant possession

5. Landlord's agreement

5.1 The Landlord agrees with the Tenant to permit the Tenant on paying the Rent and performing and observing the several obligations of the Tenant in relation to the Tenancy peaceably and quietly to hold and enjoy the Holding during the Tenancy without any interruption or disturbance from or by the Landlord or any person claiming under or in trust for the Landlord

5.2 The Landlord agrees to comply with the provisions of Schedule 2

6. Landlord's reservations

The Landlord reserves the right to go onto the Holding to do anything reasonably necessary to secure compliance with any Environmental Stewardship Schemes (including carrying out works and accompanying officials from any competent body responsible for such schemes)

7. Re-entry

At any time after the happening of any of the events specified in the following paragraphs of this clause it shall be lawful for the Landlord after giving two months' notice to the Tenant of his intention to do so to re-enter the Holding or any part thereof in the name of the whole and thereupon the Tenancy shall absolutely determine but without prejudice to any right of action or remedy of the Landlord in respect of any antecedent breach by the Tenant of any of his obligations in relation to the Tenancy

7.1 If the Tenant:-

7.1.1 is unable to pay a debt or has no reasonable prospect of being able to pay a debt within the meaning of section 268 of the Insolvency Act 1986

7.1.2 makes an application to the court for an interim order under section 253 of the Insolvency Act 1986

7.1.3 makes any assignment for the benefit of or enters into any arrangement with his creditors either by composition or otherwise

7.1.4 does anything which would entitle a petition for a bankruptcy order to be made against him

8. Custom

The rights of the parties under this agreement or otherwise in respect of the Tenancy shall not depend upon or be affected by any custom of the country

9. Early entry

9.1 Following clearance of the Tenant's crop from the holding and provided that the Landlord complies with relevant rules or regulations

of the Single Payment Scheme (including cross-compliance) the Landlord will be permitted access to the Holding before the termination of the Tenancy for the purpose of carrying out cultivations and subject to any reasonable limitations or restrictions specified by the Tenant

9.2 For the avoidance of doubt the Tenant retains possession of the Holding and remains responsible under paragraph 8 of Schedule 2 for the requirements of the Single Payment Scheme and the Landlord enters as licensee only and agrees to comply with the Tenant's reasonable requirements

10. Holdover

At the end of the Term the Landlord shall permit the Tenant free of charge to enter on the Holding in a good and husbandlike manner during the Holdover Period to harvest and remove any crop then in the ground

AS WITNESS the hands of the parties hereto the day and year first before written

SCHEDULE 1
HOLDING DESCRIPTION

ALL THAT land described below TOGETHER with the benefit of any rights which are appurtenant to the Holding or which are reasonably needed for its proper enjoyment by the Tenant and capable of grant by the Landlord SUBJECT TO any rights which are expressly excepted or reserved out of the Holding and any rights or quasi-rights which have been enjoyed until now over the Holding by any occupier of any neighbouring property or the Landlord SUBJECT ALSO TO any subsisting obligations under any covenants affecting the Holding but with any subsisting obligations under any covenants

OS Ref	**NG No**	**Total field area**	**Eligible Area**

SCHEDULE 2
THE SINGLE PAYMENT SCHEME

1. In this Schedule:-

1.1 'Single Payment Scheme' means the single payment scheme governed by Council Regulation (EC) No 73/2009 and all the relevant subsequent Commission and other regulations administrative provisions statutory guidance or codes of practice implemented by the EC and/or the UK Government ('the Regulations')

1.2 'Entitlements' mean the payment entitlements leased to the Tenant under this Agreement set out in Schedule 3

1.3 'Cross-compliance' means the statutory management requirements and the good agricultural and environmental conditions as defined under the Single Payment Scheme

1.4 the parties will do all such acts matters and things as may be necessary to formalise the transfer of the Entitlements to the Tenant by way of lease if practicable

1.5 the Tenant agrees:-

1.5.1 to comply with any statutory or regulatory requirements regarding Entitlements (including Cross-compliance) and not to allow any Entitlements to lapse or be reduced or confiscated

1.5.2 not to set aside more (if any) of the Holding than the minimum required (treating the Holding as a separate holding) under the Single Payment Scheme

1.5.3 on termination of the Tenancy (including so far as may be necessary after the end of the Tenancy) to do all such acts matters and things as may be reasonably necessary to formalise the transfer of the Entitlements to the Landlord or his nominee as soon as reasonably practical

1.6 the Tenant shall pay to the Landlord the Entitlement Charge as set out in the Particulars

1.7 if at the end of the Tenancy the amount or quality of entitlements transferred to the Landlord under paragraph 1.5.3 above is less than the amount or quality of the Entitlements leased to the Tenant by the Landlord pursuant to paragraph 1.4 (subject to adjustment on account of mandatory cuts or increases) the Tenant shall pay to the Landlord the cost of acquiring sufficient entitlements of comparable quality to restore the Entitlements and shall compensate the Landlord for any other loss suffered as a result of the reduction of the Entitlements

1.8 if the rules of the Single Payment Scheme in respect of leasing of entitlements with land change or are otherwise clarified after the date of this Agreement in such a way that the objects of this Clause cannot be achieved the parties shall in default of agreement between them refer the matter to an arbitrator and the arbitrator shall redraft the terms of this Agreement in such way as may fairly achieve the said objects and the costs thereof shall be borne by the parties as the arbitrator may direct

SCHEDULE 3
THE ENTITLEMENTS

Registered Owner	Description	Area (Ha)

SIGNED by
in the presence of:-

SIGNED by
in the presence of:-

PRECEDENT 7 FARM BUSINESS TENANCY LETTING WITH TRANSFER OF ENTITLEMENTS AND TRANSFER BACK

FARM BUSINESS TENANCY

PARTICULARS

Date of Agreement	:	....
The Landlord	:	....
The Tenant	:	....
Commencement Date	:	....
Term	:	from and including the Commencement Date
Rent	:	£[ ]
Rent Instalments	:	Equal instalments payable on the Rent days
Rent Days	:	[ ] and [ ]
Permitted Use	:	Use in the trade or business of agriculture
Entitlements	:	As described in Schedule 3
Entitlement Charge	:	A sum equal to the amount payable to the Tenant under the Single Payment Scheme in respect of the Entitlements on the assumption that the rules of the Single Payment Scheme have been fully complied with by the Tenant, payable on receipt from the Rural Payments Agency ('the Due Date')
Holdover Period	:	[ ] months following the end of the Term

THIS FARM BUSINESS TENANCY AGREEMENT is made on the Date of Agreement

BETWEEN:-

(1) [*name*] of [*address*] ('the Landlord')

and

(2) [*name*] of [*address*] ('the Tenant')

1. Particulars and definitions

1.1 The attached Particulars are incorporated in this Agreement and to the extent allowed by the context the expressions in the left-hand column of the Particulars have the respective meanings shown opposite them

1.2 In this Agreement to the extent allowed by the context the following phrases have the respective meanings set under them:-

1.2.1 'Environmental Stewardship Schemes' means national or European schemes for the conservation or enhancement of the environment under which the Landlord has assumed obligations or is subject to restrictions in respect of the Holding

1.2.2 'the Holding' means the property described in Schedule 1

1.2.3 'the Landlord' includes whoever for the time being owns the interest in the Holding which gives right to possession of it at the end of the Tenancy

1.2.4 'the Tenancy' means the tenancy hereby created

1.2.5 'the Tenant' includes the Tenant's successors in title

1.2.6 'the 1995 Act' means the Agricultural Tenancies Act 1995

2. Interpretation

In this Agreement to the extent allowed by the context:-

2.1 'the Holding' includes any part of the Holding

2.2 'the Single Payment Scheme' means the single payment scheme governed by Council Regulation (EC) No 73/2009 and all the relevant subsequent Commission and other regulations administrative provisions statutory guidance or codes of practice implemented by the EC and/or the UK Government ('the Regulations')

2.3 'Value Added Tax' includes any substituted or additional tax of a similar nature

2.4 unless stated otherwise any reference to a sum of money is to that sum exclusive of Value Added Tax and any Value Added Tax charged on it is payable in addition

2.5 any obligation not to do any act or thing includes an obligation not to cause or allow anyone else to do that act or thing

2.6 any reference to this Agreement includes its schedules

2.7 any reference to a page clause or schedule is to that page clause or schedule in this Agreement

2.8 the clause and paragraph headings are for convenient reference only and shall not affect its construction

2.9 any reference to a statute regulation order or code of practice includes any later provision substituted for or amending the same and any subsidiary provision made under powers thereby conferred

3. Letting

3.1 The Landlord lets the Holding to the Tenant for the Term

4. Tenant's covenants

The Tenant covenants with the Landlord:-

4.1 *Rent*

to pay the Rent by the Rent Instalments on the Rent Days

4.2 *Value added tax*

to pay all Value Added Tax which may be payable on any taxable supply made to the Tenant in relation to the Tenancy or on any payment by the Landlord which the Tenant is required to reimburse to the Landlord

4.3 *Repair*

4.3.1 at all times to maintain and keep in good condition and (where appropriate) neat and tidy the Holding

4.3.2 to keep clear and free from obstruction all land drains and ditches on the Holding

4.3.3 to keep all boundary features on the Holding in such state as is adequate for their purpose and (in the case of hedges) reasonably cut back

4.4 *Good agricultural and environmental practice*

to keep the Holding in good agricultural and environmental condition as defined under the Single Payment Scheme

4.5 *Environmental Stewardship Schemes*

to ensure that its activities on the Holding do not breach the requirements of any Environmental Stewardship Schemes written details of which have been provided to the Tenant by the Landlord

4.6 *Permitted use*

to use the Holding for the Permitted Use only

4.7 *Alteration*

not on the Holding to:-

4.7.1 commit waste of any kind

4.7.2 plant any trees, coppice, saplings, pollards or underwood

4.7.3 make any alteration, any addition or any tenant's improvement falling within the provision of Part III of the 1995 Act other than by any routine improvements as defined in section 19(10) of the 1995 Act

4.8 *Statutory obligations*

at the Tenant's own expense to do everything in relation to the Holding or its use which is required to comply with any statutory or other obligation relating to the occupation of the Holding and in particular those relating to:-

4.8.1 the health safety and welfare of the persons using or employed on or about or visiting the Holding

4.8.2 waste effluent manure or other matter or thing of a noxious, dangerous or offensive nature

4.9 *Not to obstruct roadway*

not to obstruct any roadway or track by which access to and from the Holding or any neighbouring property is gained

4.10 *No nuisance*

not to do or allow to remain upon the Holding anything which may be or become or cause a nuisance, disturbance, inconvenience, injury or damage to the Landlord or the owner or occupier of any neighbouring property

4.11 *Adverse rights*

to take all reasonable steps to prevent any right over the Holding being acquired and to prevent trespassers on the Holding

4.12 *Indemnity – loss*

4.12.1 not to do or omit to do on or near the Holding any act or thing which may render the Landlord liable to pay any penalty, damages, compensation, costs, charges or expenses or to carry out any work of whatever nature

4.12.2 at all times to keep the Landlord fully indemnified against all losses, claims, actions and liabilities arising (whether directly or indirectly) as a result of any act or default of the Tenant

4.13 *Dilapidations notice*

to pay to the Landlord on an indemnity basis all expenses incurred by the Landlord in the preparation and service of a notice under section 146 of the Law of Property Act 1925 or in relation to proceedings under section 146 or 147 of the Act notwithstanding that forfeiture is avoided otherwise than by relief granted by the Court

4.14 *Alienation*

not without the consent of the Landlord to assign underlet charge or part with possession of the whole or any part of the Holding or permit anyone else to occupy the whole or any part of the Holding

4.15 *Notices*

to give full particulars to the Landlord of any notice, direction, order or proposal relating to the Holding made given or issued to the Tenant within seven days of receipt

4.16 *Single Payment Scheme*

to comply with the provisions of Schedule 2

4.17 *Yield up*

on the determination of the Tenancy to yield up the Holding in such repair and so cultivated as to be in accordance with the terms of this Agreement and with vacant possession

5. Landlord's agreement

5.1 The Landlord agrees with the Tenant to permit the Tenant on paying the Rent and performing and observing the several obligations of the Tenant in relation to the Tenancy peaceably and quietly to hold and enjoy the Holding during the Tenancy without any interruption or disturbance from or by the Landlord or any person claiming under or in trust for the Landlord

5.2 The Landlord agrees to comply with the provisions of Schedule 2

6. Landlord's reservations

The Landlord reserves the right to go onto the Holding to do anything reasonably necessary to secure compliance with any Environmental Stewardship Schemes (including carrying out works and accompanying officials from any competent body responsible for such schemes)

7. Re-entry

At any time after the happening of any of the events specified in the following paragraphs of this clause it shall be lawful for the Landlord after giving two months' notice to the Tenant of his intention to do so to re-enter the Holding or any part thereof in the name of the whole and thereupon the Tenancy shall absolutely determine but without prejudice to any right of action or remedy of the Landlord in respect of any antecedent breach by the Tenant of any of his obligations in relation to the Tenancy

7.1 If the Tenant:-

7.1.1 is unable to pay a debt or has no reasonable prospect of being able to pay a debt within the meaning of section 268 of the Insolvency Act 1986

7.1.2 makes an application to the court for an interim order under section 253 of the Insolvency Act 1986

7.1.3 makes any assignment for the benefit of or enters into any arrangement with his creditors either by composition or otherwise

7.1.4 does anything which would entitle a petition for a bankruptcy order to be made against him

8. Custom

The rights of the parties under this agreement or otherwise in respect of the Tenancy shall not depend upon or be affected by any custom of the country

9. Early entry

9.1 Following clearance of the Tenant's crop from the holding and provided that the Landlord complies with relevant rules or regulations of the Single Payment Scheme (including cross-compliance) the Landlord will be permitted access to the Holding before the termination of the Tenancy for the purpose of carrying out cultivations and subject to any reasonable limitations or restrictions specified by the Tenant

9.2 For the avoidance of doubt the Tenant retains possession of the Holding and remains responsible under paragraph 8 of Schedule 2 for the requirements of the Single Payment Scheme and the Landlord enters as licensee only and agrees to comply with the Tenant's reasonable requirements

10. Holdover

At the end of the Term the Landlord shall permit the Tenant free of charge to enter on the Holding in a good and husbandlike manner during the Holdover Period to harvest and remove any crop then in the ground

AS WITNESS the hands of the parties hereto the day and year first before written

SCHEDULE 1
HOLDING DESCRIPTION

ALL THAT land described below TOGETHER with the benefit of any rights which are appurtenant to the Holding or which are reasonably needed for its proper enjoyment by the Tenant and capable of grant by the Landlord SUBJECT TO any rights which are expressly excepted or reserved out of the Holding and any rights or quasi-rights which have been enjoyed until now over the Holding by any occupier of any neighbouring property or the Landlord SUBJECT ALSO TO any subsisting obligations under any covenants affecting the Holding but with any subsisting obligations under any covenants

OS Ref	**NG No**	**Total field area**	**Eligible Area**

SCHEDULE 2
THE SINGLE PAYMENTS SCHEME

1. In this Schedule:-

1.1 'Single Payment Scheme' means the single payment scheme governed by Council Regulation (EC) No 73/2009 and all the relevant subsequent Commission and other regulations administrative provisions statutory guidance or codes of practice implemented by the EC and/or the UK Government ('the Regulations')

1.2 'Entitlements' mean the payment entitlements transferred to the Tenant under this Agreement set out in Schedule 3

1.3 'Cross-compliance' means the statutory management requirements and the good agricultural and environmental conditions as defined under the Single Payment Scheme

1.4 the parties will do all such acts matters and things as may be necessary to formalise the transfer of the Entitlements to the Tenant and in particular the Landlord will submit the necessary form RLE1 to the relevant body as soon as possible and in any event prior to 2 April 20[ ]

1.5 the Tenant agrees:-

1.5.1 to comply with any statutory or regulatory requirements regarding Entitlements (including Cross-compliance) and not to allow any Entitlements to lapse or be reduced or confiscated

1.5.2 not to set aside more (if any) of the Holding than the minimum required (treating the Holding as a separate holding) under the Single Payment Scheme

1.5.3 on termination of the Tenancy (including so far as may be necessary after the end of the Tenancy) to do all such acts matters and things as may be reasonably necessary and give such notice and submit such forms as may be required to formalise the transfer of the Entitlements to the Landlord or his nominee as soon as reasonably practical

1.6 the Tenant shall pay to the Landlord the Entitlement Charge as set out in the Particulars

1.7 if at the end of the Tenancy the amount or quality of entitlements transferred to the Landlord under paragraph 3(c) above is less than the amount or quality of the Entitlements transferred to the Tenant by the Landlord pursuant to paragraph 2 (subject to adjustment on account of mandatory cuts or increases) the Tenant shall pay to the Landlord the cost of acquiring sufficient entitlements of comparable quality to restore the Entitlements and shall compensate the Landlord for any other loss suffered as a result of the reduction of the Entitlements

1.8 if the rules of the Single Payment Scheme in respect of transferring entitlements change or are otherwise clarified after the date of this Agreement in such a way that the objects of this Clause cannot be achieved the parties shall in default of agreement between them refer the matter to an arbitrator and the arbitrator shall redraft the terms of this Agreement in such way as may fairly achieve the said objects and the costs thereof shall be borne by the parties as the arbitrator may direct

SCHEDULE 3
THE ENTITLEMENTS

Registered Owner	Description	Area (Ha)

SIGNED by
in the presence of:-

SIGNED by
in the presence of:-

PRECEDENT 8 LETTING WITHOUT ENTITLEMENTS

FARM BUSINESS TENANCY

PARTICULARS

Date of Agreement	:	....
The Landlord	:	....
The Tenant	:	....
Commencement Date	:	....

Term	:	from and including the Commencement Date
Rent	:	£[ ]
Rent Instalments	:	
Rent Days	:	and
Permitted Use	:	Use in the trade or business of agriculture
Holdover Period	:	[ ] months following the end of the Term

THIS FARM BUSINESS TENANCY AGREEMENT is made on the Date of Agreement

BETWEEN:-

(1) [*name*] of [*address*] ('the Landlord')

and

(2) [*name*] of [*address*] ('the Tenant')

1. Particulars and definitions

1.1 The attached Particulars are incorporated in this Agreement and to the extent allowed by the context the expressions in the left-hand column of the Particulars have the respective meanings shown opposite them

1.2 In this Agreement to the extent allowed by the context the following phrases have the respective meanings set under them:-

1.2.1 'Environmental Stewardship Schemes' means national or European schemes for the conservation or enhancement of the environment under which the Landlord has assumed obligations or is subject to restrictions in respect of the Holding

1.2.2 'the Holding' means the property described in the Holding Description Schedule

1.2.3 'the Single Payment Scheme' means the single payment scheme governed by Council Regulation (EC) No 73/2009 and all the relevant subsequent Commission and other regulations administrative provisions statutory guidance or codes of practice implemented by the EC and/or the UK Government ('the Regulations')

1.2.4 'the Tenancy' means the tenancy hereby created

1.2.5 'the 1995 Act' means the Agricultural Tenancies Act 1995

2. Interpretation

In this Agreement to the extent allowed by the context:-

2.1 'the Holding' includes any part of the Holding

2.2 'Value Added Tax' includes any substituted or additional tax of a similar nature

2.3 unless stated otherwise any reference to a sum of money is to that sum exclusive of Value Added Tax and any Value Added Tax charged on it is payable in addition

2.4 any obligation not to do any act or thing includes an obligation not to cause or allow anyone else to do that act or thing

2.5 any reference to this Agreement includes its schedules

2.6 any reference to a page clause or schedule is to that page clause or schedule in this Agreement

2.7 the clause and paragraph headings are for convenient reference only and shall not affect its construction

2.8 any reference to a statute regulation order or code of practice includes any later provision substituted for or amending the same and any subsidiary provision made under powers thereby conferred

3. Letting

The Landlord lets to the Tenant the Holding TO HOLD the same unto the Tenant for the Term on the terms of this Agreement YIELDING AND PAYING therefor the Rent

4. Tenant's covenants

The Tenant covenants with the Landlord:-

4.1 *Rent*

to pay the Rent by the Rent Instalments on the Rent Days

4.2 *Value added tax*

to pay all Value Added Tax which may be payable on any taxable supply made to the Tenant in relation to the Tenancy or on any payment by the Landlord which the Tenant is required to reimburse to the Landlord

4.3 *Repair*

4.3.1 at all times to maintain and keep in good condition and (where appropriate) neat and tidy the Holding

4.3.2 to keep clear and free from obstruction all land drains and ditches on the Holding

4.4 *Good agricultural and environmental practice*

to keep the Holding in good agricultural and environmental condition as defined under the Single Payment Scheme

4.5 *Environmental Stewardship Schemes*

to ensure that its activities on the Holding do not breach the requirements of any Environmental Stewardship Schemes written details of which have been provided to the Tenant by the Landlord

4.6 *Permitted use*

to use the Holding for the Permitted Use only

4.7 *Alteration*

not on the Holding to:-

4.7.1 commit waste of any kind

4.7.2 plant any trees, coppice, saplings, pollards or underwood

4.7.3 make any alteration, any addition or any tenant's improvement falling within the provision of Part III of the 1995 Act other than by any routine improvements as defined in section 19(10) of the 1995 Act

4.8 *Statutory obligations*

at the Tenant's own expense to do everything in relation to the Holding or its use which is required to comply with any statutory or other obligation relating to the occupation of the Holding and in particular those relating to:-

4.8.1 the health safety and welfare of the persons using or employed on or about or visiting the Holding

4.8.2 waste effluent manure or other matter or thing of a noxious, dangerous or offensive nature

4.9 *Not to obstruct roadway*

not to obstruct any roadway or track by which access to and from the Holding or any neighbouring property is gained

4.10 *No nuisance*

not to do or allow to remain upon the Holding anything which may be or become or cause a nuisance, disturbance, inconvenience, injury or damage to the Landlord or the owner or occupier of any neighbouring property

4.11 *Adverse rights*

to take all reasonable steps to prevent any right over the Holding being acquired and to prevent trespassers on the Holding

4.12 *Indemnity – loss*

4.12.1 not to do or omit to do on or near the Holding any act or thing which may render the Landlord liable to pay any penalty, damages, compensation, costs, charges or expenses or to carry out any work of whatever nature

4.12.2 at all times to keep the Landlord fully indemnified against all losses, claims, actions and liabilities arising (whether directly or indirectly) as a result of any act or default of the Tenant

4.13 *Dilapidations notice*

to pay to the Landlord on an indemnity basis all expenses incurred by the Landlord in the preparation and service of a notice under section 146 of the Law of Property Act 1925 or in relation to proceedings under section 146 or 147 of the Act notwithstanding that forfeiture is avoided otherwise than by relief granted by the Court

4.14 *Alienation*

not without the consent of the Landlord to assign underlet charge or part with possession of the whole or any part of the Holding or permit anyone else to occupy the whole or any part of the Holding

4.15 *Notices*

to give full particulars to the Landlord of any notice, direction, order or proposal relating to the Holding made given or issued to the Tenant within seven days of receipt

4.16 *Yield up*

on the determination of the Tenancy to yield up the Holding in such repair and so cultivated as to be in accordance with the terms of this Agreement and with vacant possession

5. Landlord's Agreement

The Landlord agrees with the Tenant to permit the Tenant on paying the Rent and performing and observing the several obligations of the Tenant in relation to the Tenancy peaceably and quietly to hold and enjoy the Holding during the Tenancy without any interruption or disturbance from or by the Landlord or any person claiming under or in trust for the Landlord

6. Landlord's reservations

The Landlord reserves the right to go onto the Holding to do anything reasonably necessary to secure compliance with any Environmental Stewardship Schemes (including carrying out works and accompanying officials from any competent body responsible for such schemes)

7. Re-entry

At any time after the happening of any of the events specified in the following paragraphs of this clause it shall be lawful for the Landlord after giving two months' notice to the Tenant of his intention to do so to re-enter the Holding or any part thereof in the name of the whole and thereupon the Tenancy shall absolutely determine but without prejudice to any right of action or remedy of the Landlord in respect of any antecedent breach by the Tenant of any of his obligations in relation to the Tenancy

7.1 If the Tenant:-

7.1.1 is unable to pay a debt or has no reasonable prospect of being able to pay a debt within the meaning of section 268 of the Insolvency Act 1986

7.1.2 makes an application to the court for an interim order under section 253 of the Insolvency Act 1986

7.1.3 makes any assignment for the benefit of or enters into any arrangement with his creditors either by composition or otherwise

7.1.4 does anything which would entitle a petition for a bankruptcy order to be made against him

8. Custom

The rights of the parties under this agreement or otherwise in respect of the Tenancy shall not depend upon or be affected by any custom of the country

9. Single Payment Scheme

The Tenant during the Term and otherwise the Landlord will comply with the requirements of the Single Payment Scheme including where appropriate cross-compliance rules and set-aside management requirements

10. Early entry

10.1 Following clearance of the Tenant's crop from the Holding and provided that the Landlord complies with relevant rules or regulations of the Single Payment Scheme (including cross-compliance) the Landlord will be permitted access to the Holding before the termination of the Tenancy for the purpose of carrying out cultivations and subject to any reasonable limitations or restrictions specified by the Tenant

10.2 For the avoidance of doubt the Tenant retains possession of the Holding and remains responsible under Clause 9 for the requirements of the Single Payment Scheme and the Landlord enters as a licensee only and agrees to comply with the Tenant's reasonable requirements

11. Holdover

At the end of the Term the Landlord shall permit the Tenant free of charge to enter on the Holding in a good and husbandlike manner during the Holdover Period to harvest and remove any crop then in the ground

AS WITNESS the hands of the parties hereto the day and year first before written

SCHEDULE
HOLDING DESCRIPTION

ALL THAT land described below TOGETHER with the benefit of any rights which are appurtenant to the Holding or which are reasonably needed for its proper enjoyment by the Tenant and capable of grant by the Landlord SUBJECT TO any rights which are expressly excepted or reserved out of the Holding and any rights or quasi-rights which have been enjoyed until now over the Holding by any occupier of any neighbouring property or the Landlord SUBJECT ALSO TO any subsisting obligations under any covenants affecting the Holding but with any subsisting obligations under any covenants

OS Ref	NG No	Total field area	Eligible Area

SIGNED by
in the presence of:-

SIGNED by
in the presence of:-

PRECEDENT 9 LETTING SPECIFYING ENTITLEMENTS

[*Letting specifying normal entitlements and/or former set-aside entitlements and transferring to Landlord without payment save in respect of any historic element.*]

FARM BUSINESS TENANCY

PARTICULARS

Date of Agreement	:	....
The Landlord	:	....
The Tenant	:	....
Commencement Date	:	....
Term	:	from and including the Commencement Date

Rent	:	£[ ]
Rent Instalments	:	Equal instalments payable on the Rent days
Rent Days	:	[ ] and [ ]
Permitted Use	:	Use in the trade or business of agriculture
Holdover Period	:	[ ] months following the end of the Term
The Entitlement Area	:	[ ] Hectares
Relevant Entitlements	:	[ ] Hectares Normal Entitlements [ ] former Hectares Set-aside Entitlements

THIS FARM BUSINESS TENANCY AGREEMENT is made on the Date of Agreement

BETWEEN:-

(1) [*name*] of [*address*] ('the Landlord')

and

(2) [*name*] of [*address*] ('the Tenant')

1. Particulars and definitions

1.1 The attached Particulars are incorporated in this Agreement and to the extent allowed by the context the expressions in the left-hand column of the Particulars have the respective meanings shown opposite them

1.2 In this Agreement to the extent allowed by the context the following phrases have the respective meanings set under them:-

1.2.1 'Environmental Stewardship Schemes' means national or European schemes for the conservation or enhancement of the environment under which the Landlord has assumed obligations or is subject to restrictions in respect of the Holding

1.2.2 'the Holding' means the property described in Schedule 1

1.2.3 'the Landlord' includes whoever for the time being owns the interest in the Holding which gives right to possession of it at the end of the Tenancy

1.2.4 'the Tenancy' means the tenancy hereby created

1.2.5 'the Tenant' includes the Tenant's successors in title

1.2.6 'the 1995 Act' means the Agricultural Tenancies Act 1995

2. Interpretation

In this Agreement to the extent allowed by the context:-

2.1 'the Holding' includes any part of the Holding

2.2 'the Single Payment Scheme' means the single payment scheme governed by Council Regulation (EC) No 73/2009 and all the relevant subsequent Commission and other regulations administrative provisions statutory guidance or codes of practice implemented by the EC and/or the UK Government ('the Regulations')

2.3 'Value Added Tax' includes any substituted or additional tax of a similar nature

2.4 unless stated otherwise any reference to a sum of money is to that sum exclusive of Value Added Tax and any Value Added Tax charged on it is payable in addition

2.5 any obligation not to do any act or thing includes an obligation not to cause or allow anyone else to do that act or thing

2.6 any reference to this Agreement includes its schedules

2.7 any reference to a page clause or schedule is to that page clause or schedule in this Agreement

2.8 the clause and paragraph headings are for convenient reference only and shall not affect its construction

2.9 any reference to a statute regulation order or code of practice includes any later provision substituted for or amending the same and any subsidiary provision made under powers thereby conferred

3. Letting

3.1 The Landlord lets the Holding to the Tenant for the Term

4. Tenant's covenants

The Tenant covenants with the Landlord:-

4.1 *Rent*

to pay the Rent by the Rent Instalments on the Rent Days

4.2 *Value added tax*

to pay all Value Added Tax which may be payable on any taxable supply made to the Tenant in relation to the Tenancy or on any payment by the Landlord which the Tenant is required to reimburse to the Landlord

4.3 *Repair*

4.3.1 at all times to maintain and keep in good condition and (where appropriate) neat and tidy the Holding

4.3.2 to keep clear and free from obstruction all land drains and ditches on the Holding

4.3.3 to keep all boundary features on the Holding in such state as is adequate for their purpose and (in the case of hedges) reasonably cut back

4.4 *Good agricultural and environmental practice*

to keep the Holding in good agricultural and environmental condition as defined under the Single Payment Scheme

4.5 *Environmental Stewardship Schemes*

to ensure that its activities on the Holding do not breach the requirements of any Environmental Stewardship Schemes written details of which have been provided to the Tenant by the Landlord

4.6 *Permitted use*

to use the Holding for the Permitted Use only

4.7 *Alteration*

not on the Holding to:-

4.7.1 commit waste of any kind

4.7.2 plant any trees, coppice, saplings, pollards or underwood

4.7.3 make any alteration, any addition or any tenant's improvement falling within the provision of Part III of the 1995 Act other than by any routine improvements as defined in section 19(10) of the 1995 Act

4.8 *Statutory obligations*

at the Tenant's own expense to do everything in relation to the Holding or its use which is required to comply with any statutory or other obligation relating to the occupation of the Holding and in particular those relating to:-

4.8.1 the health safety and welfare of the persons using or employed on or about or visiting the Holding

4.8.2 waste effluent manure or other matter or thing of a noxious, dangerous or offensive nature

4.9 *Not to obstruct roadway*

not to obstruct any roadway or track by which access to and from the Holding or any neighbouring property is gained

4.10 *No nuisance*

not to do or allow to remain upon the Holding anything which may be or become or cause a nuisance, disturbance, inconvenience, injury or damage to the Landlord or the owner or occupier of any neighbouring property

4.11 *Adverse rights*

to take all reasonable steps to prevent any right over the Holding being acquired and to prevent trespassers on the Holding

4.12 *Indemnity – loss*

4.12.1 not to do or omit to do on or near the Holding any act or thing which may render the Landlord liable to pay any penalty, damages, compensation, costs, charges or expenses or to carry out any work of whatever nature

4.12.2 at all times to keep the Landlord fully indemnified against all losses, claims, actions and liabilities arising (whether directly or indirectly) as a result of any act or default of the Tenant

4.13 *Dilapidations notice*

to pay to the Landlord on an indemnity basis all expenses incurred by the Landlord in the preparation and service of a notice under section 146 of the Law of Property Act 1925 or in relation to proceedings under section 146 or 147 of the Act notwithstanding that forfeiture is avoided otherwise than by relief granted by the Court

4.14 *Alienation*

not without the consent of the Landlord to assign underlet charge or part with possession of the whole or any part of the Holding or permit anyone else to occupy the whole or any part of the Holding

4.15 *Notices*

to give full particulars to the Landlord of any notice, direction, order or proposal relating to the Holding made given or issued to the Tenant within seven days of receipt

4.16 *Single Payment Scheme*

to comply with the provisions of Schedule 2

4.17 *Yield up*

on the determination of the Tenancy to yield up the Holding in such repair and so cultivated as to be in accordance with the terms of this Agreement and with vacant possession

5. Landlord's agreement

5.1 The Landlord agrees with the Tenant to permit the Tenant on paying the Rent and performing and observing the several obligations of the Tenant in relation to the Tenancy peaceably and quietly to hold and enjoy the Holding during the Tenancy without any interruption or disturbance from or by the Landlord or any person claiming under or in trust for the Landlord

5.2 The Landlord agrees to comply with the provisions of Schedule 2

6. Landlord's reservations

The Landlord reserves the right to go onto the Holding to do anything reasonably necessary to secure compliance with any Environmental Stewardship Schemes (including carrying out works and accompanying officials from any competent body responsible for such schemes)

7. Re-entry

At any time after the happening of any of the events specified in the following paragraphs of this clause it shall be lawful for the Landlord after giving two months' notice to the Tenant of his intention to do so to re-enter the Holding or any part thereof in the name of the whole and thereupon the Tenancy shall absolutely determine but without prejudice to any right of action or remedy of the Landlord in respect of any antecedent breach by the Tenant of any of his obligations in relation to the Tenancy

7.1 If the Tenant:-

7.1.1 is unable to pay a debt or has no reasonable prospect of being able to pay a debt within the meaning of section 268 of the Insolvency Act 1986

7.1.2 makes an application to the court for an interim order under section 253 of the Insolvency Act 1986

7.1.3 makes any assignment for the benefit of or enters into any arrangement with his creditors either by composition or otherwise

7.1.4 does anything which would entitle a petition for a bankruptcy order to be made against him

8. Custom

The rights of the parties under this agreement or otherwise in respect of the Tenancy shall not depend upon or be affected by any custom of the country

9. Early entry

9.1 Following clearance of the Tenant's crop from the holding and provided that the Landlord complies with relevant rules or regulations

of the Single Payment Scheme (including cross-compliance) the Landlord will be permitted access to the Holding before the termination of the Tenancy for the purpose of carrying out cultivations and subject to any reasonable limitations or restrictions specified by the Tenant

9.2 For the avoidance of doubt the Tenant retains possession of the Holding and remains responsible under paragraph 8 of Schedule 2 for the requirements of the Single Payment Scheme and the Landlord enters as licensee only and agrees to comply with the Tenant's reasonable requirements

10. Holdover

At the end of the Term the Landlord shall permit the Tenant free of charge to enter on the Holding in a good and husbandlike manner during the Holdover Period to harvest and remove any crop then in the ground

AS WITNESS the hands of the parties hereto the day and year first before written

SCHEDULE 1
HOLDING DESCRIPTION

ALL THAT land described below TOGETHER with the benefit of any rights which are appurtenant to the Holding or which are reasonably needed for its proper enjoyment by the Tenant and capable of grant by the Landlord SUBJECT TO any rights which are expressly excepted or reserved out of the Holding and any rights or quasi-rights which have been enjoyed until now over the Holding by any occupier of any neighbouring property or the Landlord SUBJECT ALSO TO any subsisting obligations under any covenants affecting the Holding but with any subsisting obligations under any covenants

OS Ref	**NG No**	**Total field area**	**Eligible Area**

SCHEDULE 2
THE SINGLE PAYMENT SCHEME

1. In this Schedule:-

1.1 'the Council Regulation' means Council Regulation (EC) No 73/2009

1.2 'the Single Payment Scheme' means the single payment scheme governed by Council Regulation (EC) No 73/2009 and all the relevant subsequent Commission and other regulations administrative provisions statutory guidance or codes of practice implemented by the EC and/or the UK Government ('the Regulations')

1.3 'Entitlements' has the same meaning as in the Council Regulation

1.4 'Former Set-Aside Entitlements' shall have the same meaning as in the Council Regulation

1.5 'the Entitlements' means Normal Entitlements or former Set-Aside Entitlements (as the case may be)

1.6 'the Relevant Entitlements' means the Entitlements to be transferred to the Landlord or his nominee in accordance with the terms hereof specified in the Particulars

1.7 'the Eligible Area' means the Entitlement area specified in Schedule 1

2. On quitting the Holding the Tenant will take all such steps and do all such acts as are required to procure without charge to the Landlord the transfer to the Landlord or his nominee of the Relevant Entitlements

3. The Relevant Entitlements shall be such as have a regional element only and no historic element

4. If it is not possible to transfer Entitlements with a regional element only the Tenant may transfer other Entitlements but shall be entitled to compensation in respect of the Relevant Entitlements of a sum equal to the net present value of the historic element of the Relevant Entitlements agreed between the parties and in default of agreement determined by arbitration or such other sums as shall be agreed

5. At the end of the tenancy the Tenant shall retain the payments in respect of the Relevant Entitlements for the then current harvest year or as otherwise agreed

6. The Landlord shall indemnify the Tenant in respect of any tax interest and expenses arising in respect of a disposal of Relevant Entitlements if and to the extent that the Tenant receives less compensation in respect thereof than the deemed market value ascertained for tax purposes

7. In so far as they remain to be observed and performed and are capable of taking effect after the termination of the tenancy the provisions of this clause shall remain in full force and effect notwithstanding such termination

8. The Tenant so long as the Tenancy subsists and thereafter the Landlord will ensure compliance with the rules and regulations of the Single Payment Scheme in respect of the Holding including where appropriate cross compliance rules and set-aside management requirements

SIGNED by
in the presence of:-

SIGNED by
in the presence of:-

PRECEDENT 10 NOTICE TO TERMINATE FIXED TERM TENANCY OF OVER 2 YEARS

AGRICULTURAL TENANCIES ACT 1995
SECTION 5

THE HOLDING

[*name and identification of Holding, by*

address, ordnance survey numbers or plan]

TO TENANT/LANDLORD

[*Name and Address*]

I, [*name and address of Landlord/Tenant*], give you notice that I intend to terminate the tenancy of the above Holding dated [*date*] and made between (1) [*name*] and (2) [*name*] on [*date*], being the date fixed by the tenancy agreement for the expiry of the term granted by the tenancy

Signed: ....

Dated: ....

PRECEDENT 11 LANDLORD'S NOTICE TO TERMINATE PERIODIC TENANCY

AGRICULTURAL TENANCIES ACT 1995
SECTION 6

THE HOLDING

[*name and identification of Holding, by*

address, ordnance survey numbers or plan]

TO TENANT

[*Name and address*]

I, [*name and address of Landlord*], give you notice that I intend to terminate the tenancy of the above Holding of which you are the Tenant pursuant to an agreement dated [*date*] and made between (1) [*name*] and (2) [*name*] on the [*date*], or at the expiration of the year of your tenancy which expires next after the end of twelve months from the date of service of this notice

Signed: ....

Dated: ....

PRECEDENT 12 TENANT'S NOTICE TO TERMINATE PERIODIC TENANCY

AGRICULTURAL TENANCIES ACT 1995
SECTION 6

THE HOLDING

[*name and identification of Holding, by*

address, ordnance survey numbers or plan]

TO LANDLORD

[*Name and address*]

I, [*name and address of Tenant*], give you notice that I intend to terminate the tenancy of the above Holding of which I am your Tenant pursuant to an agreement dated [*date*] and made between (1) [*name*] and (2) [*name*] on [*date*], or at the expiration of the year of my tenancy which expires next after the end of twelve months from the date of service of this notice

Signed: ....

Dated: ....

PRECEDENT 13 NOTICE EXERCISING OPTION TO TERMINATE TENANCY OR PART OF TENANCY OVER 2 YEARS

AGRICULTURAL TENANCIES ACT 1995
SECTION 7

THE HOLDING

[*name and identification of Holding, by*

address, ordnance survey numbers or plan]

TO LANDLORD/TENANT

[*Name and address*]

I, [*name and address of Tenant/Landlord*], in accordance with Clause [*break clause number*] of the Tenancy Agreement of the above Holding dated [*date*] and made between (1) [*name*] and (2) [*name*] hereby exercise the option to terminate the tenancy [as to the whole of the Holding] [as to that part of the Holding shown edged red on the plan attached to this notice] on the [*date*]

Signed: ....

Dated: ....

PRECEDENT 14 TENANT'S NOTICE DEMANDING STATUTORY REVIEW OF RENT

AGRICULTURAL TENANCIES ACT 1995
SECTION 10

THE HOLDING

[*name and identification of Holding, by*

address, ordnance survey numbers or plan]

TO LANDLORD

[*name and address*]

I, [*name and address of Tenant*], demand that the question of rent to be paid in respect of the above Holding as from [*date*] shall be referred to arbitration in accordance with Section 10 of the Agricultural Tenancies Act 1995

Signed: ....

Dated: ....

PRECEDENT 15 LANDLORD'S NOTICE DEMANDING STATUTORY REVIEW OF RENT

AGRICULTURAL TENANCIES ACT 1995
SECTION 10

THE HOLDING

[*name and identification of Holding, by*

address, ordnance survey numbers or plan]

TO TENANT

[*name and address*]

I, [*name and address of Landlord*], demand that the question of rent to be paid in respect of the above Holding as from [*date*] shall be referred to arbitration in accordance with Section 10 of the Agricultural Tenancies Act 1995

Signed: ....

Dated: ....

PRECEDENT 16 NOTICE OF PROPOSAL TO APPOINT AN ARBITRATOR

AGRICULTURAL TENANCIES ACT 1995
SECTION 28

THE HOLDING

[*name and identification of Holding, by*

address, ordnance survey numbers or plan]

TO LANDLORD or TENANT

(*as the case may be*)

[*name and address*]

I, [*name and address of Tenant/ Landlord*], propose to apply to the President of the Royal Institution of Chartered Surveyors for the appointment of an arbitrator by him to settle the dispute specified below unless within two months from the service of this notice you and I have appointed an arbitrator by agreement

[*Brief details of dispute*]

Signed: ....

Dated: ....

PRECEDENT 17 NOTICE OF REFERENCE OF DISPUTE TO A THIRD PARTY

AGRICULTURAL TENANCIES ACT 1995
SECTION 29

THE HOLDING

[*name and identification of Holding, by*

address, ordnance survey numbers or plan]

TO LANDLORD or TENANT

(*as the case may be*)

[*name and address*]

I, [*name and address of Tenant/Landlord*], in accordance with Clause [*no*] of the Tenancy Agreement dated [*date*] made between (1) [*name*] and (2) [*name*] have made a reference to [*specify third party named in Tenancy Agreement*] to settle the dispute specified below

[*Brief details of dispute*]

Signed: ....

Dated: ....

Chapter 5

OCCUPATION OF FARM BUILDINGS

INTRODUCTION

PRECEDENTS
1 Lease of farm buildings
2 Licence for storage

INTRODUCTION

It is important to be on the lookout for letting arrangements in the countryside which, in effect, are not farm business tenancies but business tenancies under the Landlord and Tenant Act 1954, Part II. The distinction can sometimes be difficult to make. For example, use as a stud farm constitutes a business use rather than an agricultural use. On the other hand, a letting for grazing purposes will be an agricultural letting rather than a business letting. Unless it is quite clear which is the predominant user, the cautious practitioner will endeavour to prepare an agreement which meets his purposes so far as possible under both statutory regimes. Most importantly, in the context of security of tenure, this may need to include service of notices on the tenant and the tenant making a declaration or swearing a statutory declaration in respect of contracting out of the security of tenure provisions of the Landlord and Tenant Act 1954, Part II.[1]

A precedent is included here for letting farm buildings which are no longer required in the agricultural business. This is an ordinary warehouse letting under the business tenancy regime and as such agreements go, relatively short. The precedent is drafted on the basis that the tenant occupies the whole building (or buildings) and there is no sharing with other tenants, and no common parts. There is no provision for service charge as it is not anticipated that the landlord will provide any services. The precedent does not include any provisions for a rent review which would usually be included in longer leases.

---

1 See the Regulatory Reform (Business Tenancies) (England and Wales) Order 2003, SI 2003/3096, art 22, Schs 1 and 2.

If the term exceeds 7 years, the lease must be registered and the prescriptive clauses are included in the precedent for such requirement.

If the letting is long term, further clauses may be thought necessary.

The landlord must bear in mind that such a letting carries with it various fiscal implications:

- *Inheritance tax*. The building will no longer be eligible for agricultural relief for inheritance tax purposes because it is no longer occupied jointly with farmland. It may perhaps be eligible for business relief but only if the requirements set out in *Farmer v IRC*[2] can be met.

- *Capital gains tax*. On a similar basis, there may be a risk of losing rollover relief.

- *Rating*. The business will be liable to rates, although one would anticipate that these would be passed on to the tenant.

It might be more appropriate if, instead of letting the building for storage purposes, the landlord was prepared to run a business storing goods on behalf of a third party. Clearly this would incur work and liability for the property owner, but he might take the view that the fiscal advantages are worth it. However, this would not assist with the rating question.

Licence for storage

Farmers often have empty barns. Where they want to let them on a long-term basis they should use the precedent for a business tenancy set out in Precedent 1. However, often the arrangement required may be a short-term simple agreement for storage of items such as caravans in the winter months.

The precedent is for a licence rather than a tenancy. The essential element is that the licensee does not have exclusive possession of the building or part of the building.

---

2 (1999) STC (SCD) 321.

PRECEDENTS

PRECEDENT 1 LEASE OF FARM BUILDINGS

[PRESCRIBED CLAUSES[3]

LR1. Date of lease

[Date]

LR2. Title number(s)

LR2.1 Landlord's title number(s)

[Insert title number(s) or leave blank if none]

LR2.2 Other title numbers

[Title number(s)] or [None]

LR3. Parties to this lease

Landlord

[Name(s)]

[Address] or [Registered office if a company]

[Company registered number]

Tenant

[Name(s)]

[Address] or [Registered office if a company]

[Company registered number]

Other parties

[None] or [name and address of a guarantor, if any]

LR4. Property

In the case of a conflict between this clause and the remainder of this lease then, for the purposes of registration, this clause shall prevail

The Property is defined in clause [ ] of this lease

LR5. Prescribed statements, etc

None

3 Leases for a term exceeding 7 years are registrable and must include prescribed clauses.

LR6. Term for which the Property is leased

The term as specified in this lease at clause [ ] in the definition of 'Term'

LR7. Premium

None

LR8. Prohibitions or restrictions on disposing of this lease

This lease contains a provision that prohibits or restricts dispositions

LR9. Rights of acquisition, etc

LR9.1 Tenant's contractual rights to renew this lease, to acquire the reversion or another lease of the Property, or to acquire an interest in other land

None

LR9.2 Tenant's covenant to (or offer to) surrender this lease

None

LR9.3 Landlord's contractual rights to acquire this lease

None

LR10. Restrictive covenants given in this lease by the Landlord in respect of land other than the Property

None

LR11. Easements

LR11.1 Easements granted by this lease for the benefit of the Property

[The easements as specified in Schedule [ ] to this lease]

LR11.2 Easements granted or reserved by this lease over the Property for the benefit of other property

The easements as specified in Schedule [ ] to this lease

LR12. Estate rentcharge burdening the Property

None

LR13. Application for standard form of restriction

[None]

LR14. Declaration of trust where there is more than one person comprising the Tenant

[Omit all inapplicable statements]

[The Tenant is more than one person. They are to hold the Property on trust for themselves as joint tenants]

[The Tenant is more than one person. They are to hold the Property on trust for themselves as tenants in common in equal shares]

[The Tenant is more than one person. They are to hold the Property on trust [Complete as necessary]

THIS LEASE is made on the [ ] day of [ ] 20[ ]

BETWEEN:-

(1) [*name*] of [*address*] ('the Landlord')

(2) [*name*] of [*address*] ('the Tenant')

and

[(3) [*name*] of [*address*] ('the Guarantor')]

OPERATIVE PROVISIONS

1. Definitions

In this lease the following expressions shall have the following meanings:-

1.1 *Annual Rent*

[*state amount*] (exclusive of VAT)

1.2 *Property*

means the [building] situate at [ ] as shown edged red on the Plan [Plan 1]

[1.3 *Environmental Legislation*

means the Environmental Protection Act 1990 the Water Resources Act 1991 the Water Industry Act 1991 the Environment Act 1995 and any other legislation relating to the control of pollution or the protection of the environment]

1.4 *Interest*

means interest at the rate of 4% above the base lending rate from time to time of [*name of the bank*] Bank plc or such other comparable rate as the Landlord may reasonably designate if the base rate shall cease to be published

[1.5 *Insurance Rent*

sums equal to the aggregate in each year of the gross cost of the premium for the insurance of the Property (other than any plate glass) for its full reinstatement cost against:

1- loss or damage by or in consequence of the Insured Risks, including costs of demolition, site clearance, professionals' fees and incidental expenses

2- loss of Annual Rent from the Property for three years and

3- any VAT and insurance premium tax payable on the above]

1.6 *Insured Risks*

means fire, explosion, lightning, earthquake, storm, flood, bursting and overflowing of water tanks, apparatus or pipes, impact by aircraft and articles dropped from them, impact by vehicles, riot, civil commotion and any other risks against which the Landlord decides to insure against from time to time and **Insured Risk** means any one of the Insured Risks

1.7 *Landlord's Adjoining Land*

means the neighbouring or adjoining property in which the Landlord has an interest and known as [*description of the Landlord's farm*] [registered at HM Land Registry with title number(s) [*insert title number(s)*] [shown edged blue on [the Plan] [Plan 2]]

1.8 *Landlord's Surveyor*

means the person or firm appointed by or acting for the Landlord

1.9 *Permitted Use*

[*specify use*]

1.10 *Plan*

the plan attached to this lease

1.11 *[Plan 1] [Plan 2]*

the plan attached to this lease and marked ['1'] ['2']

1.12 *Planning Acts*

'Planning Legislation' means the Town and Country Planning Act 1990 the Planning (Listed Buildings and Conservation Areas) Act 1990 the Planning (Consequential Provisions) Act 1990 the Planning (Hazardous Substances) Act 1990 the Planning and Compensation Act 1991 and any other legislation relating to town and country planning or the control of development

1.13 *Rent Commencement Date*

[*date*]

[1.14 *Rent Days*

[*dates*] in each year] [the first day of each month]

[1.15 *Schedule of Condition*

means a schedule annexed to this lease at Annexure [*number*]]

1.16 *Service Media*

drains, sewers, flues, gutters, gullies, ducts, shafts, watercourses, pipes, cables, wires and any other medium for the passage or transmission of soil, water, gas, electricity, air, smoke, information or other matters and includes where relevant all ancillary equipment and structures

1.17 *Term*

a term of years commencing on and including the Term Commencement Day and ending on [*date*][4]

1.18 *Term Commencement Date*

[*date*]

1.19 *VAT*

value added tax chargeable under the Value Added Tax Act 1994 or any similar replacement or additional tax

[1.20 *1954 Act*

Landlord and Tenant Act 1954]

1.21 *1995 Act*

Landlord and Tenant (Covenants) Act 1995

2. Interpretations

In this lease where the context so admits:-

2.1 references to 'the Landlord' shall include references to the person for the time being entitled to the reversion immediately expectant on the termination of the Term and references to 'the Tenant' shall include references to its successors in title and assigns

2.2 where the Landlord or the Tenant or any guarantor of the Tenant for the time being are two or more individuals obligations expressed or implied on the part of such party are joint and several

2.3 a 'person' includes a corporate or unincorporated body

2.4 references to any right exercisable by the Landlord or any right exercisable by the Tenant in common with the Landlord shall be construed as including the exercise of such right by and in common with any person authorised by the Landlord and any other person who is or who becomes entitled to use it

---

4 If the lease is to be contracted out of the protection of ss 24–28 of the 1954 Act then the definition of the term should not include references to any continuation such as 'holding over' or 'extension'; if the references are included then the lease will not be granted for 'a term of years certain' and will not be capable of being contracted out.

2.5 except for any reference to the Town and Country Planning (Use Classes) Order 1987 any reference to a statute or statutory instrument or other legislative provision includes any statutory extension, modification amendment, re-enactment of it for the time being in force and any instruments, directions, orders or permissions made under it

2.6 any covenant by the Tenant not to do any act or thing is to be construed as a covenant not to permit such act or thing to be done

2.7 any reference to approval or consent means an approval or consent which is in writing and which has been signed by or on behalf of the Landlord

2.8 words importing one gender include other genders and words importing the singular include the plural and vice versa

2.9 a reference to 'the term' is to the Term [and any statutory or agreed continuation of this lease] and a reference to 'the end of the term' is to the end of the term however it may occur

2.10 the clause, paragraph and schedule headings have been inserted for ease of reference only and do not affect the interpretation of this lease

2.11 the expression 'tenant covenant' and 'landlord covenant' each has the meaning given to it by the Landlord and Tenant (Covenants) Act 1995

3. Grant and rent

3.1 The Landlord demises the Property to the Tenant together with the rights set out in Part I of Schedule 1 but excepting and reserving to the Landlord the rights set out in Part II of Schedule 1 and subject to the Incumbrances set out in Schedule 2 to hold for the Term yielding and paying to the Landlord the following as rent:-

3.1.1 the Annual Rent by equal [monthly] [quarterly] instalments in advance on the [Rent Days] in every year of the Term and proportionately for any period of less than a year the first of such payments being a proportionate sum in respect of the period from and including the Rent Commencement Date to and including the Rent Day next following to be paid on the date hereof

[3.1.2 the Insurance Rent paid by virtue of clause 5.2 as from the Term Commencement Date][5]

3.1.3 all interest payable under this lease

3.1.4 if VAT is payable on Annual Rent the amount of such VAT[6] and

3.1.5 all other sums (including any VAT) which may become due from the Tenant to the Landlord under the provisions of this lease

---

5 If the lease is of the whole of the building the tenant may insure it himself. Where more than one tenant occupies the same building, the landlord will insure and usually recover the cost through the insurance rent.

6 VAT will be payable on rent if the landlord opted to tax.

4. Tenant's covenants

The tenant covenants with the Landlord during the Term as follows:-

4.1 *Rent*

4.1.1 to pay the rent reserved in clause 3 at the times and in the manner stated (and if required by the Landlord at the time by banker's standing order or direct debit) without any deductions whatsoever and without exercising any right of set-off

4.1.2 to pay on demand Interest from the due date until payment on all or any part of any instalment of rent hereby reserved and on all other sums covenanted to be paid by the Tenant to the Landlord which are not paid within fourteen days after the same become due

4.2 *Outgoings*

to pay and indemnify the Landlord against:-

4.2.1 all rates taxes assessments duties charges impositions and outgoings which are now or during the Term shall be charged assessed or imposed upon the Property or upon the owner or occupier of it

4.2.2 all gas electricity telecommunications and other services consumed on the Property, all charges for meters, and all standing charges and to comply with the lawful requirements and regulations of their respective suppliers and

4.2.3 any tax assessed upon the aggregate or any proportion of the income or the value of the assets of the Landlord; or any tax assessed or payable by reason of the act of the Landlord in granting this Lease

4.3 *Common items*

to pay the Landlord on demand a fair proportion of all costs payable by the Landlord for the maintenance, repair, cleaning and renewal of all Service Media, party walls, party structures, yards forecourts, ways and other items not on the Property but used or capable of being used by the Property in common with the Landlord's Adjoining Land or with the occupier of any other property such fair proportion to be conclusively determined by the Landlord's Surveyor

4.4 *VAT*

4.4.1 to pay VAT to the Landlord as rent at the standard or other appropriate rate if applicable in respect of all supplies made or deemed to be made by the Landlord to the Tenant under this lease or in respect of the Property whether the supply is taxable because of an election made by the Landlord or for any other reason

4.4.2 if the Tenant agrees in this lease to indemnify the Landlord in respect of any payments the Landlord may make the Tenant shall in addition indemnify the Landlord against any VAT paid by the Landlord in respect of that payment (unless the VAT is recoverable by the Landlord)

4.5 *Costs*

4.5.1 to pay proper and reasonable costs and expenses of the Landlord including any solicitors' or other professionals' costs and expenses (incurred both during and after the end of the Term) arising out of or in contemplation of:-

4.5.1.1 any action taken by the Landlord in relation to a breach of any of the Tenant's obligations under this lease

4.5.1.2 serving any notice in connection with this lease under section 146 or 147 of the Law of Property Act 1925 or taking any proceedings under either of those sections, notwithstanding that forfeiture is avoided otherwise than by relief granted by the Court

4.5.1.3 serving any notice in connection with this lease under section 17 of the Landlord and Tenant (Covenants) Act 1995

4.5.1.4 the preparation and service of a schedule of dilapidations in connection with this Lease and

4.5.1.5 any consent or approval applied for under this Lease, whether or not it is granted (unless the consent or approval is unreasonably withheld by the Landlord in circumstances where the Landlord is not unreasonably to withhold it)

4.5.2 where the Tenant is obliged to pay or indemnify the Landlord against any solicitors' or other professionals' costs and expenses (whether under this or any other clause of this Lease) that obligation extends to those costs and expenses assessed on a full indemnity basis

4.6 *Repairs*

to keep the Property [in no worse condition than that in which it was at the Term Commencement Date as evidenced by a Schedule of Condition agreed between the parties and annexed to this lease] [in good repair and decorative order] (damage by any Insured Risk excepted unless and to the extent that any act or omission of the Tenant renders the insurance money irrecoverable) and to yield up the same and all fixtures annexed thereto in good repair and decorative order to the Landlord on the termination of the Term howsoever determined

4.7 *Decoration*

to decorate the Property as often as is reasonably necessary and also I in the last three months before the end of the term

4.8 *Alterations and additions*

4.8.1 not to make any [structural] alterations external alterations or additions to the Property

4.8.2 not to install or alter the route of any Service Media at the Property

[4.8.3 not to make any material internal non-structural alterations to the Premises unless:-

4.8.3.1 all necessary consents from any competent authority have been obtained

4.8.3.2 the Landlord has been supplied with drawings and where appropriate specifications

4.8.3.3 the reason respective professional advisors have been paid

4.8.3.4 the Tenant has entered into such covenants as the Landlord may require as to the execution and reinstatement of the alterations

4.8.3.5 the Landlord's prior written consent has been obtained (such consent not to be unreasonably withheld)][7]

4.9 *Landlord's entry*

4.9.1 to permit the Landlord and its employees or agents at all reasonable times on prior written notice (being not less than three days except in emergency when no notice is required) to enter into inspect and view the Property for the purpose of:-

4.9.1.1 examining its condition and taking a schedule of fixtures and fittings in the Property

4.9.1.2 estimating the current value of the Property for insurance purposes

4.9.1.3 inspecting the Property for any purpose connected with any proposed action pursuant to Part II of the 1954 Act

4.9.2 if any breach of covenant, defects, disrepair, removal of fixtures or unauthorised alterations or additions are found on inspection for which the Tenant is liable, then, on notice from the Landlord, to execute to the reasonable satisfaction of the Landlord or the Landlord's Surveyor all repairs works replacements or removals required within two months (or such shorter period as shall be reasonable) after the receipt of the notice

4.9.3 if the Tenant fails to comply with a notice under the preceding sub-clause to allow the Landlord or its agents or workers to enter the Property and execute the repairs works replacements or removals

---

7 Structural alterations are rarely allowed in short tenancies but agreeing to simple internal alterations, such as light partitions (subject to restrictions) can help the tenant to run its business.

The costs incurred by the Landlord in carrying out any works pursuant to this sub-clause (and any professional fees and any VAT in respect of those costs) shall be debt due from the Tenant to the Landlord and payable on demand and any action taken by the Landlord pursuant to this clause shall be without prejudice to the Landlord's other rights, including those under clause [ ]

4.10 *Dealing with the Property*[8]

4.10.1 not to assign, mortgage or charge [part only of] the Property or underlet the whole or any part of the Property

[4.10.2 not to assign or charge the whole of the Property without the consent of the Landlord (such consent not to be unreasonably withheld or delayed) but subject to the operation of the following provisions of this sub-clause 4.10(c)

4.10.3 in addition to reasonable grounds, the Landlord may withhold its consent to an application by the Tenant for licence to assign this lease unless (for the purposes of section 19(1A) of the Landlord and Tenant Act 1927) at the at the time of assignment, the Tenant requesting consent to assign (together with any former tenant who by virtue of section 11 of the 1995 Act was not released on an earlier assignment of this lease) enters into an authorised guarantee agreement in favour of the Landlord which:-

4.10.3.1 is in respect of all the tenant covenants of this lease

4.10.3.2 imposes principal debtor liability on the assignor

4.10.3.3 is in respect of the period starting with the assignee becoming bound by the tenant covenants in this lease and ending on the date when the assignee is released from those covenants by virtue of section 5 of the 1995 Act

4.10.3.4 requires that in the event that this lease is determined by disclaimer the assignor enters into a new tenancy for a term equal to the unexpired residue of the Term and with the terms and conditions as would have applied to this lease if it had continued undetermined and

4.10.3.5 is otherwise in the form reasonably required by the Landlord

4.10.4 on an assignment, if the Landlord reasonably so requires, the Tenant procures that a person who is reasonably acceptable to the Landlord enters into a guarantee and indemnity of the tenant covenants in the form the operative provisions of which are set out in Schedule 3

---

[8] Totally barring assignment or sub-letting may not be in the landlord's interest, in particular, if the tenant's business is not yet established at the time of the grant, and there is a risk that the tenant may not be able to pay the rent. In such circumstances it may suit the landlord to consent to an assignment to a tenant with stronger covenant. Where the term of a lease is very short the assignment may not be appropriate.

4.10.5 the Landlord may withhold its consent to a proposed assignment if at the time of the assignment, there are arrears of rent or other monies due under this lease to the Landlord or if in the reasonable opinion of the Landlord, the value of the Landlord's reversion to this lease would be diminished or otherwise adversely affected by the proposed assignment (on the assumption, if not the fact, that the Landlord wished to sell the reversion immediately after the proposed assignment) and

4.10.6 nothing in this clause shall prevent the Landlord from giving consent subject to any other reasonable condition, nor from refusing consent to an assignment in any other circumstance where it is reasonable to do so]

4.11 *User*

not to use the Property for any purpose other than the Permitted Use

4.12 *Restrictions* affecting use

not to:-

4.12.1 use the Property for any illegal purpose nor for any purpose or in a manner that would cause loss, damage, injury, nuisance or inconvenience to the Landlord or any owner or occupier of neighbouring property

4.12.2 overload any structural part of any building on the Property nor any Service Media at or serving the Property

4.12.3 keep in the Property any dangerous or hazardous or specially combustible goods

4.12.4 allow any auction to be held on the Property

4.12.5 deposit any rubbish or refuse outside the Property and to ensure that all rubbish and refuse shall be regularly removed from the Property and that all goods and materials delivered to the Property shall be stored within the Property

[4.12.6 any other restriction]

4.13 *Compliance with legislation*

4.13.1 at all times to comply with the provisions of all requirements of the Planning Acts [and the Environmental Legislation] so far as those provisions relate to the Property and to observe and comply with the provisions of or imposed under any statute licence or registration regulating or permitting the use of the Property for the purpose for which it is for the time being used and at the expense of the Tenant to do all that is necessary to obtain maintain and renew all licences and registrations required by law for the use of the Property that purpose and not at any time to do or omit anything which could or may result in a contravention of them

4.13.2 not to apply for any planning permission for the Property [without the Landlord's consent[9]]

4.13.3 without prejudice to any obligation on the Tenant to obtain any consent or approval under this Lease, the Tenant shall carry out all works that are required under any law to be carried out at the Property whether by the owner or the occupier

4.13.4 forthwith upon receipt of any notice or other communication affecting the Property to send a copy of the relevant document to the Landlord and in so far as it relates to the Property, take all steps necessary to comply with the notice or other communication and take any other action in connection with it as the Landlord may require

[4.13.5 to keep the Property equipped with all fire prevention, detection and fighting machinery and equipment and fire alarms which are required under all relevant laws or required by the insurers of the Property or reasonably recommended by them or reasonably required by the Landlord and shall keep that machinery, equipment and alarms properly maintained and available for inspection]

4.14 *Encroachments obstructions and acquisition of rights*

4.14.1 not to stop up, darken or obstruct any windows or flow of light or air to the Property

4.14.2 not to grant any right or licence over the Property to any person nor permit any person to make any encroachment over the Property and immediately to notify the Landlord if any person takes or threatens to take any action to obstruct the flow of light or air to the Property

4.15 *Advertisements and signs*

4.15.1 not to place or display on the exterior of the Property [or on the windows or inside the Premises so as to be visible from the exterior of the Property] any name, writing, notice, sign, illuminated sign, display of lights, placard, poster sticker or advertisement other than a suitable sign or name plate of a size and kind first approved by the Landlord or the Landlord's Surveyor showing the Tenant's name and business

4.15.2 not to erect or permit or suffer to be erected any external aerial or mast on the Property without a consent in writing of the Landlord

---

9 If any alterations at all are allowed under the lease.

4.16 *General indemnity*

to keep the Landlord indemnified against all expenses, costs, claims, damage and loss (including any diminution in the value of the Landlord's interest in the Property and loss of amenity of the Property) arising (whether directly or indirectly) from any breach of any tenant covenants in this Lease, or any act or omission of the Tenant, [any under-tenant] or its[their respective] workers, contractors or agents or any other person on the Property with the actual or implied authority of the tenant [any of them]

[4.17 *Registration of disposition of this Lease*

within one month of any disposition or dealing with this lease to give the Landlord notice of the relevant transaction with a certified copy of the relevant documents and to pay the Landlord's solicitors for the time being such fee as they may reasonably require for the registration of such notice][10]

4.18 *Re-letting boards and viewing*

4.18.1 to allow the Landlord or its agents to enter the Property at any time during the Term to fix on any part of the Property a notice board for the sale of the interest of the Landlord and during the six months immediately preceding the termination of this lease to fix on the Property a notice board for re-letting of the Property provided that any such board does not impede the reasonable conduct of the Tenant's business on the Property

4.18.2 to permit any person authorised by the Landlord or by the Landlord's agent to view the Property at reasonable hours without interruption in connection with any such re-letting or sale

4.19 *Returning the Property to the Landlord*

at the end or earlier determination to the Term to return the Property to the Landlord in the repair and condition required by this Lease and if so requested by the Landlord, to remove from the Property all the Tenant's belongings that is to say trade fixtures and fittings and all notices, notice boards and signs bearing the name of or otherwise relating to the Tenant or its business making good to the reasonable satisfaction of the Landlord all damage to the Property from the removal of the Tenant's belongings

[4.20 *New guarantor*

4.20.1 in this clause 4.20 references to a 'new guarantor replacement event' are references in the case of an individual to death, bankruptcy, having a receiving order made against him or having a receiver appointed

---

10 Provision for registrable leases.

under the Mental Health Act 1983, and in the case of a company to passing a resolution to wind up, entering into liquidation or having a receiver appointed

4.20.2 where during the Term a guarantor replacement event occurs to any person who has entered into guarantee covenants with the Landlord pursuant to Clause 4.10 the Tenant must give notice of the event to the Landlord within twenty eight days of his becoming aware of it and if so required by the Landlord the Tenant must within twenty eight days obtain some other person reasonably acceptable to the Landlord to execute a guarantee in favour of the Landlord in such form as the Landlord may reasonably require]

4.21 *Encumbrances*

to observe and perform by way of indemnity only the obligations and restrictions comprising the Encumbrances so far as they relate to the Property and are capable of being enforced, and to keep the Landlord indemnified against liability for breach of the obligations and restrictions

5. Landlord's covenants

5.1 *Quiet enjoyment*

The Landlord covenants with the Tenant that, so long as the Tenant pays the rent reserved by and complies with its obligations in this Lease, the Tenant shall have quiet enjoyment of the Property without any lawful interruption by the Landlord or any person claiming under or in trust for the Landlord

5.2 *Insurance*

5.2.1 Subject to clause 5.2.8 the Landlord shall keep the Property (other than any plate glass) insured against:-

5.2.1.1 loss or damage by the Insured Risks for the sum which the Landlord from time to time shall consider to be the full cost of rebuilding or reinstatement (including architects surveyors and other professional fees the cost of debris removal demolition site clearance or any works that may be required by statute and incidental expenses)

5.2.1.2 the loss of the Annual Rent for [two] years or such longer period as the Landlord may from time to time reasonably consider to be sufficient for the purpose of planning and carrying out such reinstatement

5.2.2 the Landlord shall not be obliged to insure any part of the Property installed by the Tenant

5.2.3 the Landlord's obligation to insure is subject to any exclusions, limitations, excesses and conditions that may be imposed by the insurers and to insurance being available in the London insurance market on reasonable terms acceptable to the Landlord

5.2.4 the Tenant shall pay to the Landlord on demand any amount that is deducted or disallowed by the insurers pursuant to any excess provision in the insurance policy; and any costs that the Landlord incurs in obtaining a valuation of the Property for insurance purposes

5.2.5 the Tenant shall:-

5.2.5.1 not do or omit to do anything that may result in any policy of insurance of the Property or any neighbouring property of the Landlord becoming void or voidable or otherwise prejudiced

5.2.5.2 comply at all times with the requirements and recommendations of the insurers relating to the Property and the Tenant's use of it

5.2.5.3 give the Landlord immediate notice of the occurrence of any damage or loss relating to the Property arising from an Insured Risks

5.2.5.4 not effect any insurance of the Property (except any plate glass at the Property), but if it becomes entitled to the benefit of any insurance proceeds in respect of the Property (other than in respect of plate glass) pay those proceeds or cause them to be paid to the Landlord and

5.2.5.5 pay the Landlord an amount equal to any insurance money that the insurers of the Property refuse to pay by reason of any act or omission of the Tenant or any under-tenant, their workers, contractors or agents or any person at the Property with the actual or implied authority of any of them

5.2.6 the Landlord shall, subject to obtaining all necessary planning and other consents, use all insurance money received (other than for loss of rent) in connection with any damage to the Property to repair the damage for which the money has been received or (as the case may be) in the Property

5.2.7 if the Property and/or any access to the Property is damaged or destroyed by an Insured Risk so as to be unfit for occupation and use then (unless the policy of insurance in relation to the Property has been vitiated in whole or in part in consequence of any act or omission of the Tenant, [or any under-tenant] or its [their respective] workers, contractors or agents or any other person on the Property with the actual or implied authority of any of the Tenant [them] payment of the Annual Rent, or a fair proportion of it according to the nature and extent of the damage, shall be suspended until the Property has been reinstated and made fit for occupation and use or so as to make the Property accessible or until the end of two years from the date of damage or destruction, whichever is sooner

5.2.8 if following damage to or destruction of the Property, the Landlord considers that it is impossible or impractical to reinstate the Property, the Landlord may terminate this lease by giving notice to the Tenant.

On giving notice this lease shall determine but this shall be without prejudice to any right or remedy of the Landlord in respect of any breach of the tenant covenants of this Lease. Any proceeds of the insurance shall belong to the Landlord

5.2.9 provided that the Tenant has complied with its obligations in this clause the Tenant may terminate this lease by giving notice to the Landlord if following damage or destruction of the Property by an Insured Risk the Property has not been reinstated so as to be fit for occupation and use or so as to make the Property accessible or useable within [two] years after the date of damage or destruction. On giving this notice this lease shall determine but this shall be without prejudice to any right or remedy of either party to this lease in respect of any breach of the tenant covenants of this Lease. Any proceeds of the insurance shall belong to the Landlord

6. Additional provisions

6.1 *Provisions for re-entry*

The Landlord may re-enter the Property at any time after any of the following occurs:-

6.1.1 any rent is unpaid 21 days after becoming payable whether it has been formally demanded or not

6.1.2 any breach of any condition of, or tenant covenant, in this lease

6.1.3 where the Tenant or any guarantor is a corporation:-

6.1.3.1 the taking of any step in connection with any voluntary arrangement or any other compromise or arrangement for the benefit of any creditors of the Tenant or guarantor; or

6.1.3.2 the making of an application for an administration order or the making of an administration order in relation to the Tenant or guarantor; or

6.1.3.3 the giving of any notice of intention to appoint an administrator, or the filing at court of the prescribed documents in connection with the appointment of an administrator, or the appointment of an administrator, in any case in relation to the tenant or the guarantor; or

6.1.3.4 the appointment of a receiver or manager or an administrative receiver in relation to any property or income of the Tenant or guarantor; or

6.1.3.5 the commencement of a voluntary winding-up in respect of the Tenant or guarantor, except a winding-up for the purpose of amalgamation or reconstruction of a solvent company in respect of which a statutory declaration of solvency has been filed with the Registrar of Companies; or

6.1.3.6 the making of a petition for a winding-up order or a winding-up order in respect of the Tenant or guarantor; or

6.1.3.7 the striking-off of the Tenant or guarantor from the Register of Companies or the making of an application for the Tenant or the guarantor to be struck-off; or

6.1.3.8 the Tenant or guarantor otherwise ceasing to exist

6.1.4 where the Tenant or any guarantor is an individual:-

6.1.4.1 the taking of any step in connection with any voluntary arrangement or any other compromise or arrangement for the benefit of any creditors of the Tenant or guarantor; or

6.1.4.2 the presentation of a petition for a bankruptcy order or the making of a bankruptcy order against the Tenant or guarantor

If the Landlord re-enters the Property pursuant to this clause, this lease shall immediately end, but without prejudice to any right or remedy of the Landlord in respect of any breach of covenant by the Tenant or any guarantor

6.2 *Power to deal with the Landlord's Adjoining Land*

6.2.1 The Landlord may deal as he thinks fit with the Landlord's Adjoining Land and may erect or permit to be erected on such property any buildings or structures irrespective of whether they affect or diminish the light or air which may now or at any time be enjoyed by the Tenant in respect of the Property

6.2.2 Nothing herein contained shall confer on the Tenant any right to the benefit of or to enforce any covenant or agreement contained in any lease or other instrument relating to the Landlord's Adjoining Land or any other premises belonging to the Landlord

6.3 *Exclusion of warranty as to use*

Nothing in this lease or in any consent granted by the Landlord under this lease is to imply or warrant that the Property may lawfully be used under the Planning Acts for the Permitted Use

6.4 *Compensation for disturbance*

The Tenant is not entitled on quitting the Property to claim from the Landlord any compensation unless and to the extent that any statutory right to compensation precludes the operation of this clause

6.5 *Removal of property after determination of term*

6.5.1 If after the Tenant has vacated the Property following the termination of this lease any property of the Tenant remains in the Property, and the Tenant fails to remove it within fourteen days after being requested in writing by the Landlord to do so, the Landlord may as the agent of the Tenant sell such property and hold the proceeds of sale, after deducting the costs and expenses of removal storage and sale reasonably and properly incurred by it to the order of the Tenant

6.5.2 The Tenant will indemnify the Landlord against any liability incurred by it to any third party whose property has been sold by the Landlord in the bona fide mistaken belief (which is to be presumed unless the contrary be proved) that it belonged to the Tenant and was liable to be dealt with as such under this clause 6.5

6.6 *Notices, consents and approvals*

6.6.1 Except where this lease specifically states that a notice need not be in writing, or where notice is given in an emergency, any notice given pursuant to this lease shall be in writing

6.6.2 Written notice shall be delivered by hand or sent by pre-paid first class post or recorded delivery and a correctly addressed notice sent by pre-paid first class post shall be deemed to have been delivered at the time at which it would have been delivered in the normal course of the post

6.6.3 Section 196 of the Law of Property Act 1925 shall otherwise apply to notices given under this Lease

6.6.4 Where the consent of the Landlord is required under this Lease, a consent shall only be valid if it is given by deed, unless:-

6.6.4.1 it is given in writing and signed by a person duly authorised on behalf of the Landlord and

6.6.4.2 it expressly states that the Landlord waives the requirement for a deed in that particular case

6.6.5 if a waiver is given, it shall not affect the requirement for a deed for any other consent

6.6.6 where the approval of the Landlord is required under this Lease, an approval shall only be valid if it is in writing and signed by or on behalf of the Landlord, unless the approval is being given in a case of emergency or this lease expressly states that the approval need not be in writing

6.6.7 if the Landlord gives a consent or approval under this Lease, the giving of that consent or approval shall not imply that any consent or approval required from a third party has been obtained, nor shall it obviate the need to obtain any consent or approval from a third party

6.7 *Waiver*

The demand for and/or acceptance of rent by the Landlord or its agents shall not constitute a waiver of any of the covenants on the part of the Tenant contained in this lease or of the Landlord's remedies for the breach non-observance or non-performance thereof

6.8 *Rights and easements*

The operation of Section 62 of the Law of Property Act 1925 shall be excluded from this lease and the Tenant shall not be deemed to be entitled to any rights over the Landlord's Adjoining Property except those expressly set out in Schedule 1 of this Lease

6.9 *Entire understanding*

6.9.1 This lease embodies the entire understanding of the parties relating to the Property or to any of the matters dealt with by any of the provisions of this Lease

6.9.2 The Tenant acknowledges that in entering into this lease he has not relied on nor shall have any remedy in respect of, any statement or representation made by or on behalf of the Landlord

6.10 *Contracts (Rights of Third Parties) Act 1999*

No term of this lease shall be enforceable under the Contracts (Rights of Third Parties) Act 1999 by a third party under that Act the Tenant acknowledges that this lease has not been entered into in reliance wholly or partly on any statement or representation made by or on behalf of the Landlord except such statement or representation that is expressly set out in this Lease

6.11 *Exclusion of sections 24–28 of the 1954 Act*

6.11.1 Pursuant to section 38(A)(1) of the 1954 Act the parties agree that the provisions of sections 24 to 28 (inclusive) of the 1954 Act shall be excluded in relation to the tenancy created by this lease

6.11.2 The parties confirm that:-

6.11.2.1 the Landlord served on the Tenant a notice [dated [ ] in the form or substantially in the form set out in Schedule 1 to the Regulatory Reform (Business Tenancies) (England and Wales) Order 2003 ('the Order') in relation to the tenancy created by this lease [not less than fourteen days] before [this lease] [agreement for lease dated [ ]] was entered into (a certified copy of the which notice is annexed hereto)

6.11.2.2 [the Tenant] [*name of declarant*] who was duly authorised by the Tenant to do so] made [a declaration] [statutory declaration][11] dated [ ] in the form or substantially ion the form set out in Schedule 2 to the Order a certified copy of copy of which [statutory] declaration is annexed to this lease]; [and]

[6.11.2.3there is no agreement for lease to which this lease gives effect;]

---

[11] The tenant must make a statutory declaration if the landlord's statutory notice was served less than 14 days before the lease was completed (or where there is an agreement for a lease – before the date of the agreement).

6.11.3 the parties further agree that the form of this lease is in all respects material to the agreement to exclude security of tenure the same as that which was in their contemplation at the time of service of the notice

THIS DOCUMENT has been executed as a Deed and takes effect on the date stated at the beginning of it

SCHEDULE 1
PART I
Rights Granted

1. The right of free passage and running of water, soil, gas, electricity, telephone and other supplies in and through the Conducting Media now or at any time during the Term passing through over or under the Landlord's Adjoining Land and serving the Property so far as necessary for the enjoyment of the Property and in common with the Landlord and others so authorised by the Landlord and all other persons entitled thereto

[2. The right subject to temporary interruption for repair or maintenance for the Tenant to park [ ] motor vehicles in the area shown [hatched in green] on [the Plan] [Plan 1]]

[3. The right to cross over the area shown hatched in blue on [the Plan] [Plan 1] with or without vehicles for the sole purpose of gaining access to and egress from the Property [and the area shown hatched in green on [the Plan] [Plan 1]]]

[4. The right to access the Property over the area shown [coloured] on [the Plan] [Plan 1]]

[5. (*any other rights the Tenant may require*)]

PART II
Rights Excepted

1. The free and uninterrupted passage and running of water soil gas electricity telephone and other services or supplies from and to other parts of the Landlord's Adjoining Land in and through the Conducting Media which now or may hereafter during the Term be in upon through under or over the Property

2. The right to construct and to maintain in upon through under or over the Property at any time during the Term any Conducting Media for the benefit of any other part of the Landlord's Adjoining Land

3. The right at any time during the Term at reasonable times and upon reasonable notice except in case of emergency to enter (or in case of emergency to break and enter) the Premises to:-

3.1 inspect cleanse connect into lay repair remove relay replace with others alter or execute any works whatever to or in connection with the Conducting Media and services referred to in paragraphs 1 and 2 of this Schedule

3.2 carry out work or do anything whatsoever comprised within the Landlord's obligations under this lease whether or not the Tenant is liable to make a contribution

3.3 take schedules or inventories of fixtures and other items to be yielded up on the expiry of the Term and

3.4 inspect the condition and state of repair of the Property

3.5 exercise any of the rights granted to the Landlord by this Lease

4. The right for the Landlord's Surveyor or any other surveyor acting for the Landlord at convenient hours and on reasonable prior notice to enter and to inspect and measure the Property for all purposes connected with any pending or intended step under the Landlord and Tenant Act 1954

5. Full right and liberty to at any time to erect any building of any height on the Landlord's Adjoining Land in such manner as the Landlord thinks fit provided that the same shall not materially obstruct affect or interfere with the amenity of or access to the Property or the passage of light and air to the Property

SCHEDULE 2
ENCUMBRANCES

The matters contained or referred to in the [Registers of Title No [*specify*] [*if unregistered title specify root of the Landlord's title*]] (other than any subsisting financial charges) insofar as the same affect the Property

[SCHEDULE 3]

The operative provisions of the guarantee to be given by a guarantor on an assignment of this Lease

1. Guarantee

1.1 The Guarantor guarantees to the Landlord that the Tenant will pay the rents reserved by, and perform and observe all the Tenant's covenants in this Lease throughout the Term and any extension by statute of the tenancy created by this Lease, and the Guarantor will pay and make good to the Landlord on demand any losses, damages, costs and expenses suffered or incurred by the Landlord by reason of any failure of the Tenant to do so

1.2 The guarantee in clause 1.1 remains in force for so long as, and to the extent that, the Tenant is not released by law from liability for the Tenant's covenants in this Lease

1.3 The Guarantor also guarantees to the Landlord that the Tenant will observe and perform its obligations under an authorised guarantee agreement to be entered into by the Tenant under the terms of this Lease, and will pay and make good to the Landlord on demand any losses damages costs and expenses suffered or incurred by the Landlord if the Tenant fails to do so

1.4 For the purposes of this guarantee references to 'the Tenant' are to the assignee of this Lease in relation to whom the guarantee to the Landlord is given and none other

2. No waiver or release of liability

The Guarantor is not to be released from liability under these provisions by reason of:-

2.1 any forbearance the granting of time or any other indulgence on the part of the Landlord, including (but without affecting the general operation of this clause 2) any granting or extension of time or any refusal by the Landlord to accept any rents tendered by or on behalf of the assignee at the time when the Landlord is entitled (or will be entitled after the service of a notice under section 146 of the Law of Property Act 1925) to re-enter the Property; or

2.2 any variation of this Lease, whether or not made with the consent of the Guarantor, and the guarantee of the Guarantor in clause 1 is to operate in relation to this Lease as it may be varied from time to time

3. Guarantor to accept new lease upon disclaimer

3.1 If this Lease is determined by re-entry by the Landlord or is effectively determined by disclaimer, the Guarantor shall, if the Landlord by notice within three months after the date of determination so requires, take from the Landlord a lease of the Premises

3.2 The lease to be granted to the Guarantor under clause 3.1 is to be on the following terms:-

3.2.1 the term is to commence on the date of termination of this Lease and to be equal to the residue of the Term which would have remained unexpired at that date if this Lease had not then been terminated

3.2.2 the yearly rent is to be the same as would have been payable under this Lease if it had continued undetermined

3.2.3 the lease is otherwise to be on the same terms and conditions as would have applied under this Lease if it had continued undetermined and

3.2.4 the Guarantor is to succeed to the rights and assume the liability of the Tenant under this Lease as if the Lease had continued undetermined

4. Subordination of rights of the Guarantor

4.1 The provisions of clause 4.2 are to apply unless the Landlord has no subsisting claim against the Tenant for non-payment of rent or for breach of obligation under this Lease

4.2 The Guarantor may not:-

4.2.1 seek to recover from the Tenant, or any third party, whether directly or by way of set-off lien counterclaim or otherwise or accept any money or other property or security or exercise any rights in respect of any sum which may be or become due to the Guarantor on account of the failure of the Tenant to observe and perform the tenant covenants in this Lease

4.2.2 in competition with the Landlord claim prove or accept any payment in a winding-up, liquidation, bankruptcy, composition with creditors or other form of arrangement on the insolvency of the Tenant, for money owing to the Guarantor by the Tenant; nor

4.2.3 exercise any right or remedy in respect of an amount paid by the Guarantor under this Lease or any liability incurred by the Guarantor in observing performing or discharging the obligations and covenants of the Tenant

4.3 The Guarantor warrants that it has not taken, and undertakes with the Landlord that it will not without the consent of the Landlord take, any security from the Tenant in respect of this guarantee and, if any security is nevertheless taken, it is to be held on trust for the Landlord as security for the respective liabilities of the Guarantor and the Tenant

SIGNED as a Deed by)
[(*name of the Landlord*)])
in the presence of:)
[(*name of the Landlord*) acting by
a director and its secretary or by two directors]

[Director]
[Director/Secretary]

SIGNED as a Deed by)
)
in the presence of:)
Witness:
Address:
Occupation:

C/Part

SIGNED as a Deed by)
)
in the presence of:)
Witness:
Address:
Occupation:

PRECEDENT 2 LICENCE FOR STORAGE

THIS LICENCE is made the [ ] day of [ ] 20[ ]

BETWEEN:-

1 [*name*] of [*address*] ('the Owner')

and

2 [*name*] of [*address*] ('the Licensee')

NOW IT IS AGREED as follows:-

1. Definitions and interpretation

In this Licence the following expressions have the meanings given in this clause:-

1.1 *The Plan*

the 'Plan' means the plan annexed to this Licence

1.2 *The Premises*

the 'Premises' means the land and buildings shown for identification purposes only edged purple on the Plan

1.3 *The Access Route*

the 'Access Route' means the tracks of the Premises the use of which is necessary to obtain access to and egress from the Licensed Area coloured brown on the Plan

1.4 *The Building*

the 'Building' means part of the Premises shown for identification purposes only edged blue on the Plan

1.5 *The Licensed Area*

the 'Licensed Area' means the area shown for identification purposes only edged red on the Plan or such other storage space, comprising a single area of not less than [*number*] square metres within the Building as the Owner may from time to time in his absolute discretion specify

1.6 *Headings*

the clause and sub-clause headings do not form part of this Licence and must not be taken into account in the construction or interpretation

1.7 *The Licence Fee*

the 'Licence Fee' means [ ] pounds (£[ ]) per month

1.8 *The Licence Period*

the 'Licence Period' means the [ ] months from the date of this Licence or until the Licence is determined in accordance with Clause 4

1.9 *Reference to Clauses*

any reference in this Licence to a clause or sub-clause without further designation is to be construed as a reference to the clause or sub-clause of this Licence so numbered

1.10 *VAT*

'VAT' means value added tax or any other tax of a similar nature

2. The Licence

Subject to clauses 3 and 4, the Owner gives the Licensee the right, for the Licence Period in common with the Owner and all others authorised by the Owner so far as is not inconsistent with the rights given, to use the Licensed Area for [*specify purpose*] and to use the Access Route for access to and egress from the Licensed Area

3. Licensee's agreements and undertakings

The Licensee agrees and undertakes:-

3.1 to pay the Licence Fee, together with any VAT, in advance on the first day of each month

3.2 to keep the Licensed Area clean and tidy and clear of rubbish and leave it in a clean and tidy condition and free of the Licensee's equipment, goods and chattels at the end of the Licence Period

3.3 not to obstruct the Access Route or make it dirty or untidy, or leave any rubbish on it

3.4 not to use the Licensed Area or the Access Route in such a way as to cause any nuisance, damage, disturbance, annoyance, inconvenience or interference to the Premises or adjoining or neighbouring property or to the owners, occupiers or users of any adjoining or neighbouring property

3.5 not without the prior written consent of the Owner to bring onto, or permit on, the Premises any dangerous or noxious substance or thing, inflammable material or high temperature equipment

3.6 not to do anything that will or might constitute a breach of any statutory requirement affecting the Premises or that will or might wholly or partly vitiate any insurance effected in respect of the Premises from time to time

3.7 to indemnify the Owner, and keep the Owner indemnified, against all losses, claims, demands, actions, proceedings, damages, costs or expenses or other liability arising in any way from this Licence, any breach of any of the Licensee's undertakings contained in this clause, or the exercise or purported exercise of any of the rights given in clause 2

3.8 to take out his own insurance in respect of all items stored and against third party liability for a minimum cover of five million pounds (£5,000,000.00)

3.9 not to impede in any way the Owner, or his servants or agents, in the exercise of his rights of possession and control of the Premises or any part of the Premises

4. Determination

The rights granted in clause 2 are to determine (without prejudice to the Owner's rights in respect of any breach of the undertakings contained in clause 3):-

4.1 immediately on notice given by the Owner at any time following any breach by the Licensee of his undertakings contained in clause 3

and

4.2 on not less than 28 days' notice given by the Owner or the Licensee to the other party to expire on the last day of any month

5. No Assignment

The benefit of this Licence is personal to the Licensee and not assignable, and the rights given in clause 2 may only be exercisable by the Licensee

6. No Warranty

The Owner gives no warranty that the Premises are legally or physically fit for the purposes specified in clause 2

7. Exclusion of Liability

The Owner will not to be liable for damage to any property of the Licensee (or his employees) or for any losses, claims, demands, actions, proceedings, damages, costs or expenses or other liability incurred by him or them in the exercise or purported exercise of the rights granted by clause 2

AS WITNESS the hands of the parties the day and year first before written

SIGNED by
in the presence of:-

SIGNED by
in the presence of:-

Chapter 6

RESIDENTIAL TENANCIES

INTRODUCTION

PRECEDENTS

1 Assured shorthold tenancy
2 Tenant's notice proposing change from assured tenancy to assured shorthold tenancy
3 Landlord's notice proposing assured shorthold tenancy
4 Landlord's letter proposing rent increase
5 Notice proposing different terms for statutory periodic tenancy
6 Application referring a notice proposing different terms for statutory periodic tenancy to rent assessment committee
7 Notice seeking possession of property let on assured tenancy/assured agricultural occupancy
8 Landlord's notice proposing new rent under assured periodic tenancy
9 Landlord's or licensor's notice proposing a new rent or licence fees under an assured agricultural occupancy

INTRODUCTION

Because of the difficulty a landlord had in gaining possession from a tenant who was either a protected agricultural occupant[1] or an assured agricultural occupant,[2] many landlords let property to farm workers on shorthold tenancies. After some initial fears that this might be contrary to official policy it was generally accepted that these lettings were valid.

Protected shortholds were introduced by the Housing Act 1980. The tenancy had to be between one and 5 years and provided the appropriate notice was given at the start of the term, possession could be obtained by the landlord under Case 19 of the Rent Act 1977.

No protected shorthold could be granted after 15 January 1989.

---

1 Under the Rent (Agriculture) Act 1976.
2 Under the Housing Act 1988.

Assured shortholds were introduced by the Housing Act 1988. Before 28 February 1997, the commencement of the Housing Act 1996, the tenancy had to be a fixed term granted for a minimum term of 6 months but there was no longer any need for a maximum term.[3] It was also essential to serve a notice in prescribed form before the start of the tenancy informing the tenant that the tenancy was to be an assured shorthold tenancy. These requirements have been abolished by the Housing Act 1996 for most assured shortholds, but notices are still necessary if an assured shorthold is to be granted to an agricultural worker. Even if the proposed tenant does not satisfy the agricultural worker condition[4] at the start of the tenancy a notice should be served. Otherwise should the agricultural worker condition be subsequently satisfied the assured shorthold will become an assured agricultural occupancy or assured tenancy.

Although the tenancy no longer has to be for a minimum of 6 months, any order for possession may not take effect earlier than 6 months after the beginning of the tenancy.

It should be remembered that an assured shorthold is a variant of the assured tenancy and therefore the rent must be more than £250 per annum.

A tenant of an assured shorthold may demand from his landlord a written statement of the terms of the tenancy, unless he already has such a statement.[5]

If the shorthold is granted by a farm tenant the head lease should be checked to ensure that the subletting of the farm cottage is permitted.

Rent control

An assured shorthold tenant can refer the rent to a Rent Assessment Committee if he considers that the rent is significantly higher than the rent payable under similar tenancies of similar dwelling-houses in the locality.[6] The Committee has power to fix the rent at a market rent taking into account the lack of security of tenure. The tenant's right to go to the Assessment Committee arises only during the initial fixed term. Moreover, the Committee can make a determination only if there are a significant number of assured tenancies of similar dwellings in the locality.

Where the original fixed term tenancy continues as a periodic tenancy, the rent may be increased by the landlord serving a prescribed notice on the tenant. If the tenant does not agree the proposed increase he can refer the rent to the Rent Assessment Committee.[7]

---

3 Housing Act 1988, s 20.

4 A person who has worked full time in agriculture or has worked in agriculture as a permit worker for not less than 91 out of the past 104 weeks.

5 Housing Act 1996, s 97.

6 Housing Act 1988, s 22.

7 Housing Act 1988, s 13.

Obtaining possession

The landlord may obtain possession on giving 2 months' notice in writing to expire at the end of the fixed term or at the end of a completed statutory periodic tenancy.[8] Where the periodic tenancy is quarterly, 6-monthly or yearly the notice must also accord with the common law rules. Therefore the notice period would be a quarter or 6 months. A formula can be used 'at (or after) the end of that period of the tenancy which will end after the expiry of two months from the service of this notice'.[9]

The court has no discretion to refuse possession though in cases of exceptional hardship it may postpone possession for up to 6 weeks.[10]

There is an accelerated procedure for obtaining possession.[11]

8 Housing Act, s 21, as amended by the Housing Act 1996.
9 *Notting Hill Housing Trust v Roomus* [2006] EWCA Civ 407.
10 Housing Act 1980, s 89.
11 CPR 1998, SI 1998/3132, rr 55.11–55.18.

PRECEDENTS

PRECEDENT 1 ASSURED SHORTHOLD TENANCY

Date	:	.... 20[ ]
The Landlord	:	....
The Tenant(s)	:	....
The Property	:	....
		The Property includes all fixtures, fittings and electrical or mechanical appliances and is:-
		[ ] Furnished and let together with the Contents being the items listed in the attached Inventory
		[ ] Unfurnished
The Rent	:	£[ ]
		[ ] per week [ ] per calender month
		Payable in on the [ ] day of each month
		The first payment is to be made on
The Deposit	:	[ ] The deposit is £[ ]
		[√] There is no deposit
The Term	:	The tenancy is for a fixed period of starting from
The Costs Contribution	:	A sum of £[ ] (including VAT) towards the Landlord's legal costs
The Landlord's Agent	:	....
(if any)		or such other person of whom the Landlord may notify the Tenant in writing from time to time
(*including address and telephone number*)		
Occupants attaining the age of 18	:	....
		Any resident in the property over the age the age of 18 should be notified to the Landlord and named in the Tenancy agreement

AGREEMENT

1. **Letting**

The Landlord lets the Property to the Tenant for the Term at the Rent payable as set out overleaf

2. **Assured shorthold tenancy**

2.1 This Agreement creates an assured shorthold tenancy within Part I Chapter II of the Housing Act 1988. This means when the Term expires the Landlord can recover possession as set out in section 21 of that Act

[2.2 The Landlord and Tenant agree that the Tenant has before the execution of this Agreement received a notice that the tenancy created by this Agreement is to be an assured shorthold tenancy within the meaning of the Housing Act 1988 section 19A, Schedule 2A paragraph 9(2) even though the fact that the agricultural worker condition is or may be for the time being fulfilled in respect of the Property]

3. **Definitions and interpretations**

In this Agreement:-

3.1 a reference to a statute includes any amendment or re-enactment of it (whether made before or after the date of this Agreement) and any secondary legislation made under that statute

3.2 whenever there is more than one Tenant all their obligations can be enforced against all or some of them jointly or against each of them individually

3.3 any agreement by the Tenant not to do any act or thing includes an agreement not to allow anyone else to do that act or thing

3.4 the words and phrases on the preceding pages of this agreement have the meanings as shown on those pages

3.5 all sums due to be paid by the Tenant under this Lease are expressed exclusive of VAT and the Tenant in addition must pay the full amount of any VAT that is accountable in respect of those sums

3.6 the expression 'the Prescribed Information' means the information required under section 213(5) of the Housing Act 2004 as set out in the Housing (Tenancy Deposits) (Prescribed Information) Order 2007

3.7 the expression 'the Scheme Administrator' means the administrator of either a custodial or insurance Tenancy Deposit Scheme

3.8 the expression 'the TDS' means the tenancy deposit scheme

3.9 the expression 'the DPC' means the Deposit Protection Certificate issued to the Landlord by the Scheme Administrator once the tenancy has been protected under the TDS

4. Forfeiture

The Landlord is entitled to forfeit this Agreement by entering the Property whenever:-

4.1 the Tenant is 14 days late in paying the Rent, even if it was not formally demanded and/or

4.2 the Tenant has not complied with any obligation in this Agreement or has committed a breach of any term or condition of the Tenant and/or

4.3 the Tenant not less than 6 months after the start date of the tenancy is adjudicated bankrupt or has an interim receiver appointed and/or

4.4 the Court has power to make an Order for Possession of a dwellinghouse on one or more of the grounds set out in Schedule 2 to the Housing Act 1988

Forfeiture of this Agreement does not cancel any outstanding obligation which the Tenant owes to the Landlord

5. Rent reduction

During any period when the Tenant cannot live in the Property because of damage from a risk against which the Landlord has insured, any Rent which falls due from the date of the occurrence of that risk is cancelled or reduced as appropriate unless or to the extent that the insurers do not pay under the policy because of something done or not done by the Tenant

6. Insuring the Property

The Landlord is responsible for insuring the Property but the Tenant is liable for any excess of any claim made which was due to the fault of the Tenant and if the Tenant's actions result in the premium for the insurance of the Property increasing such increase may be recoverable from the Tenant

7. Notices

7.1 The Landlord's address for service of notices is as stated on page 1 of this Agreement and in addition any notice is deemed properly served by the Landlord if served by the Landlord's Agent

7.2 If details of the Landlord's Agent appear in this Agreement notices served by or on the Landlord may be served by or on the Landlord's Agent and the Landlord may elect that any Deposit is held by the Landlord's Agent

8. Schedule of Condition

If there is a Schedule of Condition either attached to this Agreement or signed by or on behalf of the parties and identified as being the Schedule of Condition (whether written or photographic) for the purposes of this Agreement then nothing in this Agreement shall require the Tenant to do any work to put any part of the Property specified in that Schedule into any better condition than that recorded in the Schedule

9. Deposit

9.1 The Landlord acknowledges receipt of the Deposit from the Tenant

9.2 At the end of the Tenancy, the Landlord shall be entitled to withhold from the Deposit such proportion of the Deposit as may be reasonably necessary to:-

9.2.1 make good any damage to the Property or the Contents (except for fair wear and tear)

9.2.2 replace any of the Contents which may be missing from the Property

9.2.3 pay any accounts for utilities, Council tax charges or any other taxes or accounts for which the Tenant may be liable, which remain unpaid

9.2.4 pay any Rent which remains unpaid and

9.2.5 pay for the Property and the Contents to be cleaned if the Tenant is in breach of its obligations under this agreement

10. Tenancy Deposit Scheme arrangements

10.1 The Deposit is protected by [ ]. The Deposit is held by the Landlord

10.2 The Landlord will provide within fourteen days of the Deposit being received the Prescribed Information

10.3 The Landlord agrees that the Deposit shall be held in accordance with the rules of the TDS

10.4 The Landlord will sign and retain one copy of the DPC

10.5 The Landlord will give a copy of the DPC to the Tenant

10.6 The Tenant will sign one copy of the DPC to acknowledge confirmation of the Deposit being protected under the TDS

10.7 The Landlord and the Tenant agree that any interest accrued from the deposit shall be paid to the [the Tenant] OR [Landlord] OR [shared equally between the Landlord or the Tenant]

10.8 The Landlord shall inform the Tenant within ten working days of the Tenancy ending if the Landlord intends to withhold all or part of the Deposit as detailed in clause 9 above

10.9 If there is a Deposit dispute the Landlord and the Tenant shall deal with the dispute in accordance with the rules of the TDS and the Housing Act 2004

11. The Tenant's obligations

The Tenant is obliged:-

11.1 *Rent*

to pay the Rent as set out above and by banker's order if so requested

11.2 *Deposit*

immediately to pay the Deposit as security in accordance with Clause 9.2 above and when requested to pay to the Landlord any sum needed to bring the Deposit up to the full amount from time to time

11.3 *Costs*

immediately to pay the Costs Contribution by way of contribution towards the Landlord's costs in connection with this agreement

11.4 *Interest*

to pay interest of [ ]% above the base rate of Barclays Bank Plc for any payments that are more than [ ] days late whether requested or not

11.5 *Outgoings*

11.5.1 immediately to notify all relevant service providers of his occupation of the property and to pay for all gas, electricity, water, oil and sewerage services supplied to or used at the Property during the tenancy and all charges for the use of any telephone and television at the Property during the tenancy and all other outgoings on the Property. Where necessary the sums demanded by the service provider will be apportioned according to use and/or the duration of the tenancy (as the Landlord shall decide). The sums covered by this clause include standing charges or other similar charges and VAT as well as charges for actual consumption plus the cost of any reconnection of any services which have been withdrawn following any act or omission by the Tenant

11.5.2 to pay the annual maintenance charge and call out charges associated with any security alarm fitted to the property

11.6 *Council Tax*

to pay all Council Tax or similar or substituted tax or payment due for the Property and/or the tenant's occupation of it

11.7 *Use of Property*

not to use the Property otherwise than for residential occupation by the Tenant and his family and not to carry on any profession trade or business at the Property

11.8 *Nuisance*

not to use or allow anyone else to use the Property or any part of it for any activities which are dangerous offensive noxious noisy (so as to be audible outside the Property) illegal or immoral or which are or may become a nuisance or annoyance to the Landlord or to any neighbours

11.9 *Notices*

to give the Landlord promptly a copy of any notice received concerning the Property or any neighbouring property

11.10 *Insurance*

not to act in any way which will or may result in the insurance of the Property being void or voidable, or in the premium for it being increased

11.11 *Repairs*

to keep the interior of the Property (other than any part for which the Landlord is responsible) and (if applicable) the Contents in good and tenantable repair and condition (fair wear and tear excepted) but the Tenant is not obliged to make good any damage against which the Landlord is insured unless the insurance monies are withheld due the act or omission of the Tenant

11.12 *Damages*

11.12.1 to report promptly to the Landlord any damage to the Property or (if applicable) the Contents caused by the Tenant and (except where covered by the Landlord's insurance) to pay for any appropriate repair or if irreparably damaged or lost to replace or (at the option of the Landlord) to pay the Landlord the cost of replacement pay the Landlord the cost of replacement

11.12.2 to pay for the steam or other appropriate cleaning of all carpets and upholstery within the Property which shall in the opinion of the Landlord (whose decision shall be final and binding on the Tenant) have been soiled during the tenancy

11.13 *Alterations or additions*

not to make any alterations or additions to the Property, including changing the colour scheme or altering or adding to the wiring or plumbing nor (without the prior written consent of the Landlord) to drill holes or drive nails into any wall or ceiling

11.14 *Contents*

not to remove any of the Contents from the Property without the prior written consent of the Landlord and to pay the replacement value of any which may have been removed during the tenancy

11.15 *Defects to the Property*

to report promptly to the Landlord all defects in the Property which it is the Landlord's duty to repair

11.16 *Keep clean and tidy*

to keep the Property clean and tidy and any garden and bushes or hedges in a neat and tidy order, including keeping the grass regularly cut and the drives, paths and garden free from weeds and fallen leaves

11.17 *Vermin*

to notify the Landlord of the presence of any vermin at the Property

11.18 *Drains and gutters*

to keep all roofs and gutters on the Property free from leaves, plants and dirt and to keep all the drains, sinks, sewers and soakaways belonging to the Property free from any obstruction

11.19 *Septic tank*

to empty any septic tank serving the Property whenever required and at least once every twelve months or (at the Landlord's discretion) to pay to the Landlord on demand the cost thereof or contribute a fair proportion of such cost

11.20 *Inspection of the Property*

11.20.1 to allow the Landlord or anyone with the Landlord's authority upon giving at least two days' notice (or without notice in an emergency) to enter the Property at reasonable times to inspect its condition, take meter readings and/or carry out any repairing obligations and for any other reasonable purpose

11.20.2 to undertake immediately any repairs which are the Tenant's obligation to undertake and of which the Landlord gives notice and if the Tenant fails to comply the Landlord may return after a period of five days (less in case of emergency) to carry out the repairs himself with the Tenant being liable to reimburse and indemnify the Landlord against any costs so incurred

11.21 *Sharing the Property*

not to assign sub-let part with possession of or share the Property or any part of it and not to take in lodgers or paying guests

11.22 *Prospective tenants*

to allow the Landlord and/or any person with written authority from the Landlord to affix a re-letting sign to the premises eight weeks prior to the end of the tenancy and to enter the Property at any reasonable time on reasonable notice to view the Property

11.23 *Keys and locks*

not to have additional keys made for any locks nor any locks changed without the prior written consent of the Landlord. If any additional or new keys are made to return them to the Landlord with all original keys at the end of the Tenancy and in the event any keys have been lost to pay to the Landlord any costs incurred by the Landlord in replacing the locks to which the lost keys belonged

11.24 *Frost damage*

to take adequate measures to prevent frost damage to the plumbing and heating installations including draining the hot water and heating system whenever that is required to prevent frost damage during a period when the Property is unoccupied during the winter months

11.25 *Smoke alarms*

to regularly check all smoke alarms in the Property (replacing batteries as necessary) to ensure they are in proper working order at all times

11.26 *Gas safety*

to ensure all ventilators are kept clear and to report immediately to Transco (or other company or body responsible for gas safety) and immediately thereafter the Landlord any brown or sooty build up on any gas appliances or other concern as to the safety of any gas appliance

11.27 *Car parking*

to park cars, motor cycles or other vehicles in the designated parking spaces (where these exist) and not to park or store on the Property any trailer, caravan or mobile home

11.28 *Refuse collection*

to ensure that all rubbish is stored in suitable bins in a suitable place and made available for collection at appropriate times

11.29 *Expenses*

to pay all charges, costs and expenses which the Landlord incurs:-

11.29.1 as a result of a breach of any of the terms of this agreement and/or

11.29.2 in preparing and serving a notice under section 146 of the Law of Property Act 1925, even if forfeiture is avoided without a court order and/or

11.29.3 in preparing and serving a schedule of dilapidations recording failure to give up possession of the Property in the appropriate state of repair when this Agreement ends

11.30 *Chimneys*

to sweep any chimneys (if used) at least once every twelve months

11.31 *Pets*

not to keep pets or other animals birds or reptiles at the Property (without the written consent of the Landlord) and then only in accordance with any conditions contained in any such consent

11.32 *Periods of vacancy*

to inform the Landlord of any periods of more than twenty one days when the Property will be uninhabited and during all periods when the Property is uninhabited to take such steps as may reasonably be necessary to protect and preserve the Property and the Contents

11.33 *Expiration of Tenancy*

to leave the Property and (if applicable) the Contents at the end of the Tenancy in the state in which this Agreement requires the Tenant to keep them and to pay for any cleaning costs and removal of rubbish that may be necessary after the tenancy

12. Landlord's obligations

The Landlord is obliged to:-

12.1 allow the Tenant to occupy the Property without interference so long as this Agreement continues and the Tenant complies with its terms

12.1 carry out such repairs to the Property as are required of the Landlord by the Landlord and Tenant Act 1985

SIGNED by
in the presence of:-

WARNING

This is a formal document designed to create legal rights and obligations. Take advice before using it.

SIGNED by
in the presence of:-

PRECEDENT 2 TENANT'S NOTICE PROPOSING CHANGE FROM ASSURED TENANCY TO ASSURED SHORTHOLD TENANCY

FORM No 8

Housing Act 1988 Schedule 2A, paragraph 7(2) as inserted by Schedule 7 to the Housing Act 1996

Tenant's notice proposing that an Assured Tenancy be replaced by an Assured Shorthold Tenancy

- Please write clearly in black ink.
- Please cross out text marked with an asterisk (*) that does not apply.
- This notice should only be used by an assured tenant. You should only use this notice to notify your landlord that you wish your assured tenancy to be replaced by an assured shorthold tenancy.
- This notice must be served by a tenant on a landlord before an assured tenancy can be replaced by an assured shorthold tenancy.
- **You should be aware that by serving this notice, you will be giving up your right to stay in the property after the first six months of the assured shorthold tenancy or, if you agree a fixed term with your landlord, after the end of the fixed term.**
- **You do not have to complete this form even if your landlord has asked you to do so. Your existing security of terms as an assured tenant will be unaffected if you do not complete it.**
- **If you are in any doubt about whether to complete this form, take it immediately to a citizen's advice bureau, housing advice centre, a law centre or a solicitor.**
- Once you are clear that you wish to issue this notice, complete the form and send it to your landlord.

---

1. To: ....

Name(s) of landlord(s)

---

2. I/We*, the tenant(s) of:

....

....

....

Address of premises

give notice that I/we* propose that the assured tenancy to which this notice relates should be replaced by a shorthold tenancy.

---

3. I/We* propose that the new shorthold tenancy should commence on:

......./ ..../ ..../

day month year

- The new shorthold tenancy cannot commence until after the date this notice is served on the landlord.

4.(a) I/We* understand that under my/our* existing tenancy, I/we* can only be required to give up possession in accordance with the grounds set out in Schedule 2 to the Housing Act 1988, whereas under the new shorthold tenancy, the landlord(s) will be able to recover possession of the premises without being required to prove a ground for possession, after the first six months of the assured shorthold tenancy, or, if there is a fixed term for longer than six months, at the end of that fixed term, subject to two months' notice.

Signed .... Date ....

....

....

To be signed and dated by the tenant. If there are joint tenants each tenant must sign.

(b) Name and address of tenant

Name(s) (*Block Capitals*): ....

....

....

Address:

....

....

....

Telephone – Daytime .... Evening ....

PRECEDENT 3 LANDLORD'S NOTICE PROPOSING ASSURED SHORTHOLD TENANCY

FORM No 9

Housing Act 1988 Schedule 2A, paragraph 9, as inserted by Schedule 7 to the Housing Act 1996

Landlord's notice proposing an Assured Shorthold Tenancy where the tenancy meets the conditions for an Assured Agricultural Occupancy

- Please write clearly in black ink.
- Please tick boxes where appropriate.
- If the agricultural worker condition in Schedule 3 to the Housing Act 1988 is met with respect to the property to which the proposed assured tenancy relates, and the landlord wishes that tenancy to be an assured shorthold tenancy, he must serve this notice on the tenant before the tenancy is entered into.

- **This notice cannot be used where the landlord has already granted to the prospective tenant (or, in the case of joint tenants, to at least one of them) a tenancy or licence under section 24 of the Housing Act 1988 (an assured agricultural occupancy).**
- **This notice does not commit the tenant to taking the tenancy.**

---

1. To: ....

....

....

Name of the proposed tenant. If a joint tenancy is being offered, enter the names of the joint tenants.

---

2. You are proposing to take a tenancy as the following address:

....

....

....

commencing on ./ ..../ ..../
day month year

---

3. This notice is to tell you that your tenancy is to be an assured shorthold tenancy.

- Provided you keep to the terms of the tenancy, you are entitled to remain in the property for at least six months after the start of the tenancy. Depending on the terms of the tenancy, once the first six months have elapsed, the landlord may have the right to seek possession at any time, subject to two months' notice.
- As an assured shorthold tenant, you have the right to apply to a rent assessment committee for the determination of a reasonable rent for the tenancy. An application to your local rent assessment committee must be made on the form headed *Application to a Rent Assessment Committee for a determination of a rent under an Assured Shorthold Tenancy* within six months of the beginning of the tenancy. You can obtain the form from a rent assessment panel or a law stationer.
- If you need help or advice about this notice, and what you should do about it, take it immediately to a citizens' advice bureau, a housing advice centre, a law centre or a solicitor.

---

4. Name and address of landlord.

To be signed and dated by the landlord or his agent (someone acting for him). If there are joint landlords, each landlord or the agent must sign unless one signs on behalf of the rest with their agreement.

Signed .... Date ....

....

....

Please specify whether:	landlord [ ]	joint landlords [ ]	agent [ ]

Name(s) (*Block Capitals*): ....
....

Address:

....
....
....

Telephone – Daytime .... Evening ....

PRECEDENT 4 LANDLORD'S LETTER PROPOSING RENT INCREASE

To [*name and address of tenant(s)*]

Dear [Sir] [*or as the case may be*]

YOUR TENANCY OF [*PROPERTY*]

It is considered that the rent of the above premises is now due for review. It is proposed that as from [*date – an appropriate date will usually be a rent day*] the rent is increased to £[. . .] per [*month/quarter/annum*]*.

If the proposal is acceptable please sign, date and return to us the enclosed copy of this letter to signify your agreement. If you pay rent by standing order, please instruct your bank to make the necessary adjustment, letting us have a copy of the instruction.

If you do not respond to this letter within [. . .] days we shall serve on you a notice under section 13(2) of the Housing Act 1988 (Landlord's Notice Proposing a new Rent under an Assured Periodic Tenancy or Agricultural Occupancy).

Your faithfully,

Agents for the Landlord [*name of Landlord*]

[To be typed on copy]

I/We agree that the rent on the above premises is increased to £[. . .] per [*month/quarter/annum*]* as from [*date*].

Signed: .... Dated: ....

Tenant(s)

**delete as appropriate.*

PRECEDENT 5 NOTICE PROPOSING DIFFERENT TERMS FOR STATUTORY PERIODIC TENANCY

SCHEDULE

FORMS PRESCRIBED FOR THE PURPOSES OF PART 1 OF THE HOUSING ACT 1988

FORM No 1

Housing Act 1988 section 6(2)

Notice proposing different terms for a Statutory Periodic Tenancy

- Please write clearly in black ink.
- Please tick boxes where appropriate and cross out text marked with an asterisk (*) that does not apply.
- This form can be used by either a landlord or a tenant to propose changes to the terms of a statutory periodic tenancy, which arises when a fixed term of an assured tenancy, an assured shorthold tenancy or an assured occupancy ends.
- This notice must be served on the landlord or tenant no later than the first anniversary of the day on which the former fixed term tenancy or occupancy ended.
- Do not use this notice if you are a landlord proposing only an increase in rent. Instead, you should use the form headed *Landlord's Notice proposing a new rent under an Assured Periodic Tenancy* OR *Landlord's or Licensor's Notice proposing a new rent or licence fee under an Assured Agricultural Occupancy*, which is available from a rent assessment panel or law stationers.

---

1. To: ....

*Name(s) of landlord(s)/tenant(s)**

Address of premises to which the tenancy relates:

....

....

....

---

2. This is to give notice that I/we* propose different terms for the statutory periodic tenancy from those of the fixed term assured tenancy which has now ended and that they should take effect from:

....

Insert date which must be at least three months after the date on which this notice is served.

---

3. Changes to the terms

(a) The existing provisions of the tenancy to be changed are:

....

....

....

Please attach relevant sections of the tenancy agreement if available.

(b) The proposed changes are:

....

....

....

Continue on another sheet if necessary.

---

4. Changes to the rent (if applicable). Go to section 5 if this does not apply.

- You should not propose a change on this form unless it is to take account of the proposed new terms at section 3. A change may be made if either the landlord or the tenant considers it appropriate.

(a)	The existing rent is:	£.................. per .... (e g week, month year)
(b)	Does the rent include council tax?	Yes [ ] No [ ]
(c)	If yes, the amount that is included for council tax is:	£ .... per .... (e g week, month year)
(d)	Does the rent include water charges?	Yes [ ] No [ ]
(e)	If yes, the amount that is included for council tax is:	£ .... per .... (e g week, month year)
(f)	The new rent which takes into account the proposed changes in terms of the tenancy will be:	£ .... per ....
(g)	Will the new rent include council tax?	Yes [ ] No [ ]
(h)	If yes, the amount that will be included for council tax is:	£ .... per .... (e g week, month year)
(i)	Will the new rent include water charges?	Yes [ ] No [ ]
(h)	If yes, the amount that will be included for water charges is:	£ .... per .... (e g week, month year)

---

5. Name and address of landlord or tenant proposing the changes

To be signed and dated by the landlord or his agent (someone acting for him) or the tenant or his agent. If there are joint landlords or joint tenants each landlord/tenant or the agent must sign unless one signs on behalf of the rest with their agreement.

Signed .... Date ....

....

....

Please specify whether:	landlord [ ]	joint landlords [ ]	agent [ ]	tenant's agent [ ]

Name(s) (*Block Capitals*): ....

....

Address:

....

....

....

Telephone – Daytime .... Evening ....

What to do if this notice is served on you

- If you agree with the new terms and rent proposed, do nothing. They will become the terms of the tenancy agreement on the date specified in section 2.
- If you don't agree with the proposed terms and any adjustment of the rent (see section 4), and you are unable to reach agreement with your landlord/tenant, or you do not wish to discuss it with him, you may refer the matter directly to your local rent assessment committee, before the date specified in section 2, using the form headed *Application referring a Notice proposing different terms for a Statutory Periodic Tenancy to a Rent Assessment Committee* which you can obtain from a rent assessment panel or a law stationer.
- The rent assessment committee will decide what, if any, changes should be made to the terms of the tenancy and, if applicable, the amount of the new rent.
- If you need help or advice about this notice and what you should do about it, take it immediately to a citizens advice bureau, a housing advice centre, a law centre or a solicitor.

PRECEDENT 6 APPLICATION REFERRING A NOTICE PROPOSING DIFFERENT TERMS FOR STATUTORY PERIODIC TENANCY TO RENT ASSESSMENT COMMITTEE

FORM No 2

Housing Act 1988 section 6(3)

Application referring a Notice proposing different terms for a Statutory Periodic Tenancy to a Rent Assessment Committee

- Please write clearly in black ink.
- Please tick boxes where appropriate and cross out text marked with an asterisk (*) that does not apply.
- This form should be used by a landlord or a tenant who has been served with a notice under section 6(2) of the Housing Act 1988, varying the terms of a statutory periodic tenancy which arises when a fixed term of an assured tenancy, an assured shorthold tenancy or an assured agricultural occupancy ends.
- When you have completed the form, please send it to your local rent assessment panel with a copy of the notice served on you proposing the new terms of the statutory periodic tenancy.

---

1. Name(s) of tenant(s):

....

....

....

---

2. Address of premises to which the tenancy relates:

....

....

....

---

3. Name(s) of landlord(s)/agent*:

....

....

....

Address of landlord(s)/agent*:

....

....

....

---

4. Details of premises.

(a) What type of accommodation is rented?

Rooms [ ] Flat [ ] Terraced House [ ]

Semi-Detached House [ ] Fully Detached House [ ] Other [ ]

(*Please specify*) ....

(b) If it is a flat or room(s), what floor(s) is it on?

Ground [ ] First [ ] Second [ ]

Other [ ] (*Please specify*) ....

(c) Give the number and type of rooms, eg living room, bathroom etc

....

....

(d) Does the tenancy include any other facilities, eg garden, garage or other separate building or land?

Yes [ ] No [ ]

(e) If yes, please give details:

....

....

....

(f) Is any of the accommodation shared with:

(i) the landlord? Yes [ ] No [ ]

(ii) any other tenant or tenants? Yes [ ] No [ ]

(g) If yes, please give details:

....

....

....

---

5. When did the statutory periodic tenancy begin?

....

---

6. Services

(a) Are any services provided under the tenancy (eg cleaning, lighting, heating, hot water or gardening etc.)?

Yes [ ] No [ ]

(b) If yes, please give details:

....

....

....

(c) Is a separate charge made for services, maintenance, repairs, landlords' costs of management or any other item?

Yes [ ] No [ ]

(d) If yes, what charge is payable? £........ per ....

(eg week, month year)

(e) Does the charge vary according to the relevant costs?

Yes [ ] No [ ]

(f) If yes, please give details:

....

....

....

---

7. (a) Is any furniture provided under the tenancy?

Yes [ ] No [ ]

(b) If yes, please give details. Continue on a separate sheet if necessary or provide a copy of the inventory.

....

....

....

---

8. What repairs are the responsibility of:

(a) the landlord? Continue on a separate sheet if necessary.

....

....

....

(b) the tenant? Continue on a separate sheet if necessary.

....

....

....

---

9. Give details (if known) of the other terms of the tenancy, e g can you assign the tenancy (pass it on to someone else) and if so is a premium (a payment which is in addition to rent and equivalent to more than two months' rent) payable on an assignment? Continue on a separate sheet if necessary.

....

....

....

---

10. (a) Is there a written tenancy agreement? Yes [ ] No [ ]

(b) If yes, please attach the tenancy agreement (with a note of any variations). It will be returned to you as soon as possible.

---

11. (a) I/We* attach a copy of the notice proposing changes to the statutory periodic tenancy and, if applicable, an adjustment of the amount of rent and apply for it to be considered by the rent assessment committee.

Signed .... Date ....

....

....

To be signed and dated by the landlord or his agent (someone acting for him) or the tenant or his agent. If there are joint landlords or joint tenants, each landlord/tenant or the agent must sign unless one signs on behalf of the rest with their agreement.

Please specify whether:	landlord	landlord's agent	agent	tenant's agent
	[ ]	[ ]	[ ]	[ ]

(b) Name and address of landlord or tenant referring to the rent assessment committee.

Name(s) (*Block Capitals*): ....

....

....

Address:

....

....

....

Telephone – Daytime .... Evening ....

PRECEDENT 7 NOTICE SEEKING POSSESSION OF PROPERTY LET ON ASSURED TENANCY/ASSURED AGRICULTURAL OCCUPANCY

FORM No 3

Housing Act 1988 section 8 as amended by section 151 of the Housing Act 1996

Notice seeking possession of a property let on an Assured Tenancy or an Assured Agricultural Occupancy

- Please write clearly in black ink.
- Please tick boxes where appropriate and cross out text marked with an asterisk (*) that does not apply.
- This form should be used where possession of accommodation let under an assured tenancy, an assured agricultural occupancy or an assured shorthold tenancy is sought on one of the grounds in Schedule 2 to the Housing Act 1988.
- Do not use this form if possession is sought on the 'shorthold' ground under section 21 of the Housing Act 1988 from an assured shorthold tenant where the fixed term has come to an end or, for assured shorthold tenancies with no fixed term which started on or after 28 February 1997, after six months has elapsed. There is no prescribed form for these cases, but you must give notice in writing.

1. To:

*Names of tenant(s)/licensees**

2. Your landlord/licensor* intends to apply to the court for an order requiring you to give up possession of

....

....

Address of premises

3. Your landlord/licensor* intends to seek possession on ground(s) . . . . . . in Schedule 2 to the Housing Act 1988, as amended by the Housing Act 1996, which read(s):

....

....

....

Give the full text (as set out in the Housing Act 1988 as amended by the Housing Act 1996) of each ground which is being relied on. Continue on a separate sheet if necessary.

4. Give a full explanation of why each ground is being relied on:

....

....

....

Continue on a separate sheet if necessary.

Notes on the grounds for possession

- If the court is satisfied that any of grounds 1 to 8 is established, it must make an order (but see below in respect of fixed term tenancies).
- Before the court will grant an order on any grounds of 9 to 17, it must be satisfied that it is reasonable to require you to leave. This means that, if one of these grounds is set out in section 3, you will be able to suggest to the court that it is not reasonable that you should have to leave, even if you accept that the ground applies.
- The court will not make an order under grounds 1, 3 to 7, 9 or 16, to take effect during the fixed term of the tenancy (if there is one) and it will only make an order during the fixed term on grounds 2, 8, 10 to 15 or 17 if the terms of the tenancy make provision for it to be brought to an end on any of these grounds.
- Where the court makes an order for possession solely on ground 6 or 9, the landlord must pay your reasonable removal expenses.

5. The court proceedings will not begin until after:

....

Give the earliest date on which court proceedings can be brought.

- Where the landlord is seeking possession on grounds 1, 2, 5 to 7, 9 or 16, court proceedings cannot begin earlier than 2 months from the date this notice is served on you (even where one of grounds 3, 4, 8, 10 to 13, 14A, 15 or 17 is specified) and not before the date on which the tenancy (had it not been assured) could have been brought to an end by a notice to quit served at the same time as this notice.
- Where the landlord is seeking possession on grounds 3, 4, 8, 10 to 13, 14A, 15 or 17, court proceedings cannot begin earlier than 2 weeks from the date this notice is served (unless one of 1, 2, 5 to 7, 9 or 16 grounds is also specified in which case they cannot begin earlier than two months from the date this notice is served).
- Where the landlord is seeking possession on ground 14 (with or without other grounds), court proceedings cannot begin before the date this notice is served.
- Where the landlord is seeking possession on ground 14A, court proceedings cannot begin unless the landlord has served, or has taken all reasonable steps to serve, a copy of this notice on the partner who has left the property.
- After the date shown in section 5, court proceedings may be begun at once but not later than 12 months from the date on which this notice is served. After this time the notice will lapse and a new notice must be served before possession can be sought.

---

6. Name and address of landlord/licensor*.

To be signed and dated by the landlord or licensor or his agent (someone acting for him). If there are joint landlords, each landlord or the agent must sign unless one signs on behalf of the rest with their agreement.

Signed .... Date ....

....

Please specify whether:	landlord	licensor	joint landlords	landlord's agent
	[ ]	[ ]	[ ]	[ ]

Name(s) (*Block Capitals*): ....

....

....

Address:

....

....

....

Telephone – Daytime .... Evening ....

What to do if this notice is served on you

- This notice is the first step requiring you to give up possession of your home. You should read it very carefully.
- Your landlord cannot make you leave your home without an order for possession issued by a court. By issuing this notice your landlord is informing you that he intends to seek such an order. If you are willing to give up possession without a court order, you should tell the person who signed this notice as soon as possible and say when you are prepared to leave.
- Whichever grounds are set out in section 3 of this form, the court may allow any of the other grounds to be added at a later date. If this is done, you will be told about it so you can discuss the additional grounds at the court hearing as well as the grounds set out in section 3.
- If you need advice about this notice, and what you should do about it, take it immediately to a citizens' advice bureau, a housing advice centre, a law centre or a solicitor.

PRECEDENT 8 LANDLORD'S NOTICE PROPOSING NEW RENT UNDER ASSURED PERIODIC TENANCY

FORM PRESCRIBED FOR THE PURPOSES OF SECTION 13(2) OF THE HOUSING ACT 1988

FORM No 4B

Housing Act 1988 section 13(2), as amended by the Regulatory Reform (Assured Periodic Tenancies) (Rent Increases) Order 2003

Landlord's Notice proposing a new rent under an Assured Periodic Tenancy of premises situated in England

The notes over the page give guidance to both landlords and tenants about this notice.

To:	....	[*Tenant(s)*]
of	....	[*Address of the premises subject to the tenancy*]
	....	
	....	
	....	
From:	....	[*Landlord(s)*][*Landlord's Agent*]* * *delete as appropriate*
	....	[*Address for correspondence*]
	....	
	....	
	....	[*Contact telephone number*]

1. **This notice affects the amount of rent you pay.** Please read it carefully.

2. The landlord is proposing a new rent of **£......per[week] [month] [year]***, in place of the existing one of £......per [week][month][year]*

* *delete as appropriate*

3. The first rent increase date after 11th February 2003 is ....

(see note 10 over the page)

4. The starting date for the new rent will be ....

(see notes 13 to 17 over the page)

5. Certain charges may be included and separately identified in your rent. (See note 11 over the page.) The amounts of the charges (if any) are:-

Charges	Amount included and separately identified *(enter 'nil' if appropriate)*	
	In the existing rent	In the proposed new rent
Council tax	£	£
Water charges	£	£
Fixed service charges	£	£

6. If you accept the proposed new rent, you should make arrangements to pay it. If you do not accept it, there are steps you should take before the starting date in paragraph 4 above. **Please see the notes over the page for what to do next.**

Signed: .... [Landlord(s)][Landlord's Agent]* (see note 12 over the page)

* *delete as appropriate*

Date: ....

Please read these notes carefully.

Guidance notes for tenants

What you must do now

1. This notice proposes that you should pay a new rent from the date in paragraph 4 of the notice. **If you are in any doubt or need advice about any aspect of this notice, you should immediately either discuss it with your landlord or take it to a citizens' advice bureau, a housing advice centre, a law centre or a solicitor.**

2. If you accept the proposed new rent, please make arrangements to pay it. If you pay by standing order through your bank, you should inform them that the amount has changed. You should also notify your Housing Benefit office if you are claiming benefit. If you are worried that you might not be able to pay your rent, you should seek advice from a citizens' advice bureau or housing advice centre.

3. If you do not accept the proposed new rent, and do not wish to discuss it with your landlord, you can refer this notice to your local rent assessment committee. **You must do this before the starting date of the proposed new rent in paragraph 4 of the notice.** You should notify your landlord that you are doing so, otherwise he or she may assume that you have agreed to pay the proposed new rent.

4. To refer the notice to the local rent assessment committee, you must use the form *Application referring a notice proposing a new rent under an Assured Periodic Tenancy or Agricultural Occupancy to a Rent Assessment Committee.* You can obtain this from a rent assessment panel, housing advice centre or legal stationer (details can be found in the telephone directory).

5. The rent assessment committee will consider your application and decide what the maximum rent for your home should be. In setting a rent, the committee must decide what rent the landlord could reasonably expect for the property if it were let on the open market under a new tenancy on the same terms. The committee may therefore set a rent that is higher, lower or the same as the proposed new rent.

Guidance notes for landlords on how to complete the notice

6. You can complete this notice in ink or arrange for it to be printed.

7. This notice should be used when proposing a new rent under an **assured periodic tenancy (including an assured shorthold periodic tenancy) of premises situated in England**. There is a different notice (Form No 4C—*Landlord's or Licensor's Notice proposing a new rent or licence fee under an Assured Agricultural Occupancy of premises situated in England*) for proposing a new rent or licence fee for an assured agricultural occupancy of premises situated in England.

8. Do not use this notice if the tenancy agreement contains a term allowing rent increases, or there is some other basis such as a separate agreement with the tenant for raising the rent. Any provision you rely on needs to be binding on the tenant. Legal advice should be sought if there is any doubt on this score.

9. You need to use a different form to propose a rent increase for a statutory periodic tenancy (the first exception mentioned in note 16) if you are seeking to adjust rent solely because of a proposed change of terms under section 6(2) of the Housing Act 1988. Seek legal advice if you think this may apply to you. You can obtain the form headed *Notice proposing different terms for a Statutory Periodic Tenancy* from a rent assessment panel or a legal stationer.

10. Unless the tenancy is a new one, or one of the exceptions mentioned in note 16 applies, you must insert in paragraph 3 of the notice the first date after 11th

February 2003, on which rent is proposed to be, or was, increased under this statutory notice procedure. That date determines the date that you can specify in paragraph 4 of the notice. See also note 15.

11. You should enter in each of the boxes in the second and third columns of the table in paragraph 5 either 'nil' or the amount of the existing or proposed charge. You should only enter amounts for council tax and water charges where the tenant does not pay these charges directly. You should only enter fixed service charges which are payable by the tenant in accordance with a term or condition which specifies that these charges will be included in the rent for the tenancy. Only enter an amount for service charges where the tenant has agreed to pay a **fixed** sum. Do **not** include in the table any **variable** service charge, ie a service charge within the meaning of section 18 of the Landlord and Tenant Act 1985, where the whole or part of the sum payable by the tenant varies or may vary according to costs.

12. You or your agent (someone acting on your behalf) must sign and date this notice. If there are joint landlords, each landlord must sign unless one signs on behalf of the rest with their agreement. The signature does not have to be hand-written if, for instance, the form is being printed or if you wish to use a laser or autosignature.

When the proposed new rent can start

13. The date in paragraph 4 of the notice must comply with the three requirements of section 13(2) of the Housing Act 1988, as amended by the Regulatory Reform (Assured Periodic Tenancies) (Rent Increases) Order 2003.

14. The **first requirement**, which applies in **all** cases, is that a minimum period of notice must be given before the proposed new rent can take effect. That period is:-

- one month for a tenancy which is monthly or for a lesser period, for instance weekly or fortnightly
- six months for a yearly tenancy
- in all other cases, a period equal to the length of the period of the tenancy – for example, three months in the case of a quarterly tenancy.

15. The **second requirement applies** in **most** cases (but see note 16 for two exceptions):-

(a) the starting date for the proposed new rent must not be earlier than 52 weeks after the date on which the rent was last increased using this statutory notice procedure or, if the tenancy is new, the date on which it started, **unless**

(b) that would result in an increase date falling one week or more before the anniversary of the date in paragraph 3 of the notice, in which case the starting date must not be earlier than 53 weeks from the date on which the rent was last increased.

This allows rent increases to take effect on a fixed day each year where the period of a tenancy is less than one month. For example, the rent for a weekly tenancy could be increased on, say, the first Monday in April. Where the period of a tenancy is monthly, quarterly, six monthly or yearly, rent increases can take effect on a fixed date, for example, 1st April.

16. The two exceptions to the second requirement, which apply where a statutory tenancy has followed on from an earlier tenancy, are:-

- where the tenancy was originally for a fixed term (for instance, 6 months), but continues on a periodic basis (for instance, monthly) after the term ends and
- where the tenancy came into existence on the death of the previous tenant who had a regulated tenancy under the Rent Act 1977.

In these cases the landlord may propose a new rent at once. However, the first and third requirements referred to in notes 14 and 17 must still be observed.

17. The **third requirement**, which applies in **all** cases, is that the proposed new rent must start at the beginning of a period of the tenancy. For instance, if the tenancy is monthly, and started on the 20th of the month, rent will be payable on that day of the month, and a new rent must begin then, not on any other day of the month. If the tenancy is weekly, and started, for instance, on a Monday, the new rent must begin on a Monday.

PRECEDENT 9 LANDLORD'S OR LICENSOR'S NOTICE PROPOSING NEW RENT OR LICENCE FEES UNDER AN ASSURED AGRICULTURAL OCCUPANCY

FORM PRESCRIBED FOR THE PURPOSES OF SECTION 13(2) OF THE HOUSING ACT 1988

FORM No 4C

Housing Act 1988 section 13(2), as amended by the Regulatory Reform (Assured Periodic Tenancies) (Rent Increases) Order 2003

Landlord's or Licensor's Notice proposing a new rent or licence fee under an Assured Agricultural Occupancy of premises situated in England

The notes over the page give guidance to both landlords and tenants about this notice.

This notice may also be used by licensors to give notice proposing a new licence fee. In that case, references in this form and in the notes over the page to landlords, tenants, tenancy and rent should be read as references to licensors, licensees, licence and licence fees, respectively.

To:	....	[*Tenant(s)*]
of	....	[*Address of the premises subject to the tenancy*]
	....	

....

....

From: .... [*Landlord(s)*][*Landlord's Agent*]*

* *delete as appropriate*

.... [*Address for correspondence*]

....

....

.... [*Contact telephone number*]

1. **This notice affects the amount of rent you pay.** Please read it carefully.

2. The landlord is proposing a new rent of **£......per[week] [month] [year]***, in place of the existing one of £......per [week][month][year]*

* *delete as appropriate*

3. The starting date for the new rent will be ....

(see notes 12 to 16 over the page)

4. Certain charges may be included and separately identified in your rent. (See note 10 over the page.) The amounts of the charges (if any) are:-

Charges	Amount included and separately identified *(enter 'nil' if appropriate)*	
	In the existing rent	In the proposed new rent
Council tax	£	£
Water charges	£	£
Fixed service charges	£	£

5. If you accept the proposed new rent, you should make arrangements to pay it. If you do not accept it, there are steps you should take before the starting date in paragraph 3 above. **Please see the notes over the page for what to do next.**

Signed: .... [Landlord(s)][Landlord's Agent]* (see note 11 over the page)

* *delete as appropriate*

Date: ....

Please read these notes carefully.

Guidance notes for agricultural occupants

What you must do now

1. This notice proposes that you should pay a new rent from the date in paragraph 3 of the notice. If you are in any doubt or need advice about any aspect of this notice, you should immediately either discuss it with your landlord or take it to a citizens' advice bureau, a housing advice centre, a law centre or a solicitor.

2. If you accept the proposed new rent, please make arrangements to pay it. If you pay by standing order through your bank, you should inform them that the amount has changed. You should also notify your Housing Benefit office if you are claiming benefit. If you are worried that you might not be able to pay your rent, you should seek advice from a citizens' advice bureau or housing advice centre.

3. If you do **not** accept the proposed new rent, and do not wish to discuss it with your landlord, you can refer this notice to your local rent assessment committee. **You must do this before the starting date of the proposed new rent in paragraph 3 of the notice.** You should notify your landlord that you are doing so, otherwise he or she may assume that you have agreed to pay the proposed new rent.

4. To refer the notice to the local rent assessment committee, you must use the form *Application referring a notice proposing a new rent under an Assured Periodic Tenancy or Agricultural Occupancy to a Rent Assessment Committee.* You can obtain this from a rent assessment panel, housing advice centre or legal stationer (details can be found in the telephone directory).

5. The rent assessment committee will consider your application and decide what the maximum rent for your home should be. In setting a rent, the committee must decide what rent the landlord could reasonably expect for the property if it were let on the open market under a new tenancy on the same terms. The committee may therefore set a rent that is higher, lower or the same as the proposed new rent.

Guidance notes for landlords on how to complete the notice

6. You can complete this notice in ink or arrange for it to be printed.

7. This notice should be used when proposing a new rent or licence fee for an **assured agricultural occupancy of premises situated in England**. There is a different notice (Form No 4B—*Landlord's Notice proposing a new rent under an Assured Periodic Tenancy of premises situated in England*) for proposing a new rent for an assured periodic tenancy (including an assured shorthold periodic tenancy) of premises situated in England.

8. Do not use this notice if the tenancy agreement contains a term allowing rent increases, or there is some other basis such as a separate agreement with the tenant for raising the rent. Any provision you rely on needs to be binding on the tenant. Legal advice should be sought if there is any doubt on this score.

9. You need to use a different notice to propose a rent increase for a statutory periodic tenancy (the first exception mentioned in note 15) if you are seeking to adjust rent solely because of a proposed change of terms under section 6(2) of the Housing Act 1988. Seek legal advice if you think this may apply to you. You can obtain the form headed *Notice proposing different terms for a Statutory Periodic Tenancy* from a rent assessment panel or a legal stationer.

10. You should enter in each of the boxes in the second and third columns of the table in paragraph 4 either 'nil' or the amount of the existing or proposed charge. You should only enter amounts for council tax and water charges where the tenant does not pay these charges directly. You should only enter fixed service charges which are payable by the tenant in accordance with a term or condition which specifies that these charges will be included in the rent for the tenancy. Only enter an amount for service charges where the tenant has agreed to pay a fixed sum. Do not include in the table any variable service charge, ie a service charge within the meaning of section 18 of the Landlord and Tenant Act 1985, where the whole or part of the sum payable by the tenant varies or may vary according to costs.

11. You or your agent (someone acting on your behalf) must sign and date this notice. If there are joint landlords, each landlord must sign unless one signs on behalf of the rest with their agreement. The signature does not have to be hand-written if, for instance, the form is being printed or if you wish to use a laser or autosignature.

When the proposed new rent can start

12. The date in paragraph 3 of the notice must comply with the three requirements of section 13(2) of the Housing Act 1988, ignoring the amendments made by the Regulatory Reform (Assured Periodic Tenancies) (Rent Increases) Order 2003.

13. The **first requirement**, which applies in **all** cases, is that a minimum period of notice must be given before the proposed new rent can take effect. That period is:-

- one month for a tenancy which is monthly or for a lesser period, for instance weekly or fortnightly
- six months for a yearly tenancy
- in all other cases, a period equal to the length of the period of the tenancy – for example, three months in the case of a quarterly tenancy.

14. The **second requirement** applies in **most** cases (but see note 15 for two exceptions). The starting date for the proposed new rent must not be earlier than the first anniversary of the date on which the rent was last increased using this statutory notice procedure or, if the tenancy is new, the date on which it started.

15. The two exceptions, which apply where a statutory tenancy has followed on from an earlier tenancy, are:-

- where the tenancy was originally for a fixed term (for instance, 6 months), but continues on a periodic (for instance, monthly) basis after the term ends and
- where the tenancy came into existence on the death of the previous tenant who had a regulated tenancy under the Rent Act 1977.

In these cases the landlord may propose a new rent at once. However, the first and third requirements referred to in notes 13 and 16 must still be observed.

16. The **third requirement**, which applies in **all** cases, is that the proposed new rent must start at the beginning of a period of the tenancy. For instance, if the tenancy is monthly, and started on the 20th of the month, rent will be payable on that day of the month, and a new rent must begin then, not on any other day of the month. If the tenancy is weekly, and started, for instance, on a Monday, the new rent must begin on a Monday.

Chapter 7

GRAZING AGREEMENTS

INTRODUCTION

PRECEDENTS
1 Deed of profit à prendre
2 Grazing agreement (licence)
3 Tack/agistment agreement for livestock to be taken in for feeding
4 Grazing agreement (tenancy) – non-business

INTRODUCTION

There are four precedents in this chapter:

- a profit à prendre agreement;
- a grazing agreement (licence);
- a tack/agistment agreement for livestock to be taken in for feeding; and
- a grazing agreement (tenancy) – non-business.

The detailed terms of each agreement will depend on the outcome of negotiations between the parties. However, the type of agreement to be used also depends to a great extent on tax considerations, particularly if there is to be only one occupier for grazing purposes of the grassland.

The issue is likely to be important principally for capital taxation reasons. Entrepreneur's relief and roll-over relief for capital gains tax will be available only if the landowner is occupying the property. Similarly, agricultural relief for inheritance tax purposes (on the basis of occupation) will be available if the agreement is appropriately structured and the landowner is prepared to do sufficient work to prove that he is indeed in occupation. This may be particularly important if the grassland is ancillary to a farmhouse and it is hoped to obtain inheritance tax relief on the house itself. A tenancy will not give any of these tax advantages.

There are various cases establishing that a licence will suffice to give the landowner occupation provided that he is effectively managing the property. See, for example, *CIR v Forsyth-Grant*.[1] The licence must not, however, give the licensee exclusive possession for if it does so, whether by the document itself or in practice, it will as a matter of law be construed as a tenancy.[2] The Irish case of *McCall*[3] is a recent example where business property relief for inheritance tax purposes was refused in this context.

Do not forget that grazing with most horses is not regarded as agricultural use for inheritance tax purposes.

A profit à prendre is the right to take something off another's land which is either part of the land or the wild animals existing on it provided that the thing taken is susceptible of ownership. Examples of profits à prendre are the profit of pasturage (the right to take grass through the mouths of cattle), herbage (a right to take grass by cutting or grazing) and vesture (a right to take all produce of land apart from mines minerals and timber). Unlike a lease a profit gives the grantee no estate in the land. There is therefore no question of exclusive possession if the grazier complies with the agreement.

Although grazing is within the definition of agriculture this tenancy will not be a farm business tenancy as the business condition is not fulfilled. There is a prohibition in the tenancy agreement against using the land for the purpose of a trade or business. The tenancy is therefore a common law tenancy which will terminate in accordance with agreed terms.

Agistment is where a farmer takes in and feeds livestock belonging to someone else, payment being made on a headage basis rather than by means of a payment per acre in respect of the area of land occupied. The same comments apply in this case, as with the Grazing Agreement, with regard to the occupation of the land by the livestock farmer and the requirements of the single payment scheme.

---

1 1943 SC 528, 25 TC 369.

2 *Street v Mountford* [1985] AC 809.

3 *McCall and Keenan v HMRC* [2009] NICA 12.

PRECEDENTS

PRECEDENT 1 DEED OF PROFIT À PRENDRE

[*In this precedent the Owner is claiming the single payment.*]

THIS DEED is made the [ ] day of [ ] 20[ ]

BETWEEN:-

(1) [*name*] of [*address*] ('the Owner')

and

(2) [*name*] of [*address*] ('the Grazier')

WHEREAS:-

1. Definitions and interpretation

In this Deed:-

1.1	**Grazing Fee**	:	means the sum of [ ] pounds (£[ ]) [plus VAT at 15% if such is applicable]
1.2	**Land**	:	means ALL THOSE pieces and parcels of land situate at [ ] full particulars of which are contained in the Schedule and are delineated and edged red on the Plan
1.3	**Payment Date**	:	means the date of this Deed
1.4	**Period of Occupation**	:	means the period beginning on [ ] and ending on [ ]
1.5	**Plan**	:	means the plan attached to this Deed
1.6	**Single Payment Scheme**	:	the single payment scheme governed by Council Regulation (EC) No 73/2009 and all the relevant subsequent Commission and other regulations administrative provisions statutory guidance or codes of practice implemented by the EC and/or the UK Government ('the Regulations')
1.7	**GAEC**	:	means good agricultural and environmental condition as defined in the Regulations
1.8	**SMR**	:	means the Statutory Management Requirements as defined in the Regulations

1.9 where any party to this Deed comprises two or more persons obligations expressed or implied to be made by or with that party are deemed to be made by or with the persons comprising that party jointly and severally

1.10 words importing one gender include all other genders; words importing the singular include the plural and vice versa and words importing persons shall be construed as importing a corporate body or a partnership and vice versa

1.11 reference to any clause, sub-clause or schedule without further designation shall be construed as a reference to the clause sub-clause or schedule in this Deed so numbered

1.12 the clause paragraph and schedule headings do not form part of this Deed and shall not be taken into account in its construction or interpretation

2. Agreement

The Owner and the Grazier have agreed as hereinafter appearing

NOW THIS DEED WITNESSES as follows:-

1. The Grazing Right

In consideration of the Grazing Fee paid by the Grazier to the Owner on the Payment Date (receipt of which the Owner hereby acknowledges) the Owner grants to the Grazier the [sole] right of herbage [together with others entitled to a like right] on the Land during the Period of Occupation

2. The Grazier's Covenants

The Grazier hereby covenants with the Owner:-

2.1 to pay the Owner the Grazing Fee on or before the Payment Date

2.2 to use the Land for the purpose of grazing [*type and number of animals – sheep, cattle or horses*] [only] [and] [not] [to cut or mow the grass]

2.3 to graze and use the Land in a good and husband like manner

2.4 not to permit any trespass on the Land

2.5 not to allow any animals other than his own to graze the Land

2.6 not to allow diseased or quarantined animals [nor confirmed fence breakers] on to the Land

2.7 to ensure that his servants or agents and any person attending or for the time being in charge of the animals present on the Land comply with The Welfare of Farmed Animals (England) Regulations 2007 or any statutory modification or re-enactment thereof for the time being in force

2.8 to indemnify the Owner in respect of all liabilities and payments including legal costs and expenses incurred by the Owner as a result of proceedings brought in respect of the welfare of animals under the Animal Welfare Act 2006 and the Welfare of Farmed Animals (England) Regulations 2007 or any statutory modification or re-enactment thereof for the time being in force

2.9 to indemnify the Owner in respect of any damage and all liabilities and costs, including legal costs and expenses incurred by or as a result of the actions or negligence of the Grazier, including any claims made by third parties resulting from any activities or negligence by the Grazier on the Land

2.10 not to do anything to cause the Land to cease to be in GAEC, and to comply with any reasonable request of the Owner to take such steps as are necessary to comply with the SMR

2.11 not to obstruct the Owner's full access to the Land at all reasonable times to carry out the Owner's obligations set out in Clause 3 of this Deed

2.12 not to assign underlet or part with possession of the whole or any part of the Land nor permit anyone else to occupy or share the occupation of the whole or any part of the Land nor share the rights granted by this Deed

3. The Owner's Covenants

The Owner hereby covenants with the Grazier:-

3.1 to mow or spray spear thistle, creeping or field thistle, curled dock, broad leaved dock and ragwort

3.2 within the provisions of the SMR to keep fertilised the Land and when and where necessary reseed and crop the grass and generally maintain the Land in a good husband like manner

3.3 to keep gates fences and ditches in good order other than damage caused by the Grazier or his servants, agents or any person attending or for the time being in charge of the livestock or livestock itself

3.4 to carry out such agricultural activity as is required to keep the Land in GAEC and to ensure compliance with the SMR

3.5 to be responsible for and control the Land in terms of management, benefit and financial risk

4. Agreements

It is hereby agreed that:-

4.1 the Owner shall have a lien upon all the Grazier's animals for the time being pastured on the Land for any sum owing or expense incurred for which under this Deed the Grazier is liable and this lien may be enforced by the sale of any animal or animals belonging to the Grazier for the time being pastured on the Land

[*Note: this may not be of great benefit in view of the need for cattle and horse passports and the tagging requirements associated with livestock*]

4.2 the Owner shall be the farmer and so the claimant and owner of entitlement to payments under the Single Payment Scheme, and the Grazier shall make no claim nor procure any third party to make such claim under the Single Payment Scheme in respect of the Land

4.3 the right of herbage hereby granted is personal to the Grazier

4.4 that if after the date of this Deed there are changes in the Regulations, or in any way the intention of the parties as far as the Single Payment Scheme is concerned as set out in this Deed would be frustrated as a result of such changes, either party may give notice to the other, in the absence of agreement between them on any appropriate changes to this Deed to achieve their objectives as set out in this Deed, to appoint a single arbitrator pursuant to the provisions of the Arbitration Act 1996 to re-draft the relevant part(s) of this Deed in such a way as may fairly achieve their objectives and the costs of such appointment shall be borne equally between them

4.5 the Owner and the Grazier shall each be entitled to terminate this Deed by giving to the other no less than 28 days' written notice to expire at any time during the Period of Occupation

5. Arbitration

If any dispute arises under this Deed either party may give notice to the other in writing specifying the dispute and requesting that agreement be reached on the identity of a single arbitrator to be appointed to determine the dispute in accordance with the provisions of the Arbitration Act 1996 and if no arbitrator is appointed by agreement within one month of the notice either party may make a written request to the President for the time being of the Royal Institution of Chartered Surveyors for him to appoint an arbitrator to determine the matter in dispute

IN WITNESS whereof the parties have set their hands to this Deed the day and year first before written

THE SCHEDULE

SIGNED as a Deed by
in the presence of:-

SIGNED as a Deed by
in the presence of:-

PRECEDENT 2 GRAZING AGREEMENT (LICENCE)

[*What follows is a short form of grazing agreement in the form of a licence whereby the licensee occupies the land for an appropriate period of the year. In other words, the owner will have the land at his disposal in accordance with the terms of the single payment scheme but the land will be grazed and in general terms kept 'clean and tidy' by the licensee who agrees with the owner not to make any claim in respect of single payment, nor to do anything which would affect the owner's right to do so and, indeed, to indemnify the owner against any deductions from his single payment arising from the licensee's activities on the land.*]

THIS AGREEMENT is made the [ ] day of [ ] 20[ ]

BETWEEN:-

(1) [*name*] of [*address*] ('the Owner')

and

(2) [*name*] of [*address*] ('the Licensee')

WITNESSES as follows:-

1. Definitions and interpretation

In this Agreement:-

1.1	**The Land**	:	means the land known as [ ] at [ ] extending to [ ] acres/hectares or thereabouts shown edged in red on the Plan and as described in Schedule 1
1.2	**The Plan**	:	means the plan annexed hereto
1.3	**The Licence Fee**	:	means the sum of [ ] pounds (£[ ])
1.4	**The Period of Occupation**	:	means the period starting on the [*date*] and ending on the [*date*]
1.5	**The Payment Date**	:	means the [*date*]
1.6	**Single Payment Scheme**	:	the single payment scheme governed by Council Regulation (EC) No 73/2009 and all the relevant subsequent Commission and other regulations administrative provisions statutory guidance or codes of practice implemented by the EC and/or the UK Government ('the Regulations')

2. Interpretation

2.1 In this Agreement the Particulars are incorporated and unless the context otherwise requires expressions in the Particulars have the respective meanings set out opposite them in the Particulars

2.2 Words importing one gender shall be construed as importing any other gender

2.3 Words importing the singular shall include the plural and vice versa

2.4 Words importing persons shall include corporations

2.5 Where there are two or more persons included in the expressions the Owner and/or the Licensee the obligations of such persons under this Agreement shall be deemed to be joint and several

2.6 Any reference in this Agreement to a statutory provision shall be deemed to include any later provision substituted for or amending the same

3. The right to graze

The Owner grants the Licensee the right to occupy the Land for grazing or mowing purposes only during the Period of Occupation

4. Licensee's Agreements

The Licensee agrees with the Owner:-

4.1 to pay to the Owner the Licence Fee in advance on the Payment Date

4.2 to pay and indemnify the Owner in respect of all water rates water charges drainage rates and any other rates taxes and outgoings in respect of the Land during the Period of Occupation

4.3 not to do or cause or permit to be done or to continue anything on or near the Land which may be or become a nuisance or inconvenience or cause injury loss damage or annoyance to the Owner or any other persons

4.4 to use the Land for the exercise of the rights granted and for no other purpose and in exercising those rights to do so in a good and husband like manner AND not allow the Land to become poached by treading during wet weather conditions and if the Owner or his agent certify that any damage is being caused the Licensee shall upon demand immediately remove his animals from the Land

4.5 to observe and comply with any and every legal requirement including any and every statute regulation by-law order direction and official code of practice governing or affecting the exercise of the rights granted

4.6 on termination of this Agreement immediately to remove any stock or things belonging to the Licensee from the Land

4.7 to indemnify the Owner against all liabilities claims damages losses costs and expenses including legal costs and expenses incurred by the Owner arising either directly or indirectly out of or otherwise related to the exercise of the rights granted or arising from any breach of the Licensee's obligations under this Agreement

4.8 not to assign underlet or part with or share possession of the Land nor to allow any other person to enter use or occupy the Land or any part of it for any purpose whatsoever which is hereby prohibited

4.9 not to permit any trespass on the Land but also not to obstruct any rights of way used or enjoyed over the Land whether granted prior to the date of this Agreement or during the Period of Occupation

4.10 not to erect any building or structure nor place any advertisement on the Land

4.11 not to claim or attempt to claim either personally or through any third party all or any part of the single payment under the Single Payment Scheme

4.12 to indemnify the Owner against any deductions from his single payment under the Single Payment Scheme arising from any action or omission on his part including but not limited to any breach of the cross compliance rules of the Single Payment Scheme and any breach of the statutory management requirements

[5. Owner's lien

The Owner shall have a lien upon all the Licensee's animals for the time being depastured on the Land for any sum owing or expense incurred for which under this Agreement the Licensee is liable and this lien may be enforced by the sale of any such animal or animals]

[*Is this really much use in view of the law relating to tagging of livestock, and passports for cattle and horses?*]

6. Owner's right of entry

The Owner and any person he authorises shall have the right to enter the Land at any time for any reasonable purposes other than for grazing animals on the Land

7. Single Payment Scheme

The provisions of Schedule 2 relating to the Single Payment Scheme apply to this Agreement

8. Termination of Agreement

If the Owner requires the whole or any part of the Land for any purpose whatsoever he shall have the right to determine this Agreement in relation to either the whole or any part of the Land on giving one calendar month's notice in writing to the Licensee whereupon the Owner shall refund to the Grazier a proportionate part of the Licence Fee

9. Owner's right to terminate on death of Grazier

The Owner shall be entitled by giving one calendar month's written notice to terminate this Agreement in the event of the death of the Licensee or on any breach by him of any of his obligations under this Agreement

10. No undertaking for new agreement

It is expressly agreed between the parties and hereby acknowledged by the Licensee that the Owner does not undertake to repeat this Agreement for another period

11. Arbitration

In the event of any dispute arising out of or in connection with this Agreement the same shall be determined in the absence of agreement between the parties by a single arbitrator to be appointed by agreement between them or in the absence of such agreement by the President for the time being of the Royal Institution of Chartered Surveyors on the application of either the Owner or the Grazier such arbitration to be conducted in accordance with the provisions of the Arbitration Act 1996

AS WITNESS the hands of the parties hereto the day and year first before written

SCHEDULE 1
THE LAND

[OS] [NG] Number	**Description**	**Acres**	**Hectares**
	Total		

SCHEDULE 2
PROVISIONS RELATING TO THE SINGLE PAYMENT SCHEME

The Owner and the Licensee agree as follows:-

1. that the Owner shall be the farmer and hence the claimant and owner of the single payment under the Single Payment Scheme and the Licensee shall make no claim nor procure any third party to make such claim under the Single Payment Scheme in respect of the Land

2. that the Owner shall carry out such (if any) agricultural activity as is required of him to keep the Land in good agricultural and environmental condition and to ensure compliance with the statutory management requirements and the Licensee shall carry out his obligations in this regard as set out in this Agreement

3. that the Owner shall be responsible for and control the Land in terms of management, benefit and financial risk

4. that if after the granting of the rights contained in this Agreement there are changes in the Regulations, or in any way the intention of the parties in connection with the Single Payment Scheme as set out in this Agreement would be frustrated as a result of such changes, either party may give notice to the other, in the absence of agreement between them on any appropriate changes to this Agreement to achieve their objectives as set out in this Agreement, to appoint a single arbitrator to re-draft the relevant parts of this Agreement in such a way as may fairly achieve their objectives and the costs of such appointment shall be borne equally between them

SIGNED by
in the presence of:-

SIGNED by
in the presence of:-

PRECEDENT 3 TACK/AGISTMENT AGREEMENT FOR LIVESTOCK TO BE TAKEN IN FOR FEEDING

[*This particular Precedent provides for the livestock owner to graze livestock on the farmer's land. Most commonly, this occurs during the winter months when dairy farmers take in sheep to graze and hence clean up their pasture land once the dairy cows have been housed for the winter.*

The difference between this and a conventional Grazing Agreement/Licence is that whereas payment under that agreement is on the basis of a certain sum of money per acre the payment under this agreement is on a headage basis with the parties agreeing that a certain sum per head per week/month be paid for each sheep/store lamb/beef animal that the livestock owner keeps on the land.

This particular Precedent has also been drawn to provide for the owner to claim single payment and it has been the introduction of the single payment scheme that

has seen these forms of agreements be committed to writing rather more than happened previously and also assuming a rather greater importance than has often been the case in the past.]

THIS AGREEMENT made the [ ] day of [ ] 20[ ]

BETWEEN:-

(1) [*name*] of [*address*] ('the Farmer')

and

(2) [*name*] of [*address*] ('the Livestock Owner')

WITNESSES as follows:-

1. Definitions and interpretation

In this Agreement:-

1.1	**Agistment Fee**	:	means the sum of [ ] pounds (£[ ]) for each head of cattle per week and the sum of [ ] pounds (£[ ]) for each head of sheep per week [plus VAT at 15% if such is applicable]
1.2	**Land**	:	means ALL THOSE pieces or parcels of land situate at [ ] full particulars of which are contained in the Schedule and for the purposes of identification are delineated and shown edged red on the Plan
1.3	**Plan**	:	means the plan attached to this Agreement
1.4	**Payment Date**	:	means the date of this Agreement and thereafter in advance every four weeks
1.5	**The First Payment**	:	means in respect of each animal when the animal is first placed on the Land by the Livestock Owner and each part of a week shall be paid for as if a complete week
1.6	**Agistment Period**	:	means the period beginning on the [*date*] and ending on the [*date*]
1.7	**Agisted Livestock**	:	means cattle and/or sheep belonging to the Livestock Owner

1.8 **Single Payment Scheme** : the single payment scheme governed by Council Regulation (EC) No 73/2009 and all the relevant subsequent Commission and other regulations administrative provisions statutory guidance or codes of practice implemented by the EC and/or the UK Government ('the Regulations')

2. Payment

The Livestock Owner shall pay the Agistment Fee to the Farmer on the Payment Date

3. Agreement to Graze

The Farmer will take in the Agisted Livestock to be agisted on the Land during the Agistment Period on the terms of this Agreement

4. Changes

The animals comprising the Agisted Livestock may from time to time be changed ('the Replacement Stock') at the option of the Livestock Owner but the Livestock Owner must not at any time exceed the number of animals comprising the Agisted Livestock and must pay for the Replacement Stock at the same rate as the original animals and those of the Agisted Stock then remaining on the Land and the Replacement Stock shall henceforth comprise the Agisted Stock

5. Dangerous and diseased animals

The Livestock Owner shall not turn out onto the Land as the Agisted Livestock or the Replacement Stock any animal he knows or believes to be dangerous or diseased

6. Risk

6.1 During the Agistment Period the Agisted Livestock shall be at the risk of the Livestock Owner unless the risk results from the misconduct of the Farmer or persons in his employment

6.2 The Farmer shall take reasonable precautions for the care and safety of the Agisted Livestock and if at any time the Farmer has knowledge or notice that any animal is suffering or apparently suffering from illness or accident he shall immediately give notice to the Livestock Owner or in the case of emergency call in a veterinary surgeon and the Livestock Owner shall be responsible for the charges of any such veterinary surgeon

7. Straying animals and claims for trespass

The Farmer shall not be responsible to the Livestock Owner for any loss or injury which may occur resulting from the escape of Agisted Livestock from the Land or any part of it and the Livestock Owner shall indemnify the Farmer against any claim which may be made against him in consequence of any escape or trespass by the Agisted Livestock except for any such claims resulting from the Farmer's own misconduct or negligence

8. Animal welfare

The Livestock Owner shall ensure compliance with all enactments concerning the welfare of livestock for the time being in force in respect of the Agisted Livestock

9. Single Payment Scheme

9.1 The parties agree that the Farmer is the person with the Land at his disposal during the Agistment Period and that the Farmer shall be the only party entitled to include the Land within any claim to entitlement to payments under the Single Payment Scheme

9.2 The Farmer agrees:-

9.2.1 to mow or spray thistles, nettles and dock, and to pull ragwort

9.2.2 in accordance with the provisions of the Single Payment Scheme to keep the Land fertilised and when and where necessary reseed and crop the grass and generally maintain the Land in a good and husband like manner

9.2.3 to keep gates fences and ditches in good order other than damage caused by the Livestock Owner or his servants, agents or any person attending or for the time being in charge of the livestock or livestock itself

9.2.4 to carry out such agricultural activity as is required to keep the Land in good agricultural and environmental condition and to ensure compliance with the statutory management requirements of the Single Payment Scheme

9.2.5 to be responsible for and control the Land in terms of management, benefit and financial risk

9.3 The Livestock Owner agrees not to include the Land in any claim under the Single Payment Scheme or other subsidy or grant scheme nor enter any management agreement in relation to the Land

9.4 The Livestock Owner agrees not to do or omit to do anything the doing or omission of which breaches the cross compliance obligations of the Farmer and agrees to indemnify the Farmer for any losses deductions or penalties suffered in respect of the Farmer's claims for single payment or entitlements arising from any act of commission or omission by the Livestock Owner relating to the Agisted Livestock

9.5 The parties further agree that if after the granting of the rights contained in this Agreement there are changes in the Regulations or in any way the intention of the parties as far as the Single Payment Scheme is concerned as set out in this Agreement would be frustrated as a result of such changes either party may give notice to the other, in the absence of agreement between them on any appropriate changes to this Agreement to achieve their objectives as set out in this Agreement to appoint a single arbitrator to re-draft the relevant parts of this Agreement in such a way as may fairly achieve their objectives and the costs of such appointment shall be borne equally between them

10. Personal Agreement and Occupation

10.1 This Agreement is personal to the parties and the benefit and/or burden of it shall not be assigned or otherwise transferred to any third party

10.2 For the avoidance of doubt this Agreement confers no rights of occupation on the Livestock Owner nor any right to come onto the Land to carry out any acts of management or husbandry save for the purposes of bringing the Agisted Livestock onto and removing it from the Land and complying with the provisions of Clause 8 above

11. Termination

This Agreement may be terminated by either party giving not less than [seven/fourteen] day's written notice to the other in the event of a serious or persistent breach by the other party of a term or condition of this Agreement

12. Arbitration Clause

If any dispute arises under this Agreement either party may give notice to the other in writing specifying the dispute and requesting that agreement be reached on the identity of a single arbitrator to be appointed to determine the dispute in accordance with the provisions of the Arbitration Act 1996 and if no arbitrator is appointed by agreement within one month of the notice either party may make a written request to the President for the time being of the Royal Institution of Chartered Surveyors for him to appoint an arbitrator to determine the matter in dispute

AS WITNESS the hands of the parties the day and year first before written

SIGNED by
in the presence of:-

SIGNED by
in the presence of:-

PRECEDENT 4 GRAZING AGREEMENT (TENANCY) – NON-BUSINESS

[*This form of agreement, often called a common law tenancy, can be used, most commonly, for grazing purposes where there is no business element and hence the landlord is not able to grant a farm business tenancy to the tenant. In short, this precedent will be most commonly employed where the tenant is a hobby farmer. There is also no reference to the single payment scheme in this precedent on the assumption that the tenant is a hobby farmer and hence will not have made an application for entitlements. If such application has been made and the tenant does have entitlements, then it may be necessary to incorporate provisions to deal with the entitlements and whatever the landlord and tenant may have agreed in connection with same. Definitions and wording from the precedents appearing in Chapter 4 will need to be incorporated if that is the case.*]

PARTICULARS

1.	**Date**	:	[*date*]
2.	**The Landlord**	:	[*name*] of [*address*]
3.	**The Tenant**	:	[*name*] of [*address*]
4.	**The Holding**	:	The property described in Schedule 1
5.	**The Term**	:	[*length of term*]
6.	**The Rent**	:	The amount specified in Schedule 5 of this Agreement
7.	**Rent instalments**	:	Equal instalments payable in advance on the Rent Days
8.	**Rent days**	:	[*date*] and [*date*] in each [*period*]
9.	**Permitted purpose**	:	The right to graze for the Tenant's own use and enjoyment the Tenant's sheep and/or cattle ('the Livestock') on the Holding
10.	**The Expert**	:	[*name*] of [*address*] and if the Expert shall die or be unwilling to act then such other expert as may be appointed on the application of either party by the President of the Royal Institution of Chartered Surveyors

THIS AGREEMENT is made on the Date set out in the Particulars

BETWEEN:-

(1) [*name*] of [*address*] ('the Landlord')

and

(2) [*name*] of [*address*] ('the Tenant')

WHEREBY IT IS AGREED as follows:-

1. **Definitions and interpretations**

In this Agreement words and phrases used have the meaning assigned to them in Schedule 4

2. **Letting**

The Landlord lets the Holding to the Tenant at the Rent for the Term [TOGETHER WITH the rights set out in Schedule 2] SUBJECT TO existing rights and other matters affecting the Holding during the Term [and EXCEPTING AND RESERVING to the Landlord the rights set out in Schedule 3]

3. **Tenant's agreements**

The Tenant agrees with the Landlord:-

3.1 *Rent*

to pay the Rent on the days and in the manner required by this Agreement

3.2 *Outgoings*

to pay all rates, taxes and outgoings whatever relating to the Holding during the Term

3.3 *Maintenance*

3.3.1 to keep all fences, hedges and gates in proper stock proof condition so as to prevent his Livestock from straying or being injured and so indemnify the Landlord against all costs claims or demands made by the owners or occupiers of any adjoining land or any other persons for damages or other money arising from the escape from the Holding of all or any of the Tenant's Livestock placed upon the Holding

3.3.2 not to remove or alter any fence hedge or other boundary on the Holding nor to destroy or damage any trees hedges or fences on the Holding and to ensure that no such damage is caused by the Tenant's Livestock

3.3.3 to keep all ditches and drains clean and free from obstructions

3.4 *Use and management of the Holding*

3.4.1 not to use or farm the Holding or any part of it for the purposes of a trade or business

3.4.2 not at any time during the Term to use the Holding other than for the Permitted Purpose set out in the Particulars

3.4.3 not to do or suffer to be done on the Holding anything which may be or become a nuisance or annoyance to the Landlord or the owners or occupiers of any adjoining land and to indemnify the Landlord against any claim caused by or arising from the Tenant's occupation of the Holding

3.4.4 not to sell off or remove from the Holding any hay or straw and not to mow the permanent pasture

3.4.5 to permit the Landlord to enter the Holding or any part of it on giving to the Tenant reasonable notice (save in the event of an emergency)

3.4.6 to comply with any other restrictions which the Landlord may reasonably require

3.4.7 not to allow the Holding to become poached by treading and if the Landlord gives the Tenant written notice that any such damage is in the Landlord's opinion being caused immediately to remove the Tenant's Livestock from the Holding

3.4.8 not to bring cause or permit to be done or brought any object or matter or thing upon the Holding by which any policy of insurance of the Landlord might be prejudiced

3.4.9 not to bring any diseased Livestock onto the Holding and to compensate the Landlord for all losses incurred as a result of any diseased Livestock being on the Holding

[3.4.10 on entering the Holding to give all necessary notices and make all necessary applications and payments to preserve any existing licence granted in respect of the Holding under the Water Resources Act 1991 or any similar enactment and subsequently to continue the licence in force during the Tenancy and not at any time by any act or omission to jeopardise the licence nor to surrender it without the consent in writing of the Landlord]

[3.4.11 on the termination of the Tenancy to make such application for the continuance or renewal of the licence as may be requisite for the time being under regulations then in force to preserve the licence for the benefit of the Landlord or his nominee without payment]

[3.4.12 not to enter into any agreement or licence for the supply of water from any spring or well pond stream watercourse or underground water supply on the Holding to any adjoining or neighbouring premises and to reimburse to the Landlord on demand all charges costs and expenses reasonably incurred by the Landlord in connection with the provision or the improved provision of water to the Holding]

3.5 *Alienation*

not to assign sublet part with or share possession or occupation of the Holding or any part of it

3.6 *Alteration of buildings*

not to alter remove or make additions to any building or other item of fixed equipment or erect any new buildings or other item of fixed equipment or make any other improvements to the Holding

3.7 *Legislation*

3.7.1 at all times immediately upon receipt of any notice order direction or other matter whatever affecting or likely to affect the Holding to produce the same for the Landlord's inspection and to permit the Landlord to take a copy

3.7.2 to comply immediately with and give sufficient effect to every notice direction or other matter whatever affecting or likely to affect the Holding served or made by any Authority

3.7.3 fully to comply with all legal obligations relating to the Holding and to keep the Landlord effectively indemnified against all actions proceedings costs expenses claims and demands in respect of any matter contravening any legal obligation

3.8 *Yield up*

on termination of this Agreement to yield up the Holding to the Landlord with vacant possession in a state of repair and condition which is consistent with the proper performance of the Tenant's agreements and obligations in this Agreement

4. Landlord's agreement

The Landlord agrees with the Tenant that if the Tenant observes and performs the Tenant's agreements and obligations in this Agreement the Tenant may peaceably hold and enjoy the Holding during the Term without any lawful interruption or disturbance from or by the Landlord or any person claiming through under or in trust for the Landlord

5. Mutual agreements

5.1 *Forfeiture*

Without prejudice to any other rights of the Landlord if:-

5.1.1 the whole or part of the Rent remains unpaid 21 days after becoming due (whether demanded or not); or

5.1.2 any of the Tenant's agreements in this Agreement are not performed or observed; or

5.1.3 the Tenant:-

5.1.3.1 proposes or enters into any composition or arrangement with his creditors generally or any class of his creditors; or

5.1.3.2 is the subject of any judgment or order made against him which is not complied with within 7 days or is the subject of any execution distress sequestration or other process levied upon or enforced against any part of his undertaking property assets or revenue; or

5.1.3.3 is the subject of a bankruptcy petition or bankruptcy order; or

5.1.3.4 is the subject of an application or order or appointment under the Insolvency Act 1986, section 253 or section 273 or section 286; or

5.1.3.5 is unable to pay or has no reasonable prospect of being able to pay his debts within the meaning of the Insolvency Act 1986, sections 267 and 268; or

5.1.4 any event occurs or proceedings are taken with respect to the Tenant in any jurisdiction to which he is subject which has an effect equivalent or similar to any of the events mentioned in clause 5.1.3, then and in any such cases the Landlord may at any time (and notwithstanding the waiver of any previous right of re-entry) upon giving the tenant two months' prior notice in writing re-enter the Holding whereupon this Agreement will absolutely determine but without prejudice to any right of action of the Landlord in respect of any previous breach by the Tenant of this Agreement

5.2 *Break clause*

5.2.1 Either party to this Agreement will be entitled on the death of the other or (in the case of more than one landlord or tenant) on the death of [any or the last] of them to terminate this Agreement by serving on the relevant deceased party's personal representatives not less than [two months' or six months' or one year's] prior written notice of his intention to do so [such notice to expire at the end of a period of the tenancy]

5.2.2 If the Landlord wishes to determine this Agreement at any time during the Term and shall give to the Tenant not less than one year's notice in writing then upon the expiration of such notice the Term shall immediately cease and determine but without prejudice to the respective rights of either party in respect of any antecedent claim or breach of covenant

5.3 *Dispute resolution*

5.3.1 In the event of any dispute arising under this Agreement between the parties the same will be determined by the Expert named in the Particulars who is to act as independent expert and not as an arbitrator and who is to be appointed on the written application of either party

5.3.2 The appointment of the Expert is to specify that his decision will be made following representations in writing by the parties and the costs of the Expert will be borne as directed by the Expert and his decision will be final and binding on all parties

5.4 *Whole agreement*

This Agreement contains the whole agreement between the Landlord and the Tenant concerning the Holding

AS WITNESS the hands of the parties hereto the day and year first before written

SCHEDULE 1
THE HOLDING

ALL THAT land situate at and known as [ ] in the County of [ ] extending to some [ ] acres or thereabouts as the same is delineated and edged red on the plan annexed hereto [TOGETHER WITH the building(s) located thereon or on some part or parts thereof]

SCHEDULE 2
RIGHTS GRANTED TO THE TENANT

[A right of way for the Tenant his agents and persons duly authorised by him at all times with or without vehicles plant machinery or animals to pass and re-pass over the track [shown coloured brown] on the attached plan] [but not at any time to park on it save in the case of emergency]

SCHEDULE 3
RESERVATIONS TO THE LANDLORD

1. All mines minerals quarries stones sand brick earth clay gravel turf petroleum and its relative hydrocarbons and all other gases and substances in or under the Holding of kind ordinarily worked or removed by underground or surface working with power to the Landlord and all persons authorised by him to enter on the Holding to search for win dress and make merchantable and carry them away from the Holding and to execute all incidental works including the right to let down the surface of the land the Tenant being paid reasonable compensation for all damage caused to him by the exercise of such power

2. The right to grant any wayleave contract easement or licence to any public or local authority or public utility company or other company or persons with a right to authorise servants and agents of such parties with or without vehicles machinery and plant to enter upon the Holding and carry out their works (subject to the payment of reasonable compensation for damage provided a claim in writing is made by the Tenant to the Landlord within a reasonable time from the occurrence of the damage) together with the benefit of all present or future such contracts agreements for easements or licenses and all rents and other money payable under them and the power to carry out on the Holding at the Landlord's cost anything required to be done under them by the Landlord

3. All timber and other trees saplings pollards and underwood with liberty for the Landlord and all persons authorised by him to enter upon the Holding in order to mark fell cut and carry them away and to replant them and to cart them from the Holding without making any payment to the Tenant for the use but making reasonable compensation to the Tenant for any damage done in the exercise of the rights reserved providing a claim in writing is made by the Tenant to the Landlord within a reasonable time

4. All the game wildfowl woodcock snipe and other wild birds listed in the Second Schedule to the Wildlife and Countryside Act 1981 their nests and eggs together with the exclusive right for the Landlord and all persons authorised by him to go upon the Holding with such equipment and persons as the Landlord authorises for the purposes of shooting fishing hawking sporting taking preserving and rearing such wildlife and also the right to hunt or permit hunting on foot or with horses and hounds

5. Subject to the Ground Game Act 1880 (as amended by s 12 of the Wildlife and Countryside Act 1981) and 1906, the right for the Landlord and all persons authorised by him to kill shoot and take away rabbits hares pigeons or any other pests

6. The right for the Landlord and all persons authorised by him with or without vehicles horses machinery and plant to enter on any part of the Holding at all reasonable times for all the above and all reasonable purposes and in particular but without prejudice to the generality of the above the right to run pipes drains conduits cables wires or other works (either already existing or any new ones) for the benefit of any other part of the Landlord's estate and the right to carry out works for the benefit of any other part of the same estate the Tenant being paid reasonable compensation for damage caused to them by the exercise of the right for the benefit of any other part of the same estate

SCHEDULE 4
DEFINITIONS AND INTERPRETATION

1. Definitions

1.1 'Authority' means any statutory public local or other authority or any court of law or any government department or any of their duly authorised officers

1.2 'Enactment' means

1.2.1 any Act of Parliament or

1.2.2 any European Community Legislation or decrees having effect of law in the United Kingdom

1.3 'Holding' means the land (including all buildings and fixed equipment on the land) set out in the Particulars as more fully described in Schedule 1

1.4 'Particulars' means the Particulars page set out at the beginning of this Agreement

1.5 'Rent' means the sum(s) of rent set out in the Particulars payable by the instalments set out in the Particulars

1.6 'Term' means the term of the agreement set out in the Particulars

2. Interpretation

In this Agreement unless the context otherwise requires:-

2.1 words importing any gender include every gender

2.2 words importing the singular number include the plural number and vice versa

2.3 words importing persons include firms companies and corporations and vice versa

2.4 references to a numbered clause or schedule are references to the relevant clause in or schedule to this Agreement

2.5 references in any schedule to this Agreement to numbered paragraphs relate to the numbered paragraphs of that schedule

2.6 where any obligation is undertaken by two or more persons jointly they are to be jointly and severally liable in respect of that obligation

2.7 any obligation on any party not to do or omit to do anything is to include an obligation not to allow that thing to be done or omitted to be done

2.8 any party who agrees to do something will be deemed to fulfil that obligation if that party procures that it is done

2.9 the headings to the clauses schedules and paragraphs of this Agreement will not affect the interpretation

2.10 any sum payable by one party to the other is to be exclusive of Value Added Tax which will where it is chargeable be paid in addition to the sum in question at the time when the sum in question is due to be paid

2.11 any reference to an Enactment includes the reference to that Enactment as amended or replaced from time to time and to any subordinate legislation or bye-law made under that Enactment

SCHEDULE 5
RENT

In this Schedule:-

1. 'the Index' means the Index of retail prices published by HM Government

2. 'the Relevant Percentage' means such percentage as is equal to the percentage increase (if any) in the figure at which the Index stands at

the relevant date over the figure for the Index last published prior to the date hereof PROVIDED THAT in the event of any change after the date hereof in the reference base used to compile the Index the figure taken to be shown in the Index after such change shall be the figure which would have been shown in the Index had the reference base current at the date hereof been retained or in the event of it becoming impossible or impracticable by reason of any change after the date hereof in the methods used to compile the Index or for any other reason whatsoever to calculate the said percentage increase under the foregoing provision then such percentage increase shall be that calculated to give the same purchasing power at the relevant date as at the date hereof the amount of such percentage to be agreed between the parties hereto or in default of agreement to be determined at their joint expense by such independent person as they shall agree or in default as shall be appointed by the President of the Institute of Chartered Accountants in England and Wales such determination to be made as an expert and to be final and binding on all persons concerned

3. during the Term the rent in every year commencing on the date hereof and subsequent anniversaries thereof (and proportionately for that part of a year) shall be [ ] pounds (£[ ]) together with the Relevant Percentage of such sum calculated at the commencement of each such year

SIGNED by
in the presence of:-

SIGNED by
in the presence of:-

Chapter 8

ALTERNATIVE LAND USES

INTRODUCTION

PRECEDENTS

1 Metal detecting licence
2 Film contract
3 Permissive path agreement
4 Statement and declaration for purposes of s 31(6) of the Highways Act 1980, as amended by the Countryside and Rights of Way Act 2000
5 Agreement for deposit of archives with County Records Office

INTRODUCTION

This chapter contains a selection of items which may be of interest to the rural practitioner. There is almost no limit to the various situations which may need to be covered. This is a small selection based on the authors' experience.

Metal detecting licence

Although metal detecting was once frowned on by the archaeological establishment it is now accepted that without enthusiastic amateurs many important finds would not have been made.

The licence seeks to protect both the interests of the landowners and the finds themselves. The landowner and the licensee should obtain a copy of the Code of Practice for Responsible Metal Detecting in England and Wales. This is available from the Department of Culture Media and Sport, 2–4b Cockspur Street, London SW1Y 5DH or online.[1]

1 www.culture.gov.uk.

Film contract

Rural land is often in demand for films or television. Many film companies will have their own form of contract. These should be studied carefully to check that the safeguards in the precedent given in this section are included.

Permissive path agreement

Before the advent of the Countryside and Rights of Way Act 2000, landowners were encouraged to enter into voluntary agreements with regard to access in the hope that this would avoid compulsory access to open country and common land. This tactic did not work. There is now access to mapped open country and registered common land and proposals for access to coastal land.

Although there may be less enthusiasm in general now for permissive paths, there may be occasions when such an agreement would avoid having a creation order imposed or would make sense for other reasons, such as to generate good will or to encourage people to visit retail outlets which would otherwise be inaccessible.

Statement and declaration under Highways Act 1980, s 31(6)

Where members of the public have used a route for 20 years without force, secrecy or permission they may be able to claim a public right of way unless the landowner can establish that there was no intention to dedicate such a right.

The recent House of Lords decision in *R (on the application of Godmanchester Town Council (Appellants) v Secretary of State for the Environment, Food and Rural Affairs (Respondent); R (on the application of Drain) (Appellant) v Secretary of State for the Environment, Food and Rural Affairs (Respondent)*[2] has made it extremely difficult for a landowner to prove that he or she had no intention to dedicate a right of way. It is therefore important for landowners, before there is any significant public use, to rebut such an intention by lodging a statement and declaration under the Highways Act 1980, s 31(6). This should be done routinely when rural land is held or acquired.

The statement and declaration should be made by the owner. 'Owner' is defined as the person who is for the time being entitled to dispose of the fee simple of the land.[3] The declaration originally had to be made within 6 years from the date of deposit or within 6 years from the date on which the last declaration was made. This period has been altered to 10 years by the Countryside and Rights of Way Act 2000[4] (CROW Act 2000). As it is the declaration which rebuts the deemed dedication, it should be lodged as soon as the highway

---

2 [2007] UKHL 28.
3 Highways Act 1980, s 31(7).
4 Section 57, Sch 6.

authority gives an official date for the lodging of the statement. This may be several days after the posting of the statement by the landowner or his or her advisers. Landowners should remember to lodge a further declaration before 10 years have elapsed from the original declaration.

It is important to stress that this procedure rebuts any intention of dedication in the future only. It cannot take away any claims which are based on past user. Nor will it overcome any claims based on documentary evidence.

The appropriate council is required to keep a register of deposited maps and statements and declarations lodged with that council. The register must be available for inspection free of charge at all reasonable hours.[5]

Deposit of archives with County Records Office

The title deeds to rural property are often of considerable historical interest. Many landowners have therefore agreed to deposit them at the County Records Office where they will be available for research. The agreement contains provisions to protect the interests of the owner of the documents while they are on deposit.

[5] Highways Act 1980, s 31A, inserted by the CROW Act 2000, Sch 6, para 4.

PRECEDENTS

PRECEDENT 1 METAL DETECTING LICENCE

THIS LICENCE is made the [ ] day of [ ] 20[ ]

BETWEEN:-

(1) [*name*] of [*address*] ('the Owner')

and

(2) [*name*] of [*address*] ('the Licensee')

WHEREBY IT IS HEREBY AGREED as follows:-

1.1 The Owner agrees that on payment of the sum of [ ] pounds (£[ ]) ('the Licence Fee') the Licensee may use the property described below ('the Property') for the purposes described below (but only for those purposes) for the period of [*period in weeks/months*] starting on the (*date*) and ending on the (*date*) ('the Licence Period')

1.2 The Property is [*description*] as more particularly identified on the plan annexed hereto ('the Plan')

1.3 The Licensee is entitled to obtain access to the Property over the accessway coloured brown on the Plan but by no other route

1.4 While this Licence continues the Licensee is entitled:-

1.4.1 to search for, examine and dig for items of buried metal or other material, whether antique or modern, or any other item or objects including those specified as follows:-

[*specify any as appropriate in the circumstances*]

1.4.2 to use equipment agreed by the Owner for searching or digging

1.4.3 to enter the Property with the agreed mechanical/electronic devices for searching for and/or digging on the Property but for no other purposes whatsoever

1.5 Either party can end this Licence at any time during the Licence Period by giving to the other at least one month's notice taking effect at the end of any month

1.6 Notwithstanding the grant of this Licence the Owner is entitled to possession of the Property and has a right of access to it at all times and for all purposes

1.7 This Licence is personal to the Licensee and cannot be transferred. The Licensee is not permitted to allow any other person to use the Property

2. THE Licensee agrees with the Owner:-

2.1 to pay the Licence Fee to the Owner on the signing of this Agreement

2.2 that the proceeds of sale of any find and any reward paid by the Secretary of State pursuant to the Treasure Act 1996 shall be divided as follows:-

2.2.1 to the Owner [ ] per cent ([ ] %)

2.2.2 to the Licensee [ ] per cent ([ ] %)

2.3 to indemnify the Owner against all injuries damages costs claims and demands arising from the activities authorised by this Licence and to produce evidence of existing insurance or where there is none or it is for inadequate cover to insure on terms agreed by the Owner

2.4 to fill any holes, replacing soil and grass carefully, repair or restore any fences, buildings or enclosures dismantled, opened or altered as a result of the activities of the Licensee

2.5 to protect any livestock upon the Property and prevent their escape

2.6 to protect all wildlife whether plant or animal upon the Property

2.7 to protect any archaeological deposits below plough soil depth

2.8 to prevent fire

2.9 to bear full responsibility for any damage to the Property including the livestock wildlife plant and animals thereon and including any injuries suffered by the Licensee or any other person as a result of the Licensee's activities thereon whether arising during or following termination of the Licence Period

2.10 to report immediately to the Owner any finds of whatsoever nature

2.11 to report the find of any items which may or which on reasonable grounds are likely to be treasure within the meaning of the Treasure Act 1996 as varied by the Treasure (Designation) Order 2002 (SI 2002 No 2666) to the Coroner for the district in which the Property is situate within [14] days of the date of the find and to supply evidence to the Owner that he has done this

2.12 to observe the provisions of the Treasure Act 1996 as varied by the Treasure (Designation) Order 2002 (SI 2002 No 2666) and the Treasure Act Code of Practice Second revision (or any subsequent revision) in relation to any find

2.13 to comply with the provisions of the Portable Antiquities Scheme

2.14 to report any archaeological object to the Owner and where the Owner so agrees to the County Sites and Monuments Record

2.15 to report any bombs or live ammunition to the Police having first marked the area on the Property where they were located

2.16 to respect the country code as issued by Natural England or the Countryside Council for Wales (as the case may be)

2.17 not to use the Property, or any part of it, nor allow any one else to do so for any activities which are dangerous, offensive, noxious, noisome, illegal or immoral, or which are or may become a nuisance or annoyance to the Owner or occupier or to the owner or occupier of any adjoining property

2.18 to carry and produce on request an identification card [*and membership card of the National Council for Metal Detecting and a Certificate of Public Liability Insurance*]

3. THE Owner and the Licensee further agree:-

3.1 that all objects found other than treasure are the property of the Owner and will be recorded or disposed of according to the Owner's wishes

3.2 the value of any objects will be determined by agreement between the Owner and the Licensee or by sale for the highest available price or by appraisal by at least two independent persons jointly selected. Thereafter payment of any monies due to either party must be made upon the sale or within [ ] days of any valuation or sale

3.3 if there shall be any breach or non-observance by the Licensee of any of the terms and conditions of this Licence the Owner may revoke this Licence and thereupon the Licence and all the rights hereby granted shall cease and be determined with or without notice by the Owner and immediately on being requested to do so the Licensee will vacate the Property

3.4 that where the excavation is not limited to the disturbance of plough-soil the Licensee will discuss with the Owner the need to consult the appropriate County Sites and Monuments Records and should the Owner so direct the Licensee will undertake such consultations

3.5 that the Licensee will not search for or dig for items on areas of permanent pasture [without first obtaining the Owner's written consent]

3.6 that where the Owner agrees the Licensee will submit all finds to the County Sites and Monuments Records for identification and recording by archaeologists

3.7 that the Owner and the Licensee except where disclosure is required by an order of the Court or in order to comply with the requirements of any statutory or other competent authority will not disclose to any third party any details of this Licence nor the existence nature or position of any find without the prior written consent of the other and will use all reasonable endeavours to prevent disclosure or publication of such details or information

4. Dispute resolution

4.1 Any disputes between the parties arising out of this Licence shall be referred to an independent expert ('the Expert') in accordance with the following provisions:-

4.1.1 either party may give written notice to the other of his intention to refer the matter to the Expert pursuant to the provisions of this clause

4.1.2 the parties may agree on the identity of the Expert in which event they shall jointly appoint him

4.1.3 in the absence of such agreement within two months of the serving of notice under clause 4.1.2 above, either party may apply to the President for the time being of the Law Society for the appointment of a suitably qualified person to act as Expert

4.1.4 the expert's decision shall be final and binding on the parties

4.1.5 the fees and expenses of the Expert and (as the case may be) of the President of the Law Society shall be borne as the Expert shall decide or in the absence of such decision equally

4.1.6 in the event that the Expert shall die or shall become unwillingly to act or shall fail to deliver a decision within four months of his appointment the procedure set out in this clause shall be repeated and may be repeated as often as is necessary to achieve a resolution of the dispute

AS WITNESS the hands of the parties hereto the day and year first before written

SIGNED by
in the presence of:-

SIGNED by
in the presence of:-

PRECEDENT 2 FILM CONTRACT

THIS AGREEMENT is made the [ ] day of [ ] 20[ ]

BETWEEN:-

(1) [*name*] of [*address*] ('the Owner')

(2) [*name*] of [*address*] ('the Company')

WHEREBY IT IS HEREBY AGREED as follows:-

1 The Owner agrees that the Company may use the areas of the property ('the Permitted Areas') described in the First Schedule for the purposes specified in the Second Schedule on the following dates [*dates*] between the hours of [*times*]

2. The Company agrees with the Owner:-

2.1 to pay the fees calculated in accordance with the formula set out in the Third Schedule

2.2 to pay for all electricity, telephone calls and other expenses incurred during the period covered by this Agreement

2.3 to supply in advance a list of all persons authorised (the 'Authorised Persons') by the Company to enter the Permitted Areas

2.4 to ensure that the Authorised Persons carry and produce on request an identification card which has on it an up to date photograph of the Authorised Person and is signed by a duly authorised officer of the Company

2.5 to comply at all times with the Lighting Specification in the Fourth Schedule

2.6 not to use the property, or any part of it, nor allow anyone else to do so for activities which are dangerous, offensive, noxious, noisome, illegal or immoral, or which are or may become a nuisance or annoyance to the Owner or any occupier of the Owner's property or to the owner or occupier of any adjoining property, so that this clause shall not preclude the exercise by the Company of the rights granted in the Second Schedule

2.7 not to smoke in any of the Permitted Areas

2.8 not to do any damage to the Permitted Areas or the contents thereof

2.9 to compensate the Owner for any damage to property or contents, the value of the loss to be determined by an expert appointed by agreement between the parties or in default of agreement by the President of the Royal Institution of Chartered Surveyors

2.10 to ensure that no food or drink is consumed except in the designated parts of the Permitted Areas

2.11 to use only [*specify area*] for relaxation and not any other part or parts of the Permitted Areas

2.12 to indemnify the Owner against all injuries damage claims and demands arising from the activities authorised by this Agreement

2.13 to remove all equipment and leave the Owner's property in a clean and tidy condition on the last of the dates specified in clause 1 above

3. IT IS FURTHER AGREED:-

3.1 the wording of the credit for the use of the Permitted Areas shall be [*set out desired wording*]

3.2 the copyright of the photographs/films will be vested in [*include name(s)*]

3.3 the photographs/films shall be used for [*specify agreed uses*] but not for any other purpose whatsoever

3.4 if there shall be any breach or non-observance by the parties of the terms and conditions of this Agreement the party not in breach may revoke the Agreement forthwith but without prejudice to any of the rights and liabilities of either party accruing before such breach

SIGNED by the parties hereto the day and year first before written

FIRST SCHEDULE
(PERMITTED AREAS AND ADDRESS OF THE OWNER'S PROPERTY)

SECOND SCHEDULE
(RIGHTS GRANTED)

[*eg including parking and catering facilities in designated areas*]

THIRD SCHEDULE
(CALCULATION OF FEES)

FOURTH SCHEDULE
(LIGHTING SPECIFICATION)

SIGNED by
in presence of:-

SIGNED by
in the presence of:-

PRECEDENT 3 PERMISSIVE PATH AGREEMENT

PARTICULARS

The Date	:	[*date*]
The Landowner	:	[*name*] of [*address*]
The Tenant	:	[*name*] of [*address*]
The District Council	:	[*name*]
The Permissive Path	:	[*description*]
Width	:	[*metres*]
The Whole Path	:	[*description*]
Permitted Purposes	:	[*description*]
Limitations on Use	:	[*description*]
The Term	:	the period from the [*date*] to the [*date*]
Payment	:	£[ ]

THIS AGREEMENT is made on the date set out in the Particulars

BETWEEN:-

(1) [*name*] of [*address*] ('the Grantor')

and

(2) [*name*] of [*address*] ('the Council')

WHEREBY IT IS AGREED as follows:-

1. The Particulars

The details in the Particulars apply to and are part of this Agreement

2. The Grant

In consideration of the Payment and the covenants given by the Council in clause 8 below the Grantor grants to the Council and its servants and agents the following rights:-

2.1 to use and allow the public to use the Permissive Path for the Permitted Purposes subject to the Limitations on Use

2.2 to enter on the Permissive Path and the land on either side for a distance of [*one*] metre for the purposes of clearing undergrowth and maintaining the Permissive Path

2.3 to execute the works described in the Schedule hereto

2.4 to enter on the Permissive Path and do work which is necessary to comply with clause 7.2 below where the Grantor has failed to do so

3. Duration

The rights will subsist throughout the Term unless this Agreement is terminated in accordance with clause [*4.2 or*] 4.3 below

4. Termination

4.1 This Agreement will terminate on the expiration of the Term

[4.2 The Grantor may terminate this Agreement at any time on or after the [*fifth*] [*tenth*] anniversary of the Date by giving three months' prior notice in writing to the Council and at the expiration of a valid notice this Agreement shall absolutely determine without prejudice to the rights and liabilities of either party against the other for any antecedent breach of covenant]

4.3 The Agreement will terminate in the event of there being a breach of covenant by the Grantor or the Council which is not remedied within twenty-one days after the service by one party of written notice on the other specifying the breach and the steps which are necessary to remedy the same but without prejudice to the rights and liabilities of either party against the other for any antecedent breach of covenant

5. Diversion

5.1 Either the Council or the Grantor may request that the Permissive Path be diverted

5.2 Where the request is granted any costs incurred in making the diverted path suitable for use for the public and any costs incurred in order to publicise to the public the change in the path should be borne by the party making the request

6. Temporary closure

The Grantor may close all or part of the Permissive Path on a temporary basis for agricultural or other land management operations or sporting events or where it is reasonable to do so for security reasons provided that the conditions set out below are observed:-

6.1 save in an emergency prior written notice of at least two weeks is given to the Council

6.2 signs giving notice of the intended closure and reasons for it and the date of the re-opening of the Permissive Path are posted at each end of the path where it is to be closed

6.3 such signs are removed when the Permissive Path is re-opened in accordance with the date given in the notice referred to in clause 6.2 above

6.4 the Permissive Path is not closed for more than [*ten*] days in any year

7. Covenants by the Grantor

The Grantor covenants:-

7.1 not to dispose of his land or that part of his land over which the Permissive Path runs without using his best endeavours to ensure that the new landowner or tenant enters into an agreement with the Council similar in terms to this one

7.2 to keep the Permissive Path clear of growing crops and other obstructions and in a reasonably tidy state

7.3 to reimburse the Council the costs of any work undertaken by the Council where the Grantor has failed to comply with clause 7.2 above, reasonable notice having been given to him to do so

[7.4 not to keep bulls more than ten months old or any animals known to be dangerous or unpredictable on any of the Grantor's land over which the Permissive Path runs]

7.5 not to do or permit to be done anything whereby the public liability insurance policy maintained by the Council in accordance with clause 8.15 may become void or voidable or whereby the insurer may refuse payment of any claim in whole or in part or whereby the rate of premium may be increased

7.6 to notify the Council forthwith of any notice circumstances or events which may affect the said insurance policy or give rise to a claim thereunder

7.7 to repay to the Council on demand all expenses incurred as a result of a breach of the covenants in clause 7.5 and 7.6 above

8. Covenants by the Council

The Council covenants:-

8.1 forthwith to execute the works described in the Schedule hereto and thereafter to maintain such works in good repair and condition

8.2 on the termination of the Agreement or permanent closure of the Permissive Path to remove all signs and other works which have been executed by the Council

8.3 to keep the surface of the Permissive Path free from all natural vegetation and to trim back vegetation growing from the sides or above the Permissive Path as may be required to keep it in a suitable condition for the Permitted Purposes

8.4 to keep the Permissive Path free from litter and rubbish

8.5 to erect and maintain any gates stiles footbridges and other furniture relating to the Permissive Path in an appropriate condition

8.6 not to cut down or maim or injure any tree or sapling other than as provided for in clause 8.3 above without the previous consent in writing of the Grantor

8.7 to put up and maintain coloured waymarks of the size and colour and in such a position as may be agreed in writing with the Grantor

8.8 to erect notices at either end of the Whole Path stating that the path is available for use by the public on a permissive basis only and stating the Permitted Purposes and the Limitations on Use which must be observed

8.9 to erect notices warning users of the Permissive Path of any dangers on or near the same

8.10 to take all reasonable steps to ensure that dogs are kept on a lead where this is specified in the Particulars as one of the Limitations on Use

8.11 not to assign or part with possession or control of any of the rights hereby granted

8.12 not to do anything in connection with the rights hereby granted which may be or become a nuisance or annoyance or cause damage to the Grantor or to the Tenant or any occupiers of the property or to the owners tenants or occupiers of any adjoining or neighbouring property

8.13 to indemnify and keep indemnified the Grantor from and against all costs charges expenses claims and demands and damages of any description in any way arising or connected with any act or default on the part of the Council or their duly authorised officers servants agents or other persons authorised or impliedly authorised in relation to the rights hereby granted

8.14 to ensure that the Council's public liability policy covers the use by the public of the Permitted Path and to maintain such cover throughout the Term unless the policy shall be vitiated by any act of the Grantor or by anyone acting with the express or implied authority of the Grantor

8.15 to pay all the proper costs and expenses incurred by the Grantor in connection with the preparation of this Agreement

9. Disputes

9.1 If there shall be a dispute between the parties concerning their rights and obligations under the terms of this Agreement the parties may refer the dispute for determination by an independent expert appointed in accordance with the sub-clauses set out below

9.2 After such a dispute has arisen the parties may agree in writing to refer the dispute to an independent expert appointed by them both whose decision shall be final and binding upon them

9.3 The procedure to be adopted by the independent expert shall be determined by him but shall provide an opportunity for the parties to state their case orally or in writing as the independent expert may direct

9.4 The independent expert shall have the discretion to award the costs of the appointment failing which the costs shall be borne equally by the parties

9.5 In the event of the parties failing to agree to refer the dispute to an independent expert or on the person to be appointed within one month of the dispute arising either party may apply to the President of the Royal Institution of Chartered Surveyors for the appointment of an expert to determine the dispute

10. No deemed dedication

For the avoidance of doubt it is hereby declared that nothing herein contained shall amount to or be construed as a permanent grant demise or dedication of or agreement for a public right of way

AS WITNESS the hands of the parties hereto the day and year first before written

SCHEDULE
(WORKS WHICH THE COUNCIL HAVE AGREED TO UNDERTAKE)

SIGNED by
in the presence of:-

SIGNED by
in the presence of:-

PRECEDENT 4 STATEMENT AND DECLARATION FOR PURPOSES OF S 31(6) OF THE HIGHWAYS ACT 1980, AS AMENDED BY THE COUNTRYSIDE AND RIGHTS OF WAY ACT 2000

DEPOSIT OF STATEMENT AND MAP
SECTION 31(6) OF THE HIGHWAYS ACT 1980

To [*name*] Council

1 I am the owner within the meaning of the above section of the land known as [*name of farm, etc.*] ('the Land') more particularly delineated on the map accompanying this statement and thereon edged red

*2 The ways coloured brown on the said map have been dedicated as highways for mechanically propelled vehicles

*3 The ways coloured orange on the said map have been dedicated or designated as restricted byways

*4 The ways coloured green on the said map have been dedicated as bridleways

*5 The ways coloured purple on the said map have been dedicated as footpaths

6 No [*other*] ways over the land have been dedicated as highways and I have no intention of dedicating [any] [further] highways

* Delete as applicable

Signed (landowner)	....
Name (of landowner)	....
Address	....
	....
	....
Date	....
Signed (witness)	....
Name (of witness)	....
Address	....
	....
	....

Occupation ....

STATUTORY DECLARATION
SECTION 31(6) OF THE HIGHWAYS ACT 1980

I [*landowner*] DO SOLEMNLY AND SINCERELY DECLARE as follows:-

1 I am the owner of the land known as [*name of farm, etc*] ('the land') more particularly delineated on the map accompanying this declaration ('the declaration map') and thereon edged red

2 On the [*date*] I [*or my predecessor in title (name)*] deposited with [*name*] Council, being the appropriate Council, a statement accompanied by a map ('the statement map') delineating the land by red edging which stated that [the ways coloured brown on the statement map had been dedicated as highways for mechanically propelled vehicles] [the ways coloured green on the statement map had been dedicated as bridleways] [the ways coloured orange on the statement map had been dedicated or designated as restricted byways] [the ways coloured purple on the statement map had been dedicated as footpaths] and no [other] ways had been dedicated as highways over the land

*3 On the [*date*] I [*or my predecessor in title (name)*] deposited with [*name*] Council, being the appropriate Council, a statutory declaration dated [*date*], stating that no additional ways [other than those marked in the appropriate colour on the declaration map] had been dedicated as [highways for mechanically propelled vehicles] [restricted byways] [bridleways] [footpaths] over the land since the deposit of the statement referred to in clause 2 above

4 No additional ways have been dedicated over the land since the statement dated [*date*] referred to in clause 2 above [since the date of the statutory declaration referred to in clause 3 above] [other than those [highways with vehicular status] [restricted byways] [bridleways] [footpaths] marked in the appropriate colour on the declaration map] and at the present time I have no intention of dedicating any more public rights of way over the land

* *Delete if not applicable*

AND I MAKE this solemn declaration on the [*date*] conscientiously believing it to be true and by virtue of the Statutory Declarations Act 1835

Declared at [*address*] .... (signature of landowner) ....

....

Before me

....

[Commissioner for Oaths or a
Justice of the Peace or Solicitor]

PRECEDENT 5 AGREEMENT FOR DEPOSIT OF ARCHIVES WITH COUNTY RECORDS OFFICE

THIS AGREEMENT is made the [ ] day of [ ] 20[ ]

BETWEEN:-

(1) [*name*] of [*address*] ('the Bailor')

and

(2) [*name*] of [*address*] ('the Bailee')

WHEREAS IT IS AGREED as follows:-

1. The Bailment

The Bailor agrees to deposit the documents ('the Documents') listed in the Schedule on the terms and conditions set out in this Agreement ('the Bailment')

2. Title to Documents

The Bailor shall at all times retain title to the Documents

3. Continuation of Bailment

The Bailment shall continue until either the Bailor or the Bailee terminates this Agreement in accordance with clause 6.2 or 6.3 below

4. Obligations of Bailee

The Bailee agrees with the Bailor as follows:-

4.1 to provide the Bailor with a receipt and description of the Documents

4.2 to provide the Bailor free of charge with copies of any lists of the Documents made by the Bailee in accordance with its current practice

4.3 to store the Documents taking no less care than it does in the storage of its own documents and to take reasonable precautions against damp, fire, vermin and illegal access

4.4 to maintain its public liability insurance policy in respect of damage to its archive collections arising from the negligence of the Bailee or its employees and to ensure that the Documents are covered by such a policy

4.5 not without the Bailor's written consent:-

4.5.1 to allow the Documents or extracts from them to be published

4.5.2 to remove the Documents from the Record Office for exhibitions, lecture or for any other purpose whatsoever

4.5.3 to produce the Documents for any official or legal purpose

5. Rights of Bailee

5.1 The Bailee may photograph or microfilm the Documents or any of them for purposes of security or where the Documents are to be returned under clause 6.1 and 6.2 below for record purposes and the copyright and ownership of such photographs or microfilms shall remain vested in the Bailee but subject to clauses 4.5.1 and 4.5.2 above

5.2 The Bailee may mark the Documents with a finding reference for safety and identification purposes

5.3 With the consent of the Bailor the Bailee may undertake any necessary conservation work on the Documents

5.4 The Bailee shall be entitled to publish the lists referred to in clause 4.2 above and the copyright of such lists shall be vested absolutely in the Bailee

5.5 The Bailee may make any of the Documents more than thirty years old and with the consent of the Bailor any of the Documents of less than thirty years old available to the public for inspection free of charge under supervision during advertised opening hours and in accordance with search room regulations but not otherwise

5.6 Subject to the requirements of the Copyright Acts 1911–1988 the Bailee may provide on request by a member of the public, photographic copies of the Documents for the purposes of private study and research only, provided that the originals would not be damaged by such photocopying and any further reproduction of such photocopies is prohibited

6. Withdrawal and termination

6.1 On the giving of prior written notice of at least five working days the Bailor may withdraw the Documents for any reason for a period not exceeding three months or longer with the consent of the Bailee

6.2 On the giving of prior written notice of at least one calendar month the Bailor may terminate the Agreement and withdraw the Documents provided that if the Bailee has with the consent of the Bailor expended labour and materials on the Documents while in the custody of the Bailee the Bailor shall repay to the Bailee a sum equivalent to the cost of the labour and materials so expended

6.3 On the giving of prior written notice of at least one calendar month the Bailee may return the Documents to the Bailor

SIGNED by the parties hereto the day and year first before written

SCHEDULE

SIGNED by
in the presence of:-

SIGNED by
in the presence of:-

Chapter 9

SPORTING RIGHTS

INTRODUCTION

PRECEDENT

1 Sporting rights – long-term agreement

INTRODUCTION

Farmers need to maximise their assets and, accordingly, in appropriate cases, shooting may be let off. This may not be possible if the farmer is a tenant and the rights are reserved to the landlord, or if a freeholder and the rights are reserved out of the conveyance or transfer to the farmer. The tenancy agreement/deeds should be checked carefully before such an agreement is entered into.

The agreement in this chapter refers to land primarily destined for shooting purposes. If it is ancillary to farmland, the landowner might wish to impose more restrictions on the lessee. The advantage of the landowner remaining in possession for agricultural purposes is that he will be entitled to agricultural property relief under the inheritance tax legislation.

Where sporting rights are enjoyed separately from the land over which they exist no rates are payable.[1]

Sporting rights are a form of profit à prendre. The precedent grants an exclusive right for a term of years. Were it a *non exclusive* grant (shooting being shared with the grantor) *in fee simple* it would amount to a right in common with the danger that the land could then be registered as common land which would carry a right of public access.

Where a profit à prendre in gross has been granted in fee simple or for a term of years of which more than 7 years are unexpired the interest may be voluntarily registered with its own title.[2] If the title to the land over which the shooting

1 Local Government and Rating Act 1997, s 2.

2 Land Registration Act 2002, s 3(1), (2) and (3).

right is to be exercised is registered the grantee of the profit in fee simple or for a term exceeding 7 years must be registered as proprietor of the profit and a notice of the right must be entered on the register of title of the grantor.[3]

3 Land Registration Act 2002, Sch 2, para 6.

PRECEDENT

PRECEDENT 1 SPORTING RIGHTS – LONG-TERM AGREEMENT

THIS DEED made the [ ] day of [ ] 20[ ]

BETWEEN:-

(1) [*name*] of [*address*] ('the Grantor')

and

(2) [*name*] of [*address*] ('the Grantee')

WITNESSES as follows:-

1. Definitions

In this Deed the following expressions have the meanings set out below unless the context otherwise requires:-

1.1	**'the Plan'**	:	means the plan annexed hereto
1.2	**'the Land'**	:	means the Grantor's property details of which are set out in the First Schedule
1.3	**'the Rights of Access'**	:	means the rights of access referred to in the First Schedule
1.4	**'the Buildings'**	:	means the [secure store] coloured blue on the Plan
1.5	**'the Shooting Rights'**	:	means the exclusive right to shoot kill and carry away game from the Land, but excludes fishing rights and deer stalking rights
1.6	**'Game'**	:	means the game specified in the Second Schedule
1.7	**'Ground Game'**	:	means the Ground Game as defined in the Ground Game Act 1880 and the Ground Game (Amendment) Act 1906
1.8	**'Wildfowl'**	:	means the wildfowl specified in the Third Schedule
1.9	**'Deer'**	:	means the deer specified in the Fourth Schedule
1.10	**'Vermin'**	:	means the vermin specified in the Fifth Schedule
1.11	**'the Term'**	:	means the period of [ ] years from the (*date*)

1.12	**'the Fee'**	:	means the sum of [ ] (plus any VAT payable on that sum) and such sum as may be substituted therefore in accordance with the provisions of this Deed such sum to be paid in accordance with the terms hereof the first payment to be made on the (*date*) and subsequent payments to be made as hereinafter provided
1.13	**'the Return Date'**	:	means the 15th day of March in each year of the Term
1.14	**'Estate Management'**	:	means the right of the Grantor to carry out on the Land all normal acts (including but without prejudice to the generality hereof the creation and demolition of buildings construction of roads and paths and the laying of pipes and cables) which a prudent landowner would do to ensure good management of the Land
1.15	**'the Review Dates'**	:	mean the [*date*] and the [*date*]
1.16	**'the Interest Rate'**	:	means 4% above the base lending rate from time to time of the [ ] Bank plc
1.17	**'DEFRA'**	:	means the Department for the Environment Food and Rural Affairs (or any successor department)

2. Interpretation

Unless the context otherwise requires:-

2.1 words importing one gender shall be construed as importing any other gender

2.2 words importing the singular shall include the plural and vice versa

2.3 words importing persons shall include corporations

2.4 where there are two or more persons included in the expressions the Grantor and/or the Grantee the obligations of such persons under this Deed shall be deemed to be joint and several

2.5 the clause headings in this Deed shall not affect the construction of this Deed

2.6 any reference in this Deed to a statutory provision, Code of Conduct, Code of Practice, or anything of a similar nature shall be deemed to include any later provision substituted for or amending the same

3. Grant Exceptions and Reservations

In consideration of the Grantee's covenants set out in the Sixth Schedule and of payment of the Fee the Grantor grants to the Grantee the Shooting Rights over and upon the Land and the right to kill and take Ground Game and to shoot and trap all vermin TOGETHER WITH the Rights of Access and the right to enter on and pass over the Land for the exercise of the Shooting Rights and for the purpose of preserving and rearing the Game SUBJECT TO the provisions of the Sixth Schedule but RESERVING to the Grantor the rights of good Estate Management and the rights of all persons in occupation of any parts of the Land as tenants of the Grantor whether such rights arise under a tenancy or by general law TO HOLD the same unto the Grantee for the Term paying throughout the Term (subject as hereinafter provided) the Fee PROVIDED THAT on the Review Dates the Grantor may increase the Fee hereby reserved the Fee, the new fee payable following the relevant Review Date being the greater of the Fee and the sum calculated by reference to the Retail Price Index applicable at the Review Date such sum thereafter to be called 'the Fee'

4. Covenants by the Grantee

The Grantee covenants with the Grantor to comply with the covenants and conditions set out in the Sixth Schedule

5. Covenants by the Grantor

Subject to the Grantee complying with the covenants and conditions set out in the Sixth Schedule the Grantor covenants with the Grantee that it will comply with the covenants and conditions set out in the Seventh Schedule

6. Provisos

It is agreed that the provisions contained in the Eighth Schedule shall apply to this Deed

IN WITNESS whereof the parties have set their hands to this Deed the day and year first before written

FIRST SCHEDULE
THE LAND

ALL THAT farmland known as [ ] Farm, in the County of [ ] extending to some [ ] acres or thereabouts as the same is more particularly described below and delineated and edged red on the Plan

TOGETHER WITH a right of access to and egress from the Land over and along the roads marked brown on the Plan with or without vehicles

AND a right to park vehicles on the areas marked orange on the Plan

TOGETHER ALSO WITH the right to maintain a [secure store] in the location marked blue on the Plan

OS Number	**Acreages**

SECOND SCHEDULE
GAME

Such of the following (and no other) as are for the time being lawful to kill or take:-

Pheasant, partridge, woodcock, common snipe, and golden plover and any such species as are from time to time included in the quarry list contained in the Wildlife and Countryside Act 1981 ('the 1981 Act') Schedule 2 Part 1

THIRD SCHEDULE
WILDFOWL

Such of the following (and no other) as are for the time being lawful to kill or take:-

Canada geese, pinkfoot geese, greylag geese, whitefront geese, mallard, tufted duck, wigeon, teal, common pochard, pintail, gadwall, shoveler duck, golden eye, and geese and duck of such species as are from time to time included in the quarry list contained in the 1981 Act

FOURTH SCHEDULE
DEER

The Grantee shall NOT kill or take any of the following:-

Red deer, fallow deer, sika deer, roe deer, muntjac deer, Chinese water deer and any cross or hybrid between these species

FIFTH SCHEDULE
VERMIN

Such of the following (and no other) as are for the time being lawful to kill or take:-

Crows, rooks, jackdaws, magpies, jays, herring gulls, lesser black backed gulls, greater black backed gulls, feral pigeons, collared doves, wood pigeons, weasels, stoats, mink, foxes, rabbits, hares and grey squirrels and any such species as are from time to time included in the quarry list contained in the 1981 Act or listed on the General Licence WLF 18 under section 16(1)(j) and (k) and section 16(5) of the 1981 Act

SIXTH SCHEDULE
GRANTEE'S COVENANTS AND CONDITIONS

1. To pay the Fee and any VAT thereon by equal half yearly instalments payable in advance on the [*date*] and the [*date*] in each year

2. To indemnify the Grantor against all rates taxes assessments duties impositions and outgoings charged assessed or imposed on the Land or the owner or occupier thereof in respect of the Buildings or Shooting Rights AND any VAT chargeable in respect of any payment made by the Grantee under this Deed or in respect of any payment made by the Grantor on behalf of the Grantee

3. Not to reduce any payment by making any deduction from it or by setting any sum off against it

4. To pay interest on all monies due under this Deed which are not paid within seven days after they fall due

5. To keep in good condition and repair throughout the Term the Buildings

6. To share the occupation and use of the Buildings with the Grantor

7. To comply with the Regulations set out in the Ninth Schedule and ensure that all persons authorised by the Grantee to enter the Land and Buildings do so

8. To comply with the Code of Good Shooting Practice referred to in the Tenth Schedule and ensure that all persons authorised by the Grantee to enter the Land and Buildings do so

9. To comply with the terms of the Rearing and Release Policy set out in the Eleventh Schedule

10. To comply with the bio-security measures set out in the Twelfth Schedule and ensure that all persons authorised by the Grantee to enter the Land and Buildings do so

11. As far as possible to prevent:-

11.1 the creation of any new public or private right of way over the Land

11.2 anyone acquiring any other indefeasible right over any part of the Land and Buildings

11.3 any trespass on the Land and Buildings and to inform the Grantor of any persistent trespassing

12. To help the Grantor to take legal proceedings in the Grantee's name against trespassers and poachers to include but without prejudice to

the generality hereof the laying of information giving evidence and signing notices requiring people to leave or not enter the Land and Buildings

13. To indemnify the Grantor against all injuries, damage, proceedings, costs, expenses, claims and demands arising from:-

13.1 the exercise of the Shooting Rights granted by this Deed

13.2 any damage caused by Game on the Land

14. Not to assign underlet transfer mortgage charge part with or share with any other person or organisation the Sporting Rights granted by this Deed which is prohibited

15. To effect proper insurance against all liability arising out of the exercise of the Shooting Rights granted by this Deed and any damage caused by the Game on the Land including employer's and public liability cover and all other matters as the Grantor shall from time to time reasonably deem necessary with a reputable insurance company or underwriter, in such sum as the Grantor from time to time approves, and ensure that the Grantor's name is endorsed upon such policy as landowner and grantor of the Shooting Rights over the Land and must produce the insurance policy and evidence of payment of the last premium to the Grantor when reasonably requested so to do

[16. To pay all solicitors' and agents' fees which the Grantor incurs in preparing and granting this Deed, including all disbursements and other expenses, and the stamp duty land tax (if any) payable on this Deed]

SEVENTH SCHEDULE
GRANTOR'S OBLIGATIONS

1. Subject to its observance of the provisions of this Deed the Grantee may peaceably enjoy the Shooting Rights without interference from the Grantor or any person claiming under it

2. To ensure as far as practical that the method of Estate Management of the Land is conducted in such a way as to prevent harm to the Game

3. To take all reasonable steps to assist the Grantee in the apprehension and prosecution of persons found poaching or trespassing in search of game or the eggs of game

EIGHTH SCHEDULE
PROVISOS

1. The Grantor is entitled to obtain an independent report from the Game Conservancy Trust advisory service (or such other industry body as the Grantor shall deem appropriate for such purposes) at the Grantee's expense at any time it considers the Grantee is or may be in breach of the provisions of this Deed and may require the Grantee at

the Grantee's expense to comply with all or some of the recommendations (as the Grantor shall decide) contained in such report

2. The Grantor is entitled to terminate this Deed after giving the Grantee twenty one days' prior written notice whenever the Grantee:-

2.1 is twenty one days late in paying the Fee or any other monies payable under the terms of this Deed even if not formally demanded

2.2 has not complied with any obligation in this Deed

2.3 is adjudicated bankrupt or an interim receiver is appointed in respect of the property of the Grantee

2.4 executes any assignment for the benefit of or makes any arrangement with his creditors

3. The forfeiture of this Deed does not cancel any outstanding obligation which the Grantee owes the Grantor

4. The Grantor may determine the grant of the Shooting Rights granted over the whole or any part of the Land on giving 3 months previous notice in writing to the Grantee

5. The Grantor shall be at liberty to enter on all or any part of the Land and Buildings at any time for any purpose without making any abatement in the Fee hereby reserved

6. At the end or sooner determination of the Term the Grantee must peaceably yield up the Shooting Rights and the Buildings to the Grantor

7. If any dispute arises between the parties in relation to this Deed it shall be referred for the determination of a single arbitrator under the Arbitration Act 1996

8. The identity of the arbitrator is to be agreed in writing by the parties or failing such agreement within one month is to be nominated by the President for the time being of the Royal Institution of Chartered Surveyors on the application of either party

9. The Law of Property Act 1925 as amended by the Recorded Delivery Act 1962 applies to any notice served under this Deed

10. No terms of this Deed may be enforced solely by virtue of the Contracts (Rights of Third Parties) Act 1999

11. It is agreed that the provisions of the Landlord and Tenant Act 1954 are excluded from this Deed

NINTH SCHEDULE
THE REGULATIONS

The Grantee further agrees with the Grantor:-

1. to exercise the rights granted by this Deed in a proper and sporting manner and fully in accordance with The Code of Good Shooting Practice (the Code) referred to in the Tenth Schedule and in particular to ensure that all employees, servants and guests comply with the letter and spirit of the Code

2. to comply with all laws and byelaws relating to the Land and Buildings and in particular the provisions of the 1981 Act, the Environment Protection Act 1990, the Animal Welfare Act 2006 and the Foot-and-Mouth and Blue Tongue Disease Orders and any modifications or re-enactments relating thereto to ensure the natural flora and fauna are protected and preserved and the laws and byelaws from time to time current relating to the possession and use of poisonous substances

3. to exercise the Shooting Rights by legal methods only and during the proper seasons and not, without the written consent of the Grantor, to use any guns other than shotguns not exceeding 8 bore in calibre or airguns

4. not to cause any damage to the Land and Buildings woods plantations crops hedges fences banks and gates access roads footpaths and other property of the Grantor or its tenants and to pay full compensation for such damage as may be caused by the exercise of the Shooting Rights to the Grantor or any agricultural tenant for such damage as may be caused

5. to employ at his own expense a competent [part time] gamekeeper to supervise the Land and protect the Game and prevent poaching and trespassing on the Land throughout the Term

6. to take all reasonable steps:-

6.1 to keep down Ground Game so as to prevent damage to the Land woods crops plantations or standing timber belonging to the Grantor or its tenants

6.2 to control pests and Vermin

7. to prevent poaching or trespassing on the Land and Buildings throughout the Term

8. to protect preserve and keep a proper stock of Game on the Land

9. on or before the Return Date in each year of the Term to give the Grantor particulars of the numbers and types of Game and Wildfowl including Ground Game and Vermin killed during the year to that date

10. to preserve the nests eggs and young of the Game from destruction and interference

11. not do anything that is or may become a nuisance to the Grantor or its agricultural tenant or to the owner or occupier of any adjacent land or property

12. not to allow any persons to shoot over the Land in use by the Grantor as an active deposit ground or on which bank raising or other works of construction or maintenance are in progress without the express permission of the Grantor

13. not to object to any planning permission for which the Grantor may have applied or may apply during the Term

14. not to exercise the rights and privileges hereby granted so as to prejudicially affect or interfere with the Grantor or endanger any person undertaking works for the Grantor

15. to give immediate notice to the Grantor of any claims made by third parties in respect of the Shooting Rights or the Buildings and to permit the Grantor to take proceedings in the name of the Grantee (but at the expense of the Grantor) against any such claims

16. not to use any spring trap or snares except in accordance with the Pests Act 1954 and the Spring Traps Approval Order 1995

17. not to erect or suffer to be erected any constructions or buildings or release pens without the Grantor's prior written consent

18. not to use lead shot on that part of the Land on the north side of the canal or within the area covered by the Special Protection Area or Site of Special Scientific interest or within 100 metres of the boundary of the said land

19. not to obstruct any roads or access ways providing access to or egress from the Land or Buildings

20. no one person permitted to exercise the Shooting Rights by the Grantee shall obtain or attempt to obtain a bag of more than [25] birds in any one day's shooting

21. at the end or sooner determination of the Term to peaceably yield up the Shooting Rights and the Buildings to the Grantor

TENTH SCHEDULE
CODE OF PRACTICE

The Code of Good Shooting Practice which is overseen by a Steering Committee comprising representatives of the British Association for Shooting and Conservation, Countryside Alliance, Campaign for Shooting, Country Land and Business Association, Devon and Cornwall Association of Sporting Shoots Game Farmers' Association, National Game Dealers' Association, National Gamekeepers' Organisation, Scottish Gamekeepers' Association, The Game Conservancy Trust, and The Scottish Landowners' Federation

ELEVENTH SCHEDULE
REARING AND RELEASE POLICY

1. The Grantee may not release on the Land in any one year more than the following numbers of the following species:-

[ ] Pheasants

[ ] Partridge

[ ] Duck

2. The Grantee must not exceed the carrying capacity of the Land as advised by the Game Conservancy and should follow the Code of Practice with regard to stocking and rearing

3. No pheasants shall be released after the 1 September in any year and no catching up for re-release during the same season shall take place

4. All partridge should be released from their pens before shooting commences on the Land. Caller birds shall not be held except during the release itself

TWELFTH SCHEDULE
BIO-SECURITY MEASURES

1. Whenever so required by DEFRA or when it is practical so to do (as the case may be) the Grantee shall ensure that:-

1.1 all shoot participants (guns, beaters, etc) clothing and footwear are clean prior to commencement of a shoot

1.2 all shoot participants walk through a disinfectant foot dip before and after a shoot and where applicable when passing from one holding to another on a shoot

1.3 all vehicles are clean and free of mud prior to entering a shoot

1.4 only vehicles that have had their wheels sprayed with approved disinfectant will be allowed to drive around a shoot and where applicable when passing from one holding to another on a shoot

1.5 all dogs are kept under effective control at all times and are not allowed to stray

1.6 all shoot participants avoid all contact with livestock before, during and after a shoot

1.7 no waste food or litter is left on a shoot

1.8 it maintains records of all participants in a shoot

1.9 in any infected area it complies with any and all licensing requirements

1.10 it complies with any other regulations or recommendations relating to bio-security that may be imposed and/or issued either by DEFRA, the Grantor or any shoot organiser employed by the Grantee

SIGNED as a Deed by
in the presence of:-

SIGNED as a Deed by
in the presence of:-

Chapter 10

MILK QUOTA

INTRODUCTION

PRECEDENTS

1 Supplemental deed apportioning milk quota between tenanted and owned holdings
2 Deed regulating ownership of quota between the Landlord, Tenant and Producer
3 Declaration of trust

INTRODUCTION

Since the first edition of this book the economic importance of milk quotas has fallen away dramatically, although they are still going to be an item, and something for advisers to know about and advise on, until the present regime comes to an end in 2014.

The dramatic fall in value during the last few years has meant that the buoyant market which existed previously has certainly gone and such a market as there is is now very small indeed.

Although the permanent and short-term transfer market is only a shadow of its former self both in terms of values and numbers of transactions taking place (and hence the omission of any documentation relating to these matters from this second edition), the ongoing existence of milk quota is something which needs to be remembered when dealing with sale and purchase of agricultural property, the granting of new tenancies and termination of existing tenancy agreements, as well as in the context of partnership agreements and other similar arrangements. Where appropriate, the authors have made reference to milk quota in other parts of this work and provided suggested wording for the Precedents concerned.

In addition to the provisions which have been suggested for the above documents, the authors have also included in this chapter the three precedents which, together with the earlier documents, they believe to be the ones which readers are going to meet most usually in practice at the present time and in the foreseeable future.

In connection with the disposal of all or part of a dairy farm, or land which has been used for milk production purposes during the last 5 years, the quota implications must still be borne in mind. The apportionment of milk quota rules considered in the cases of *Puncknowle Farms v Kane*[1] and *Posthumous v Oostervoud*[2] need to be borne in mind. Basically, milk quota attaches to land (unlike entitlements, as will have been seen in Chapter 1), and therefore the starting position is that when dairy land comes to be sold the milk quota should be transferred with it. That does not always happen, of course, and if it is the intention to sell bare land and for the seller to retain the milk quota care needs to be taken to ensure that correct provisions are incorporated into the contract for sale, as will be evident from the precedents which appear in Chapter 2. It is simply not enough to include in the contract, 'No milk quota attaches to the Land'. The correct provisions need to be incorporated into the contract documentation.

Moving from one holding to another is also, potentially, fraught with difficulty – see *Holdcroft v Staffordshire County Council*[3] for something of a horror story.

In dealings between landlord and tenant care needs to be taken if a client wishes to acquire additional milk quota to ensure that this is permissible under the terms of the tenancy agreement. It would also be sensible, at that stage, to arrange for the landlord to enter into a form of deed as in Precedent 2 below to avoid there being any arguments, at the end of the tenancy, with regard to the fact that the quota which the tenant acquires is purchased quota and compensatable as such under the terms of Sch 1 to the Agricultural Act 1986.

---

1 [1985] 3 All ER 790.
2 [1992] 2 CMIR 336.
3 [1994] 28 EG 131.

PRECEDENTS

PRECEDENT 1 SUPPLEMENTAL DEED APPORTIONING MILK QUOTA BETWEEN TENANTED AND OWNED HOLDINGS

THIS DEED is made the [ ] day of [ ] 20[ ]

BETWEEN:-

(1) [*name*] of [*address*] ('the Tenant')

(2) [*name*] of [*address*] ('the First Landlord')

and

(3) [*name*] of [*address*] ('the Second Landlord')

WHEREAS:-

In this Deed:-

1.1	**The First Tenancy Agreement**	:	means the First Tenancy dated [*date*] made between [*name*] [*name*]
1.2	**The Second Tenancy Agreement**	:	means Agreement for the Tenancy dated [*date*] made between [*name*] [*name*]
1.3	**The First Land**	:	means the Land let by the First Tenancy Agreement
1.4	**The Second Land**	:	means the Land let by the Second Tenancy Agreement
1.5	**The Tenants Land**	:	means all that land belonging to the Tenant and comprised within the Holding other than the First Land and the Second Land
1.6	**The Holding**	:	means that land farmed by the Tenant comprising the First Land, the Second Land, and the Tenants Land
1.7	**The Relevant Quota**	:	means the milk quota registered in the name of the Tenant with the RPA amounting to [ ] litres with a butterfat base of [ ]%
1.8	**DEFRA**	:	means the Department for the Environment Food and Rural Affairs or any successor department
1.9	**RPA**	:	means the Rural Payment Agency or any successor body carrying out similar functions

1.10 **The Regulations** : means the Dairy Produce Quotas Regulations 2005

2. This Deed is supplemental to the First Tenancy Agreement and the Second Tenancy Agreement

3. Subject to the provisions of this Deed the parties have agreed to the apportionment of the Relevant Quota throughout the Holding

NOW THIS DEED WITNESSES as follows:-

1. the Relevant Quota is apportioned to the First Land, the Second Land and the Tenants Land in accordance with the Regulations and as agreed between the parties the details of such agreed apportionment being set out in the Schedule

2. the First Landlord and the Second Landlord each acknowledges the Tenant's right to compensation in full in respect of such parts of the Relevant Quota as attach to the First Land and the Second Land held of them by the Tenant in accordance with the provisions of the Agriculture Act 1986 on the termination of the First Tenancy Agreement and the Second Tenancy Agreement respectively and that nothing in this Deed shall affect those rights

3. The Tenant covenants with the First Landlord:-

3.1 not to transfer charge lease or otherwise remove from the First Land the whole or any part of that part of the Relevant Quota which presently attaches to it without the previous consent in writing of the First Landlord

3.2 to take all such steps as are necessary to ensure that that part of the Relevant Quota which presently attaches to it continues to attach to the First Land (and to no other property) for the duration of the First Tenancy Agreement to the intent that the same will be transferable to the First Landlord or to an incoming tenant on the termination of the First Tenancy Agreement

3.3 not to do or omit anything the doing or omission of which would cause a loss forfeiture or reduction in that part of the Relevant Quota which attaches to the First Land

3.4 to carry out during the continuance of the First Tenancy Agreement any obligation imposed on the Tenant as occupier or producer with regard to notices compliance with directions payment of levies maintenance of records or otherwise in connection with milk quota under the Regulations and/or the Agriculture Act 1986

3.5 to supply the First Landlord with copies of any notices served on the Tenant which affect or may affect that part of the Relevant Quota which attaches to the First Land

3.6 to co-operate with the First Landlord on the termination of the First Tenancy Agreement to ensure that the First Landlord or a person nominated by him will be registered by the RPA in respect of that part of the Relevant Quota which attaches to the First Land

4. The Tenant covenants with the Second Landlord:-

4.1 not to transfer charge lease or otherwise remove from the Second Land the whole or any part of that part of the Relevant Quota which presently attaches to it without the previous consent in writing of the Second Landlord

4.2 to take all such steps as are necessary to ensure that that part of the Relevant Quota which presently attaches to it continues to attach to the Second Land (and to no other property) for the duration of the Second Tenancy Agreement to the intent that the same will be transferable to the Second Landlord or to an incoming tenant on the termination of the Second Tenancy Agreement

4.3 not to do or omit anything the doing or omission of which would cause a loss forfeiture or reduction in that part of the Relevant Quota which attaches to the Second Land

4.4 to carry out during the continuance of the Second Tenancy Agreement any obligation imposed upon the Tenant as occupier or producer with regard to notices compliance with directions payment of levies maintenance of records or otherwise in connection with milk quota under the Regulations and/or the Agriculture Act 1986

4.5 to supply the Second Landlord with copies of any notices served on the Tenant which affect or may affect that part of the Relevant Quota which attaches to the Second Land

4.6 to co-operate with the Second Landlord on the termination of the Second Tenancy Agreement to ensure that the Second Landlord or a person nominated by him will be registered by the RPA in respect of that part of the Relevant Quota which attaches to the Second Land

[5. If at any time any of the Relevant Quota is charged to any person firm or company to secure any form of indebtedness and the First Landlord or the Second Landlord (as the case may be) pursuant to such charge pays compensation in respect of that quota to the chargee such payment shall be deemed to satisfy pro tanto any claim for compensation which might otherwise be due to the Tenant in respect of the same quota litreage]

6. So far as relates to that part of the Relevant Quota not attaching to either the First Land or the Second Land the First Landlord and the Second Landlord respectively covenant with the Tenant that the First Landlord and the Second Landlord (as the case may be) will sign all necessary forms and do all necessary things to enable the Tenant to transfer that part of the Relevant Quota attaching to the Tenant's Land to one or more third parties in the event of the Tenant wishing to transfer the same

7. The parties covenant each with the other not to make any further representation(s) to the RPA DEFRA or other statutory authority nor to make any further claims either personally or through any third party or parties in respect of the Relevant Quota

8. If notwithstanding the terms of this Deed any claims are made outside the provisions of this Deed in respect of the Relevant Quota or any part or parts thereof consequent upon the termination of the First Tenancy Agreement the Second Tenancy Agreement or either of them or otherwise resulting in an apportionment of the Relevant Quota in favour of the First Landlord or the Second Landlord or the Tenant pursuant to a decision of an Arbitrator appointed under the Regulations or otherwise the First Landlord the Second Landlord or the Tenant (as the case may be) will pay to the other party or parties all lawfully recoverable losses thereby incurred by the other party or parties including all costs incurred by the other party in the acquisition of milk quota of comparable butterfat base and usage to replace such quota

9. To the extent that quantification of the losses referred to in clause 8 is or may be impossible whether by reason of milk quota having no independent value or otherwise the recoverable losses shall be determined by reference to any reduced value of the Holding (or any part of it) consequent upon the loss of the milk quota from it

10. In the event of there being an additional allocation of or reduction in milk quota imposed by DEFRA as Agent for the EU or otherwise howsoever such additional allocation or reduction (as the case may be) shall be borne rateably by all parts of the Relevant Quota

11. The provisions of this Deed shall so far as is legally possible benefit and bind the successors in title of the parties hereto their assigns and any insolvency practitioner appointed in respect of the affairs of any of them

12. References in this Agreement to the Agriculture Act 1986 and to the Dairy Produce Quotas Regulations 2005 shall apply to:-

12.1 such legislation and/or subordinate legislation as amended

12.2 any substituted legislation and/or subordinate legislation

13. In this Deed where the context so admits words importing the singular only shall include the plural and vice versa and words importing the masculine gender only shall include the feminine and vice versa and where two or more persons are included in the words 'the Tenant' 'the First Landlord' or 'the Second Landlord' any agreements or obligations entered into by them shall be deemed to be entered into jointly and severally

14. If any dispute arises under this Deed either party may give notice to the other in writing specifying the dispute and requesting that agreement be reached on the identity of a single arbitrator to be

appointed to determine the dispute in accordance with the provisions of the Arbitration Act 1996 and if no arbitrator is appointed by agreement within one month of the notice either party may make a written request to the President for the time being of the Royal Institution of Chartered Surveyors for him to appoint an arbitrator to determine the matter in dispute

IN WITNESS whereof the parties have set their hands to this Deed the day and year first before written

SCHEDULE

First Land	[ ] Litres
Second Land	[ ] Litres
Tenant's Land	[ ] Litres

SIGNED as a Deed by
in the presence of:-

SIGNED as a Deed by
in the presence of:-

SIGNED as a Deed by
in the presence of:-

PRECEDENT 2 DEED REGULATING OWNERSHIP OF QUOTA BETWEEN THE LANDLORD, TENANT AND PRODUCER

THIS DEED is made on this [ ] day of [ ] 20[ ]

BETWEEN:-

(1) [*name*] of [*address*] ('the Landlord')

(2) [*name*] of [*address*] ('the Tenant')

and

(3) [*name*] of [*address*] ('the Producer')

WHEREAS:-

1. This Deed is SUPPLEMENTAL to a tenancy agreement ('the Tenancy Agreement') made the [*date*] between the Landlord of the one part and the Tenant of the other part [as subsequently amended] relating to [*holding details*] ('the Holding')

2. The Landlord is the landlord and the Tenant is the tenant of the Holding under the Tenancy Agreement

3. The Producer is registered with the Rural Payments Agency ('the RPA') under Producer Number [*no*] as a producer of milk on the Holding the Producer's registered quota prior to entering into this Deed being [ ] litres

4. The Producer wishes to purchase further milk quota to be added to the milk quota registered in his name

NOW THIS DEED WITNESSES as follows:-

1. If the Producer purchases milk quota after the date of this Deed he shall notify the Landlord of such milk quota purchased ('the Purchased Quota')

2. It is hereby agreed and declared that the Purchased Quota shall be deemed to be and for all purposes shall be treated as Transferred Quota within the meaning of paragraph 2(2) Schedule 1 to the Agriculture Act 1986 and that in respect of the Purchased Quota the Producer shall be deemed to be and shall be treated as the tenant for the purposes of section 13 of the said Act but not further or otherwise

3. If the Producer wishes to dispose of all or any part of the Purchased Quota during the term granted by the Tenancy Agreement the Landlord will facilitate such disposal by giving all necessary consents (including permitting the Tenant to grant a short-term farm business tenancy notwithstanding the terms of the Tenancy Agreement) and/or taking any other action required by the RPA the Department for the Environment Food and Rural Affairs ('DEFRA') or any other statutory authority to facilitate such disposal

4. The entirety of the proceeds of sale of the Purchased Quota will be payable to and belong to and retained by the Producer

5. The provisions of this Deed are without prejudice to the rights of the parties in respect of the milk quota now registered in the name of the Producer in respect of the Holding on the termination of the Tenancy Agreement and shall bind so far as legally possible the respective successors in title of the parties and the expressions the Landlord the Tenant and the Producer shall be construed appropriately

6. The Landlord and the Producer shall co-operate with each other in the signing and lodging of all necessary documents including any required by DEFRA, the RPA or other statutory authority to ensure so far as possible that there shall be no transfer of Purchased Quota from the Producer to the Landlord

7. If notwithstanding the terms of this Deed the whole or any part of the Purchased Quota is transferred to or becomes vested in the Landlord howsoever or if the Landlord receives all or any part of the proceeds of sale of the Purchased Quota the Landlord shall pay to or reimburse the Producer the full value thereof

8. If any dispute arises under this Deed either party may give notice to the other in writing specifying the dispute and requesting that agreement be reached on the identity of a single arbitrator to be appointed to determine the dispute in accordance with the provisions of the Arbitration Act 1996 and if no arbitrator is appointed by agreement within one month of the notice either party may make a written request to the President for the time being of the Royal Institution of Chartered Surveyors for him to appoint an arbitrator to determine the matter in dispute

IN WITNESS whereof the parties hereto have set their hands to this Deed the day and year first before written

SIGNED as a Deed by
in the presence of:-

SIGNED as a Deed by
in the presence of:-

SIGNED as a Deed by
in the presence of:-

PRECEDENT 3 DECLARATION OF TRUST

THIS DEED is made the [ ] day of [ ] 20[ ]

BETWEEN:-

(1) [*name*] of [*address*] ('the Tenant')

and

(2) [*name*] of [*address*] ('the Producer')

WHEREAS:-

1. The Tenant is the tenant of certain land known as [*holding description*] ('the Holding') which is farmed by the Producer

2. The Producer is registered with [*name*] under Producer Number [*no*] as a producer of milk on the Holding

3. The Producer's registered milk quota at the date of this Deed is [ ] litres with a butterfat base of [%] ('the Registered Quota')

4. Having regard to the reductions which have been applied to milk quotas generally the milk quota now remaining of that originally allocated to the Producer ('the Allocated Quota') comprises [ ] litres with a butterfat base of [%] and the milk quota subsequently purchased by the Producer ('the Purchased Quota') comprises [ ] litres with a butterfat base of [%]

[5. No change has been made to the butterfat level of the Registered Quota to date] [The present butterfat base is [ ]%]

6. It has been agreed that the Purchased Quota which was paid for by the Producer belongs beneficially to the Producer absolutely [and a deed was completed on the [*date*] with the landlord of the Holding at that time protecting the Producer's position in that regard]

NOW THIS DEED WITNESSES as follows:-

1. It is hereby agreed and declared that the Producer is beneficially entitled to the Purchased Quota

2. If the Producer wishes to dispose of all or any part of the Purchased Quota the Tenant will do everything in his power to achieve the same including obtaining the Landlord's consent to a sub-letting and to sign any relevant forms

3. The entirety of the proceeds of sale of the Purchased Quota shall be payable to and belong to and be retained by the Producer

4. The parties agree that the Allocated Quota belongs to the Tenant and the Producer agrees to sign all relevant forms to facilitate any dealing therewith at the Tenant's request or the payment of compensation in respect thereof to the Tenant

5. The entirety of the proceeds of sale of the Allocated Quota or compensation received in respect thereof shall be payable to and belong to and be retained by the Tenant

6. The provisions of this Deed shall so far as legally possible bind the respective successors in title of the parties hereto

7. If notwithstanding the terms of this Deed the Allocated Quota or any sale proceeds or compensation payable in respect of it are transferred or paid to the Producer the Producer shall compensate the Tenant to the extent that the Producer has been enriched thereby

8. To the extent that the Purchased Quota the proceeds thereof or compensation payable in respect of it are received by the Tenant the Tenant shall compensate the Producer to the extent that the Tenant has been enriched thereby

9. If any dispute arises under this Deed either party may give notice to the other in writing specifying the dispute and requesting that agreement be reached on the identity of a single arbitrator to be appointed to determine the dispute in accordance with the provisions of the Arbitration Act 1996 and if no arbitrator is appointed by agreement within one month of the notice either party may make a written request to the President for the time being of the Royal Institution of Chartered Surveyors for him to appoint an arbitrator to determine the matter in dispute

IN WITNESS whereof the parties have set their hands to this Deed the day and year first before written

SIGNED as a Deed by
in the presence of:-

SIGNED as a Deed by
in the presence of:-

Chapter 11

HORSES

INTRODUCTION

PRECEDENTS
1 Grazing agreement (licence)
2 Grazing agreement (alternative form)
3 Livery agreement – DIY
4 Livery agreement – landowner dealing with feeding and exercising of horses
5 Livery agreement (shorter form)

INTRODUCTION

The Town and Country Planning Act 1990[1] provides that 'the use of any land for the purpose of agriculture or forestry … and the use for any of those purposes of any building occupied together with land so used' does not constitute development and does not require planning permission. The use of land for grazing by horses is agricultural user within the meaning of the Town and Country Planning Act 1990,[2] even though the animals are not livestock. However, this exemption applies only where the horses are grazing the land primarily in order to be fed and does not cover the position where they are being kept on that land for recreation or for exercise.

Planning permission should be obtained for the erection of stables on any land unless the development is within the curtilage of a dwelling house and so permitted by the Town and Country Planning General Permitted Development Order 1995.[3] Where stables have been erected without permission and have stood for more than 4 years without any action being taken or notice being served by the local planning authority, the stables are protected from enforcement action.

Where a horse enterprise is run, for example, as livery or a riding school, planning permission is required for the erection or change of use of any

1 Section 55(2)(e).
2 Section 366(1).
3 SI 1995/418.

buildings and of the land. The enterprise as a whole establishes a non-agricultural activity on all of the land. On this basis, even the grazing of the land is not agricultural as it is on land which forms part of a unit used for a non-agricultural purpose.

Grazing by most horses is not agricultural use for inheritance tax purposes.

Precedent 1 is a precedent for a grazing licence. The owner retains possession and the use of the land is restricted to grazing. As exclusive possession is not given to the occupier no tenancy is created.

Precedent 2 grants the grazier exclusive possession and so creates a tenancy. The use of any building is additional to the grazing so planning permission is required. As no business use is permitted, a farm business tenancy is not created.

The livery agreements are a form of contract of bailment. They give the owner of the horse, the bailor, no interest in the land.

In Precedent 3 the Livery Operator merely provides stabling for the horse.

In Precedent 4 the Livery Operator has a more active role which includes feeding and exercising the horses.

In Precedent 5 a shorter form of agreement is provided.

PRECEDENTS

PRECEDENT 1 GRAZING AGREEMENT (LICENCE)

PARTICULARS

Date of Agreement:

The Owner	:	
The Grazier	:	
The Land	:	the land situate at and known as [ ] at [ ] extending to [ ] [acres]/[hectares] or thereabouts as the same is more particularly described in Schedule 1 and delineated and edged red on the Plan
The Plan	:	the plan annexed to this Agreement
The Grazing Fee	:	£[ ] payable on [*date*] [by [ ] equal instalments of £ [ ] on [*date*] and [*date*]]
The Grazing Period	:	the period starting on [*date*] and expiring on [*date*]
Rights	:	[the right to keep no more than [ ] horses the property of the Grazier on the Land for grazing purposes only] [and] [the right to mow the Land [once] [twice] during the Grazing Period and to take away the grass] [TOGETHER WITH the right in common with the Owner to obtain access to the Land along the route shown coloured brown on the Plan]
Horses	:	means horses, ponies, mules and donkeys belonging to the Grazier
The Single Payment Scheme ('SPS')	:	the single payment scheme governed by Council Regulation (EC) No 73/2009 and all the relevant subsequent Commission and other regulations administrative provisions statutory guidance or codes of practice implemented by the EC and/or the UK Government ('the Regulations')

Where these words and phrases are used in this Agreement they shall have the meanings defined above

THIS AGREEMENT is made on the Date of Agreement

BETWEEN:-

(1) [*name*] of [*address*] ('the Owner')

and

(2) [*name*] of [*address*] ('the Grazier')

WHEREBY IT IS AGREED as follows:-

1. The Rights

The Owner grants to the Grazier the following:-

1.1 the Rights for the Grazing Period in accordance with the terms of this Agreement

1.2 the right to enter onto and to use the Land only to the extent necessary to exercise the Rights and for no other purposes and on the terms set out in this Agreement

2. Ownership of the Land

For the avoidance of doubt, the Land remains within the possession and control of the Owner throughout the Grazing Period

3. Grazier's agreements

The Grazier agrees with the Owner as follows:-

3.1 to pay the Grazing Fee in advance on [*insert date*]

3.2 by way of additional consideration to pay or indemnify the Owner in respect of a reasonable proportion in accordance with use of all water rates water charges drainage rates and any other rates taxes and outgoings in respect of the Land during the Grazing Period

3.3 not to do or cause or permit to be done or to continue anything on or near the Land which may be or become a nuisance or inconvenience or cause injury loss damage or annoyance to the Owner or any other person

3.4 to use the Land for the exercise of the Rights and for no other purpose

3.5 to observe and comply with any and every legal requirement including any and every statute regulation by-law order direction and official code of practice governing or affecting the exercise of the Rights including the keeping of the Horses on the Land

3.6 not to provide supplementary feeding without the prior consent in writing of the Owner other than the provision of mineral blocks

3.7 to use best endeavours to ensure that the Horses graze evenly over the Land and that no part of the Land is overgrazed nor becomes poached during wet weather and if the Owner or his agent confirms that damage is being caused to remove the Horses immediately

3.8 not bring on to or allow to remain on the Land any dangerous or diseased Horse and to remove as soon as possible any infected or troublesome Horse from the Land

3.9 not to bring more than the number of Horses referred to in the Rights onto the Land

3.10 not to use the Land for the purposes of any trade or business

3.11 not to undertake any cultivations or drainage works nor use any fertilisers or manures, pesticides or herbicides on the Land except as set out in 3.12 below

3.12 keep the Land clear of thistles and other noxious weeds if necessary by the application of such herbicides as are approved by the Owner using a weed-wiper or spot treatment only PROVIDED THAT the Grazier obtains the approval of the Owner before carrying out any such operation and having obtained such approval carries out the operation strictly in accordance with its terms

3.13 not to destroy or damage any trees, hedges, fences, gates, watercourses, ponds, ditches, historical or archaeological features on the Land and ensure that no such damage is caused by the Horses

3.14 not do or cause or permit to be done or omit to do or to be done anything by which any policy of insurance of the Owner would or might be prejudicially affected

3.15 not permit or allow his Horses to stray from the Land and to indemnify the Owner against any claims arising from such straying

3.16 to insure to an adequate level of cover (such cover to be for a sum of not less than [two/five] million pounds) against liability to third parties for loss or damage arising in relation to the Holding with an Insurance Company approved by the Owner (such approval not to be unreasonably withheld) [in the joint names of the Owner and the Tenant] [in the name of the Tenant but with a note of the Owner's interest endorsed on the policy]

3.17 on termination of this Agreement forthwith remove all his Horses from the Land

4. Termination

4.1 The Owner may terminate this Agreement at any time by giving the Grazier [one] month's notice in writing sent to the Grazier's address as set out in the Particulars, and on the expiry of such notice this Agreement will terminate immediately and the Grazier will remove the Horses and all (if any) other things belonging to him from the Land

4.2 If this Agreement is terminated in accordance with paragraph 4.1, the Owner will refund to the Grazier a proportion of any Grazing Fee paid in advance such proportion to be calculated on a pro rata basis for each day of the Grazing Period in respect of which it was paid remaining unexpired following termination

4.3 This Agreement shall terminate immediately if any of the following happen namely:-

4.3.1 an administration order is made against the Grazier

4.3.2 a receiver or manager is appointed in respect of the Grazier's affairs or the whole or any part of the Grazier's property or undertaking

4.3.3 the Grazier is the subject of a Bankruptcy Petition or Bankruptcy Order

4.3.4 the Grazier dies or becomes incapable by reason of mental or physical illness of discharging his obligations hereunder

4.3.5 the Grazier enters into any arrangement or composition with his creditors (including for the avoidance of doubt any voluntary arrangement within the meaning of Part I or Part VIII of the Insolvency Act 1986)

4.3.6 the interest of the Grazier under this Agreement is taken in execution

4.3.7 the Grazier commits any serious or persistent breach of this Agreement and after the Owner has given written notice to the Grazier of such breach or breaches the Grazier fails within such reasonable period as the Owner may specify to rectify such breach or breaches (if capable of rectification)

5. Personal Licence

This Agreement is personal to the Grazier and neither the Agreement nor the Rights conferred by it are capable of being assigned or otherwise dealt with

6. Single Payment Scheme

[The Owner has not registered the Land for the Single Payment Scheme pursuant to the Regulations and the Grazier agrees that he will not use the Land as part of his Holding (as defined in the Regulations) in connection with any claim the Grazier may make for single payment under the Single Payment Scheme during the Grazing Period]

[The parties agree that the Owner shall be the farmer and so the claimant and the one entitled to receive the single payment under the Single Payment Scheme and the Grazier shall make no claim under the Single Payment Scheme in respect of the Land]

[The Owner shall be entitled to retain all payments he received from the Rural Payments Agency (or other authority) under the Single Payment Scheme relating to the Land]

7. Arbitration

If any dispute arises under this Agreement either party may give notice to the other in writing specifying the dispute and requesting that agreement be reached on the identity of a single arbitrator to be appointed to determine the dispute in accordance with the provisions of the Arbitration Act 1996 and if no arbitrator has been appointed by agreement within one month of the notice either party may make a written request to the President for the time being of the Royal Institution of Chartered Surveyors for him to appoint an arbitrator to determine the matter in dispute

AS WITNESS the hands of the parties hereto the day and year first before written

SCHEDULE
THE LAND

[OS] [NG] Number	**Description**	**[Acres]**	**[Hectares]**
	Total		

SIGNED by
in the presence of:-

SIGNED by
in the presence of:-

PRECEDENT 2 GRAZING AGREEMENT (ALTERNATIVE FORM)

THIS AGREEMENT is made the [ ] day of [ ] 20[ ]

PARTICULARS

1.1	**the Owner**	:	
1.2	**the Grazier**	:	
1.3	**the Land**	:	ALL THAT land situate at [ ] as the same is delineated and edged red on the plan annexed hereto ('the Plan')
1.4	**the Period of Occupation**	:	the period beginning on the [*date*] and ending on the [*date*]
1.5	**the Grazing Fee**	:	the sum of [ ] pounds (£[ ])
1.6	**the Payment Date**	:	the [*date*]

2. Interpretation

2.1 Unless the context otherwise requires:-

2.1.1 words importing one gender shall be construed as importing any other gender

2.1.2 words importing the singular shall include the plural and vice versa

2.1.3 words importing persons shall include corporations

2.1.4 where there are two or more persons included in the expressions the Owner and/or the Grazier the obligations of such persons under this Agreement shall be deemed to be joint and several

2.2 The clause headings in this Agreement shall not affect the construction of this Agreement

2.3 Any reference in this Agreement to a statutory provision shall be deemed to include any later provision substituted for or amending the same

3. Right to graze

The Grazier shall have the right to occupy the Land for grazing purposes only during the Period of Occupation and may also but without any further payment use the Landlord's building(s) on the Land to provide his horses and ponies with shelter

4. Grazier's covenants

The Grazier shall:-

4.1 pay to the Owner the Grazing Fee on or before the Payment Date and any other payments referred to in this Agreement within seven days of being requested to make such payment by production of an invoice by the Owner

4.2 use the Land for the purpose of grazing by horses and ponies belonging to him only and to no-one else ('the Horses') and for no other purpose

4.3 graze and use the Land in a good and husband like manner

[4.4 keep the building(s) on the Land in no worse condition that at the start of the Period of Occupation]

4.5 not allow the Land to be entered upon or in any way used by any vicious or diseased Horses. In the event of the Grazier admitting any such Horses on to the Land in contravention of this sub-clause the Owner may treat the Horses as trespassers and impound them at the expense of the Grazier who shall be liable for all loss or damage caused by them either by straying or in any other way

4.6 not allow the Land to become poached by treading during wet weather conditions and if the Owner or his agent shall certify that any damage is being caused then the Grazier shall upon demand immediately remove the Horses

4.7 not permit any trespass on the Land but shall not obstruct any rights of way used or enjoyed over the Land prior to the granting of this Agreement

4.8 keep all fences ditches and watercourses in proper stock proof condition so as to prevent the Horses straying or being injured and indemnify the Owner against all costs claims and demands made by the owners or occupiers of adjoining land or any other person for damages or other money arising from the escape from the Land of all or any of the Horses placed by the Grazier upon the Land

4.9 preserve all trees or bushes on the Land from injury by the Horses or otherwise

4.10 not sub-let the grazing rights granted by this Agreement nor place upon the Land any agisted animals

4.11 not erect any building or structure, nor place any advertisement on the Land

5. Owner's Agreement

The Owner agrees that the Grazier shall have a right:-

5.1 to take a supply of water through the supply pipe ('the Pipe') in the approximate position shown by a green line on the Plan the Grazier to pay for all water so consumed and to bear all costs incurred in maintaining and repairing the Pipe

5.2 of way for himself his agents and those authorized by him at all times with or without cars, horseboxes, caravans or animals over the accessway shown coloured brown on the Plan [subject to the Grazier maintaining and repairing the same]

6. Owner's lien

The Owner shall have a lien upon all the Horses for the time being grazing on the Land for any sum owing or expense incurred for which under this Agreement the Grazier is liable and this lien may be enforced by the sale of the Horses [*Note: the difficulty in exercising a lien because of the requirement for each horse to be accompanied by its own passport at the time of sale.*]

7. Owner's right of entry

The Owner and any person he authorises shall have the right to enter the Land at any time for any reasonable purpose(s) other than for grazing animals there

8. Termination of agreement

8.1 If the Owner requires the whole or any part of the Land for any purpose whatsoever other than for grazing animals on it he shall have the right to determine this Agreement in relation to either the whole or any part of the Land upon giving [one month's] written notice to the Grazier whereupon the Owner shall refund to the Grazier a proportionate part of the Grazing Fee

8.2 The Owner may by written notice determine this Agreement immediately on the death of the Grazier or on any breach by the Grazier of any of his obligations under this Agreement

9. No undertaking for new agreement

It is expressly agreed between the parties and hereby acknowledged by the Grazier that the Owner does not undertake to repeat this Grazing Agreement for another period

AS WITNESS the hands of the parties hereto the day and year first before written

SIGNED by
in the presence of:-

SIGNED by
in the presence of:-

PRECEDENT 3 LIVERY AGREEMENT – DIY

THIS AGREEMENT is made the [ ] day of [ ] 20[ ]

BETWEEN:-

(1) [*name*] of [*address*] ('the Livery Operator')

and

(2) [*name*] of [*address*] ('the Licensee')

NOW IT IS AGREED as follows:-

1. Agreement for Livery

The Livery Operator agrees to provide board and stabling for the horse specified in Schedule 1 belonging to or in the possession of the Licensee ('the Horse') on the terms and conditions appearing below

2. Warranty

The Licensee warrants that the Horse is in a general state of good health and free from any infectious disease at the commencement of the Period of Livery and that it is of a reasonable demeanour and temperament

3. **Period of Livery**

The stabling and board will be provided for the Horse by the Livery Operator from [Monday] the (*date*) to [Sunday] the (*date*) and thereafter from week to week until terminated by either party giving one week's notice to the other to expire on the [Sunday] next following the giving of the notice ('the Period of Livery')

4. **Livery Fee**

The Licensee must pay the Livery Operator the sum of [ ] pounds (£[ ]) in respect of the Period of Livery, such sum being payable in advance on [Monday] of each week, unless otherwise agreed, together with such sum as is due and owing from the Licensee to the Livery Operator in respect of all hay, straw, shavings, feedstuffs, supplements and other items provided by the Livery Operator during the previous week, such payment to be made in arrears on [Monday] of each week in respect of the previous week ('the Livery Fee')

5. **Increase of Livery Fee**

The Livery Operator must serve at least [four] weeks written notice on the Licensee to increase the Livery Fee

6. **VAT**

Neither the Livery Fee nor any other sums payable to the Livery Operator in this Agreement include VAT which shall be added in the event of the above payments being liable to VAT at the rate of [15%]

7. **Interest**

The Licensee must pay interest on any sums due under this Agreement for more than 7 days at the rate of 3% above the base lending rate from time to time of NatWest Bank plc

8. **Provision of stabling and grazing by the Livery Operator**

8.1 *Permission to stable the Horse*

The Livery Operator permits the Licensee to keep the Horse during the Period of Livery at [ ] ('the Property')

8.2 *Stable and Paddock*

The Licensee shall occupy such stable at the Property as allocated from time to time by the Livery Operator and shall also be permitted to use such field(s) as may from time to time be notified to the Licensee as being available for the grazing and exercising of the Horse

8.3 *Washing Down*

If the Licensee wishes to wash down the Horse, the Licensee must use the washing down area designated for this purpose by the Livery Operator

8.4 *Water*

The Livery Operator will ensure that a proper supply of fresh water is available at all times, both in the stable and in any field provided under this Agreement

8.5 *Storing forage*

The Licensee will purchase from the Livery Operator at a price or prices to be agreed from time to time between the parties all hay, straw, shavings, feedstuffs, supplements and other consumable items required by the Licensee, which will be stored by the Licensee, at the Licensee's risk and expense, in the area designated for this purpose by the Livery Operator

8.6 *Storing tack*

The Licensee may keep tack in any tack room ('the Tack Room') provided by the Livery Operator at the Licensee's risk and expense

8.7 *Exercise areas*

The Licensee may use any exercise area ('the Exercise Area') provided by the Livery Operator, but is responsible for clearing all droppings after use of the Exercise Area and for leaving the same clean and tidy, in all respects, after use

9. Obligations of the Licensee towards the Horse

The Licensee is wholly and solely responsible for the welfare of the Horse. In particular the Licensee must:-

9.1 feed the Horse and ensure that it has clean and fresh water at all times

9.2 maintain clean and adequate bedding for the Horse at all times

9.3 groom, exercise and turn out the Horse as necessary and appropriate

9.4 observe the worming programme required by the Livery Operator from time to time, details of which are available from the Livery Operator and can also be found posted in the Tack Room

9.5 ensure that the Horse's teeth are inspected regularly

9.6 ensure that the Horse has all necessary inoculations, as from time to time required by the Livery Operator and

9.7 see that the Horse is regularly attended by the registered farrier, details of whom appear in Schedule 2, and when the Horse is so attended, the

Licensee shall ensure that the farrier makes use of the area designated for this purpose and that all hoof clippings, shoes and nails are removed once the Horse has been attended to

10. General obligations of the Licensee

10.1 *Tidiness*

The Licensee must keep all the Licensee's property on the Property tidy, and must clear all mess caused or created in connection with the rights granted under this agreement. All manure must be placed in such location or locations as from time to time provided by the Livery Operator

10.2 *Insurance*

The Licensee must maintain a policy of insurance with reputable insurers in respect of all liability arising from the presence of the Licensee, the Licensee's tack and any other property of the Licensee bought onto the Property and particularly (but without prejudice to the generality hereof) any damage caused by the Horse in connection with its presence and use at or from the Property together with public liability insurance for a minimum sum of £[five/ten] million in respect of each incident. The Licensee must also provide the Livery Operator with a copy of such policy of insurance when requested to do so and ensure that the interest of the Livery Operator is endorsed on the policy

10.3 *Riding hat*

The Licensee, and any other person accompanying or riding the Horse, must at all times wear a riding hat that conforms with the current BSI standards when riding the Horse at or from the Property

10.4 *Emergency contact numbers*

The Licensee has provided the Livery Operator with the emergency contact number or numbers specified in Schedule 3 below, and must update these from time to time as necessary

11. Visiting the Property

11.1 *Permission to visit*

The Licensee may visit the Property with his servants agents and invitees to see and inspect the Horse at all reasonable times without giving notice to the Livery Operator save that no person under the age of 16 years shall visit unless accompanied by a person who is over the age of 16 years

11.2 *Parking*

Vehicles must be parked in such areas as from time to time designated by the Livery Operator, when visiting the Property, and all vehicles are parked at the Licensee's risk. The Livery Operator accepts no responsibility for any loss or damage caused or suffered whilst vehicles are parked at the Property

11.3 *Prohibited areas*

Under no circumstances must any persons, vehicles or animals go on any part of the Property, other than the stable, exercise area, and field(s) and any other area designated by the Livery Operator, which is hereby expressly prohibited

11.4 *Dogs*

No dogs are allowed on to the Property

11.5 *Smoking*

There shall be no smoking at the Property other than in the Tack Room

11.6 *Trespass*

Except as expressly permitted under this agreement, and excepting any public rights of way, the Licensee must not ride the Horse on any other land belonging to the Livery Operator and must not trespass on any other person's land when riding the Horse from the Property during the Period of Livery

11.7 *Regulations*

The Livery Operator may make such other regulations as the Livery Operator deems necessary from time to time governing the access of the Licensee to the Property

12. The Livery Operator's obligations

12.1 *Insurance of the stables*

The Livery Operator must maintain a policy of insurance with reputable insurers in respect of the stables and other designated areas at the Property and public liability insurance in respect of the Livery Operator's business conducted at the Property

12.2 *Liability*

The Livery Operator is not liable for any loss, injury, illness or death suffered by the Licensee, the Licensee's agents, servants or any property of the Licensee, the Licensee's agents servants or invitees

13. Overriding welfare of the Horse

13.1 *Right to call out a vet*

Without imposing any obligation on the Livery Operator for the welfare of the Horse, the Livery Operator may call out a veterinary surgeon to attend the Horse if, in the Livery Operator's judgment, this is necessary. Any veterinary surgeon called out must have experience in treating horses, and if possible be the veterinary surgeon nominated by the Licensee in Schedule 4 below, or such other person as may be advised by the Licensee in writing from time to time. The Livery Operator may act upon the advice of the veterinary surgeon in any such case. In this regard, the Livery Operator is to act as agent for the Licensee, and is not to incur any personal liability for veterinary fees

13.2 *Right to take other action*

The Livery Operator may take such other action as in the Livery Operator's best judgment may be necessary for the welfare of the Horse, and may charge the Licensee the reasonable costs incurred as a result

13.3 *Costs*

If any costs are incurred by the Livery Operator in connection with either of the above rights, the Licensee must pay these on demand by the Livery Operator

14. Personal nature of the Agreement

This Agreement is personal to the Licensee, and relates only to the Horse. Under no circumstances may the benefit of this Agreement be transferred by the Licensee to a third party, nor may it be used for the benefit of anything but the Horse

15. Destruction

15.1 *Removal*

If the horse dies, the Licensee must arrange for its carcase to be removed from the Property as soon as reasonably possible. If the death of the Horse occurs without the knowledge of the Licensee and the Licensee cannot be contacted, the Livery Operator may arrange for its collection. In this regard the Livery Operator is to act as agent for the Licensee, and shall not incur any personal liability for the costs of the removal. Any costs that are paid by the Livery Operator are to be re-payable to the Livery Operator by the Licensee on demand

15.2 *Emergency contact number*

The Licensee has given to the Livery Operator the telephone number or numbers specified in Schedule 3 below for use in an emergency, and must update these from time to time as necessary

16. Early termination

16.1 *Termination on breach*

Notwithstanding the Period of Livery, this Agreement may be terminated forthwith by either party giving notice to the other for breach of any obligation or warranty under this Agreement by the other

16.2 *Termination on death of the Horse*

If the Horse dies, whether naturally or unnaturally, this Agreement is to terminate automatically on the removal of the carcase from the Property

16.3 *Apportionment*

If this Agreement is terminated under either of the above provisions, in either case, without prejudice to the date of termination in relation to any payments made or due, any payment or repayments must be made accordingly

17. Obligations at the end of this Agreement

The Licensee must remove the Horse, all tack, personal possessions and rubbish from the Property on the termination of this agreement

18. Interpretation

In this Agreement unless the context otherwise requires words importing the masculine gender only shall be deemed to include the feminine gender and words importing the singular number only shall be deemed to include the plural number and where there are two or more persons included in the expression 'the Livery Operator' or 'the Licensee' agreements expressly to be made by the Livery Operator and/or the Licensee shall be deemed to be made by such persons jointly or severally

AS WITNESS the hands of the parties hereto the day and year first before written

SCHEDULE 1
THE HORSE

[*insert details*]

SCHEDULE 2
REGISTERED FARRIER

[*insert details*]

SCHEDULE 3
EMERGENCY NUMBERS

[*insert emergency contact numbers of the Licensee at the commencement of the Agreement*]

SCHEDULE 4
THE VETERINARY SURGEON

[*name of nominated veterinary surgeon at the commencement of the Agreement*]

SIGNED by
in the presence of:-

SIGNED by
in the presence of:-

PRECEDENT 4 LIVERY AGREEMENT – LANDOWNER DEALING WITH FEEDING AND EXERCISING OF HORSES

THIS AGREEMENT is made on the [ ] day of [ ] 20[ ]

BETWEEN:-

(1) [*name*] of [*address*] ('the Livery Operator')

and

(2) [*name*] of [*address*] ('the Horse Owner')

NOW IT IS AGREED as follows:-

1. Agreement for Livery

The Livery Operator agrees to provide board and stabling for the horse specified in Schedule 1 below and belonging to the Horse Owner ('the Horse') on the terms and conditions appearing below

2. Warranty

The Horse Owner warrants that the Horse is in a general state of good health and free from any infectious disease at the commencement of the Period of Livery and that Horse is of a reasonable demeanour and temperament

3. Period of Livery

The stabling and board must be provided for the Horse by the Livery Operator [from [*date*] until [*date*] [*or*] from month to month beginning on [*date*] until terminated by either party giving at least one month's written notice to the other to expire on the same day of the month as the commencement of this agreement] ('the Period of Livery')

4. Livery Fee

4.1 *The Livery Fee*

The Horse Owner must pay the Livery Operator [the sum of £[ ] in respect of the Period of Livery, [such sum being payable on or before the commencement of the Period of Livery [*or such other arrangement as is agreed, e g by monthly instalments*]] [*or*] the monthly fee of [ ] in advance on the [ ] day of each month, commencing on [*date*]] ('the Livery Fee')

4.2 *Increase of Livery Fee*

The Livery Operator must serve at least [six weeks (*or as required*)] written notice on the Horse Owner increasing the Livery Fee. The notice must take effect on the day of the month on which the Livery Fee is due. The Livery Fee may not be increased during the first twelve months of the Livery Period, nor may it be increased more than once every [ ] in any twelve month period thereafter

4.3 *Deposit*

The Horse Owner must pay a deposit of [ ] pounds (£[ ]) on the execution of this Agreement, to be held by the Livery Operator as security for the performance of the obligations of the Horse Owner under this Agreement

4.4 *VAT*

The above sums do not include VAT, which is to be added in the event of any of the above payments being liable to VAT

4.5 *Interest*

The Horse Owner must pay interest on any sums due under this agreement for more than fourteen days at the rate of [3% (*or as required*)] above the base lending rate of [*NatWest Bank plc*]

5. Location

During the Livery Period the Horse must be kept in the stables provided by the Livery Operator ('the Stables') located on the Livery Operator's property situate at and known as [*description of address of location of property*] ('the Property'), as delineated by [red edging (*or as the case may be*)] on the plan attached to this agreement ('the Plan')

6. General obligations of the Livery Operator towards the Horse

6.1 *General care*

The Livery Operator must provide the Horse with proper food, stabling, [use of a paddock] and attendance, and must do all that is reasonably necessary to keep the Horse in sound health and condition. In particular, the Livery Operator must:-

6.1.1 ensure that the Stables provide dry, well ventilated, well insulated, and draught free stabling with adequate drainage, in a sheltered position, with adequate lighting and well protected electrical fittings, a good water supply, a solid, durable and non-slip floor, of a suitable height for the Horse, and kicking boards fitted in the interior

6.1.2 provide proper food and bedding for the Horse, taking into account any special requirements or allergies of the Horse, including those notified by the Horse Owner in writing in advance of the signing of this agreement and

6.1.3 be responsible for worming the Horse, the inspection of the Horse's teeth

If any special requirements come to light during the Period of Livery which involve additional cost to the Livery Operator then these costs are to be payable by the Horse Owner upon demand, in addition to the Livery Fee

6.2 *Exercise*

The Livery Operator must exercise the Horse as often as is reasonably necessary, taking into account any riding by the Horse Owner under clause 11.1 below. However for this purpose the Horse may only be ridden by the Livery Operator, his employees and such other persons connected with the Livery Operator as may be permitted in writing by the Horse Owner from time to time

6.3 *Protection from disease*

The Livery Operator must take all reasonable steps to prevent the Horse contracting any contagious disease which any other horse or beast kept at or near the Stables is known or suspected to have

6.4 *Protection of Mares*

Where a Horse is a mare, the Livery Operator must take all reasonable steps to prevent the Horse being served by any stallion whilst the Horse is 'in season'

7. Farrier

7.1 *Attendance of a Farrier*

The Livery Operator must ensure that the Horse is attended by a registered farrier at least once every six weeks, and at such other times as may be necessary

7.2 *Costs*

[The Livery Fee includes the cost of the farrier, excepting the cost of any special call out. (*or*) The Livery Fee does not include the cost of such a farrier and the Horse Owner must pay all such reasonable costs on demand by the Livery Operator]

8. Veterinary Surgeon

8.1 *Call-out of a veterinary surgeon*

The Livery Operator must summon a veterinary surgeon for the Horse whenever this is reasonably necessary. In this regard the Livery Operator is to act as agent for the Horse Owner, and shall not to incur any personal liability for veterinary fees

8.2 *Qualifications*

The vet must be a member of the Royal College of Veterinary Surgeons, and must have experience in treating horses

8.3 *Veterinary fees*

The Horse Owner must pay all veterinary fees on demand by the Livery Operator, in addition to the Livery Fee

9. Schooling and training

[*Insert provisions as to any schooling and training of the Horse to be done whilst it is in livery*]

10. Tack

10.1 *Provision of tack*

The Horse Owner must provide his own tack for the Horse

10.2 *Tack Room*

The tack must be kept in the Tack Room, as delineated by [brown colouring (*or as required*)] on the Plan

10.3 *Use of the Horse Owner's tack*

The Horse Owner's tack [may not (*or*) may] be used in the permitted exercising of the Horse by the Livery Operator

[10.4 *Cleaning of tack*

The Livery Operator must clean [and maintain] the tack]

11. Riding of the Horse by the Horse Owner

11.1 *Notice of rides*

The Horse Owner (and anyone authorised by him in writing) may ride the Horse [at any reasonable hour of the day or week (*or stable times*)], providing that at least 24 hours' notice is given to the Livery Operator, whether in writing or orally, of the time and approximate duration of the proposed ride

11.2 *Tacking*

The Livery Operator is to be responsible for tacking and un-tacking the Horse before and after any such rides, and for doing all necessary post-ride grooming, unless otherwise agreed between the parties

11.3 *Riding hats*

The Horse Owner, and any other person authorised by him to ride the Horse under clause 11.1, must at all times wear a riding hat that conforms with the current BSI standards when riding the Horse from the Stables

11.4 *Access to land*

The Horse Owner may ride [*specify areas of land on the Livery Operator's Property which may be used for riding, including means of access and egress from the Stables to the highway, with reference to the Plan if appropriate*]

11.5 *Forbidden areas*

The Horse Owner must not ride the Horse on any other land belonging to the Livery Operator, except on any public rights of way on that land extending to horses, and must not trespass on any other person's land in the locality when riding the Horse from the Stables during the Period of Livery

12. Visiting the Stables

12.1 *Rights to visit in addition to visits under clause 11.1*

The Horse Owner may visit the Stables with his family friends agents and invitees to see and inspect the Horse at all reasonable times without giving notice to the Livery Operator[. The Horse Owner himself must always accompany any of his visitors]

12.2 *Access road*

The Horse Owner may only use the roadway shown [coloured yellow (*or as required*)] on the Plan for gaining access to and from the Stables

12.3 *Parking*

Vehicles must be parked in the area shown [hatched yellow (*or as required*)] on the Plan when visiting the Stables

12.4 *Restriction of vehicles and animals*

Under no circumstances may any vehicles or animals be brought into the stable area shown [hatched green (*or as required*)] on the Plan

12.5 *Regulations*

The Livery Operator may make such other regulations as he deems necessary from time to time governing the access of the Horse Owner to the Stables. Any such further regulations must be notified to the Horse Owner in writing

13. Insurance

13.1 *The Livery Operator's obligation*

The Livery Operator must maintain a policy of insurance with reputable insurers for the building and contents in respect of the Stables [including the tack belonging to the Horse Owner], and public liability in respect of the livery business conducted at the Stables

13.2 *The Horse Owner's obligation*

The Horse Owner must maintain a policy with reputable insurers for public liability arising from damage caused by the Horse in connection with its presence and use at or from the Stables, and in the event of any shortfall in any payment made under the policy the Horse Owner shall make that up from his own resources

14. Rates

The Livery Operator shall be liable for and must pay all rates and local taxes payable in respect of the Stables, and no sum additional to the Livery Fee is payable by the Horse Owner in this respect

15. Personal Agreement

This Agreement is personal to the Horse Owner, and relates only to the Horse. Under no circumstances may the benefit of this Agreement be transferred by the Horse Owner to a third party, nor may it be used for the benefit of anything but the Horse

16. Death and destruction of the Horse

16.1 *Destruction*

If the Horse suffers a serious injury or contracts a serious illness, the Livery Operator may use his best judgement as to whether the Horse should be destroyed. If it is practicable, the Livery Operator must use his best endeavours to contact the Horse Owner in advance of this, and must take the advice of a veterinary surgeon

16.2 *Death*

If the Horse dies, the Horse Owner must arrange for its carcass to be removed from the Stables as soon as reasonably possible. If the death of the Horse occurs without the knowledge of the Horse Owner, and the Horse Owner cannot be contacted, then the Livery Operator may arrange for its collection. In this regard the Livery Operator is to act as agent for the Horse Owner, and is not to incur any personal liability for the costs of the removal. However, any costs that are paid by the Livery Operator are to be repaid by the Horse Owner immediately on demand

16.3 *Emergency contact number*

The Horse Owner has given the Livery Operator the telephone number or numbers specified in Schedule 2 for use in an emergency, and must update these from time to time as necessary

17. Early termination

17.1 *Termination on breach*

Notwithstanding the Period of Livery, this Agreement may be terminated forthwith by either party giving the other written notice for breach of any obligation or warranty under this agreement by the other

17.2 *Termination on death of the Horse*

If the Horse dies, whether naturally or unnaturally, this agreement is to terminate automatically on the removal of the carcass from the Stables

17.3 *Apportionment of the Livery Fee*

In either case, without prejudice to any claim of either party against the other, the Livery Fee is to be apportioned according to the date of termination in relation to any payments made or due, and any payment or repayments must be made accordingly

18. Obligations

The Horse Owner must remove the Horse from the Stables and all (if any) tack from the Tack Room on the termination of this Agreement, and the Livery Operator must repay the Deposit to the Horse Owner after making any proper deductions to include (but without prejudice to the generality of this provision) the cost of making good any damage caused by the Horse, and the removal and/or disposal of any property or mess left by the Horse Owner

AS WITNESS the hands of the parties the day and year first before written

SCHEDULE 1
THE HORSE
[*insert details*]

SCHEDULE 2
EMERGENCY NUMBERS
[*insert emergency contact numbers of the Horse Owner at the commencement of the agreement*]

SIGNED by
in the presence of:-

SIGNED by
in the presence of:-

PRECEDENT 5 LIVERY AGREEMENT (SHORTER FORM)

THIS AGREEMENT is made the [ ] day of [ ] 20[ ]

BETWEEN:-

(1) [*name*] of [*address*] ('the Proprietor')

and

(2) [*name*] of [*address*] ('the Horse Owner')

NOW IT IS AGREED as follows:-

The Proprietor agrees to accept for livery at [*name of stables*] ('the Stables') the horse [*name of horse*] ('the Horse') belonging to the Owner on the following terms:-

1. The livery service to be provided by him comprises the reasonable feeding, exercise, grooming, cleaning and provision of accommodation for the Horse ('the Livery Service')

2. The agreed livery fees ('the Livery Fees') for the provision of the Livery Service are [ ] pounds (£[ ]) per [week] [month] plus VAT which are to be paid [weekly] [monthly] in advance on the first day of each month

3. The Livery Service excludes any veterinary fees, farriers fees, clipping, travelling costs, foaling mares or any other special services, all of which will be charged as extra items in accordance with the Proprietor's price list for the time being

4. The Horse and all tack and equipment belonging to the Horse Owner must be fully insured by the Horse Owner

5. Any professional attendances upon the Horse, by a veterinary surgeon or farrier, which the Proprietor or stable manager (if any) deems to be necessary must be paid for by the Horse Owner. The Horse Owner's preferred veterinary surgeon will be called first but in the event of his unavailability, the Horse Owner agrees that the Proprietor or his stable manager (as the case may be) may call any other suitable veterinary surgeon and agrees to pay the costs arising and to be bound by any professional advice given at the time

6. The Proprietor reserves the right to allocate grooms and accommodation as he thinks fit

7. The exercising of the Horse by the Proprietor's staff at the Stables will be under the direction of the Proprietor or his yard manager (if any) as the Proprietor shall from time to time decide. No responsibility shall be accepted by the Proprietor for any personal injury or loss or damage to property (including the Horse, any equipment or vehicle belonging to the Horse Owner or anyone with him) except where caused by the negligence of the Proprietor or his staff in the course of their employment

8. Notice of any riding or other arrangements for the Horse must be given by no later than 3 pm on the afternoon before the day concerned

9. This Agreement may be terminated by the giving of four weeks' written notice by either party to the other

10. If pursuant to an agreement between the Proprietor and the Horse Owner the Horse is sold by the Proprietor the sale will attract a commission payment to the Proprietor of [10%/20%] of the selling

price ('the Commission') the Commission to be deducted by the Proprietor before the remainder of the selling price is paid to the Horse Owner

SIGNED by the parties on the day and year first before written

SIGNED by
in the presence of:-

SIGNED by
in the presence of:-

Chapter 12

PARTNERSHIPS, CONTRACTING AGREEMENTS AND SHARE FARMING

INTRODUCTION

PRECEDENTS

1 Family partnership agreement
2 Declaration of trust
3 Deed of dissolution
4 Joint venture partnership agreement
5 LLP agreement between partnerships
6 Agreement for joint ownership of sugar beet harvester
7 Contracting agreement
8 Livestock share farming agreement
9 Arable share farming agreement

INTRODUCTION

This chapter deals with what may loosely be called joint ventures, namely partnerships, contract farming agreements and share farming.

Traditional partnerships

A partnership is a 'business carried on with a view to profit'.[1] Before the coming into force of the Agricultural Tenancies Act 1995 on 1 September 1995 partnerships were often entered into in order to avoid the security provisions of the Agricultural Holdings Act 1986. As the non-landowning partners have a non-exclusive licence to occupy the property no tenancy is created.[2] The disadvantage of a traditional partnership is that each partner is liable for all the partnership debts and is jointly and severally liable in tort for any loss or damage resulting from wrongful actions or omissions of any of the partners. There is no limited liability.

---

1 Partnership Act 1890, s 1.
2 *Harrison-Broadley v Smith* [1964] 1 All ER 867.

However, partnerships continue to be used for family, commercial or financial reasons. The contractual arrangements governing partnerships vary greatly. There is very little statutory interference. For example, a landowner may enter into partnership with an established farmer and make an agreement as to the contribution of capital, payment of a salary and sharing of profits and losses. The partners may be given expressly or by implication a non-exclusive licence to enter the land. Or land may be let to the partnership.

A family partnership agreement is set out in full at Precedent 1. In this document there is an option for declaring milk quota or entitlements partnership assets. This avoids the sort of disputes which arose in *Faulks v Faulks*[3] and *Davis v H & R Ecroyd*[4] where milk quota was held on the facts not to be a separate partnership asset.

Also included is a declaration of trust at Precedent 2 declaring that the specified land is to be a partnership asset. This is most likely to occur in family partnerships. There is a deed of dissolution of a partnership at Precedent 3.

It is of course now possible to have a limited liability partnership.

The partnership and LLP agreements in this chapter at Precedents 4 and 5 were designed for machinery joint ventures where a number of farmers join together to run a machinery fleet. There is also an agreement for joint ownership of a single machine at Precedent 6.

Limited liability partnerships

Limited liability partnerships (LLPs) were introduced by the Limited Partnership Act 2000. According to the explanatory notes which accompanied the Act:

> 'The essential feature of an LLP is that it combines the organisational flexibility and tax status[5] of a partnership with limited liability for its members. This limited liability is possible because an LLP is a legal person separate from its members.'

The Act provides powers to apply the provisions of company law and insolvency law, with appropriate modifications, to LLPs. These powers will be used to put safeguards in place for those dealing with LLPs.[6]

In order to establish an LLP the members must subscribe to an incorporation document, which must be delivered to the Registrar of Companies together with a statement that there has been compliance with the requirement that at

---

3 [1992] 15 EG 82.

4 [1996] 2 EGLR 5.

5 For detailed provisions relating to tax see Limited Liability Partnership Act 2000, ss 10, 11 and 12.

6 See Limited Partnership Regulations 2001, SI 2001/1090.

least two people, associated for the purpose of carrying on a lawful business with a view to profit, have subscribed their names to the incorporation document.[7]

Contract farming agreements

Under farm contract farming agreements, the contractor is employed under a contract for the provision of services, usually on the basis of receiving a modest fixed fee or basic income and a share of the profits. The farmer remains the farmer for tax purposes and the contractor acts as his agent in dealings with both DEFRA and the RPA.

Share farming

The concept of share farming was developed in order to avoid creating either a tenancy which would attract the security provisions of the Agricultural Holdings Act 1986 or a partnership with the consequence of a partner being liable not only for his own contracts and misdemeanours but also for those of his co-partners. Now that new tenancies are governed by the Agricultural Tenancies Act 1995 which confers no security of tenure, other than that created by the contract itself, and the liability of a partner may be reduced by establishing a LLP, the main reasons for setting up share farming arrangements have been removed. All the same there may be farming, financial or taxation grounds for doing so.

If a tenant wants to enter into a share farming agreement it is essential that he checks the terms of his tenancy to ensure that he will not be in breach of any covenant not to share possession.

The essence of a share farming agreement is that the land owner and the share farmer both have independent businesses which work the land together. Both of those businesses occupy the land. Whatever arrangement is entered into, it is essential that the agreement records what is to happen in practice and that the parties stick to those arrangements. Otherwise, there is a risk that a partnership or a tenancy will in fact have been created.

7 Limited Partnership Act 2000, s 2.

PRECEDENTS

PRECEDENT 1 FAMILY PARTNERSHIP AGREEMENT

PARTICULARS

Date of Agreement:

Parties:

(1) [*name*] of [*address*] ('the First Partner')

(2) [*name*] of [*address*] ('the Second Partner')

(3) [*name*] of [*address*] ('the Third Partner')

Name of firm:

Business:

Commencement Date:

Place of Business:

Bankers:

Cheque signing: [*All cheques may be signed by either the First Partner the Second Partner or the Third Partner*] [*All cheques must be signed by the First Partner the Second Partner and the Third Partner*] [*All cheques below the value of (say) £500 may be signed by either the First Partner the Second Partner or the Third Partner but all cheques above that value must be signed by at least two of the First Partner the Second Partner and the Third Partner*]

The profit and loss-sharing proportions: [*The First Partner – [ ]%, the Second Partner – [ ]%, and the Third Partner – [ ]% or such other percentages as they shall from time to time agree*]

Account Date: [the [ ] day of [ ] in each year]

Interest Rate: [2/3%] above the base rate of [ ] Bank plc

Payment Period:

Partnership Accountants:

1. Definitions and interpretation

1.1 The terms used in the Particulars shall unless the context otherwise requires have the same meaning throughout this Agreement

1.2 'The Partners' means the First Partner the Second Partner and the Third Partner or such of them as remain for the time being partners under the terms of this Agreement and 'the Partnership' means the Business carried out by the Partners pursuant to the terms of this Agreement

1.3 'The Business' means the business referred to in the Particulars and carried on by the Partners pursuant to the terms of this Agreement

1.4 'The Account Date' means the account date specified in the Particulars or such other date as may be agreed by the Partners from time to time

1.5 'The Accounting Period' means the period of twelve months ending on the Account Date or such other period as the Partners agree from time to time

1.6 'An Outgoing Partner' means a Partner who dies or gives notice under clause 15 of this Agreement or is adjudicated bankrupt or enters into an arrangement or composition for the benefit of his creditors or applies for an interim order under the Insolvency Act 1986 or becomes a patient under the Mental Health Act 1983 and so far as legally possible includes the personal representatives, trustee or receiver of such a Partner or his estate

1.7 'Determining Event' means in relation to an Outgoing Partner his death or the expiry of such notice or such adjudication arrangement composition or application or becoming such a patient (as the case may be)

1.8 'The Remaining Partners' means the Partners other than the Outgoing Partner

1.9 'The Option' means the option contained in clause 16 of this Agreement

1.10 'The Purchasing Partners' means such of the Remaining Partners as exercise the Option

1.11 'The Option Date' means:-

1.11.1 in the case of a deceased Partner the date of the grant of representation to his estate

1.11.2 in the case of a Retiring Partner the date of the expiry of the notice given under clause 15 of this Agreement

1.11.3 in the case of a Partner who becomes bankrupt the date of appointment of a trustee of his estate

1.11.4 in the case of a Partner entering into an arrangement or composition with his creditors the date of such arrangement or composition

1.11.5 in the case of a Partner applying for an interim order under the Insolvency Act 1986 the date of such application and

1.11.6 in the case of a Partner becoming a patient as aforesaid the date of appointment of a receiver of his estate

1.12 'The Payment Period' means the period of [ ] [months/years]

1.13 'The Relevant Accounting Period' means the Accounting Period during which the Determining Event occurs

1.14 The Single Payment Scheme' means the single payment scheme governed by Council Regulation (EC) No 73/2009 and all the relevant subsequent Commission and other regulations administrative provisions statutory guidance or codes of practice implemented by the EC and/or the UK Government ('the Regulations')

1.15 'Entitlements' means such entitlements to the single income payment and any other entitlement to financial support of whatever kind deriving in any way from the Regulations and allocated (and/or subsequently transferred) to one or more of the Partners

1.16 References to the singular include the plural and vice versa, references to one gender include the others and references to legislation include legislation by which it is amended or replaced

1.17 The clause headings are for convenience only and do not form part of this Agreement

2. Commencement

[The Partners shall carry on the Business and the Partnership shall commence on the Commencement Date] [The Partners have been carrying on their business in partnership since or about the [*date*] but having not previously set out the terms of their agreement in writing have decided to enter into this Partnership Agreement to confirm the agreement between them]

3. Determination

The Partnership shall continue until determined in manner hereinafter provided

4. Place of Business

The Business shall be carried on at the Place of Business and/or at such other places as the Partners shall from time to time decide

5. Bankers

The bankers of the Partnership shall be the Bankers or such other bank as the Partners from time to time decide

6. Cheque signing

All cheques on the Partnership bank account shall be signed as set out in the Particulars

7. Capital

The capital for the time being of the Partnership shall belong to the Partners in the proportions in which it [is deemed to have been] [has been] contributed by them as set out in the accounts of the Partnership

8. Partners' salaries

The Partners or any of them may be paid such sum by way of salary as the Partners from time to time agree such salary to be deemed to accrue or to have accrued from day to day. Any such salary shall be treated as a working expense of the Partnership before the profits and losses are ascertained

9. Profits and losses

The profits and losses of the Partnership shall be divided among and be borne by the Partners as set out in the Particulars

[There shall be no revaluation of freehold land/ tenancies/ milk quota/ herd/ Entitlements unless all the Partners agree]

10. Accounts

As at the Account Date in each year an account shall be taken by the Partnership Accountants of all assets and liabilities of the Partnership and of its profits and losses for the Accounting Period ending on that date

A copy of such accounts shall be provided to and signed by each Partner who shall be bound by them unless some error is discovered in them within three months from the signing of them in which case such error shall be rectified and any appropriate adjustments made

11. Drawings

Each of the Partners may draw out of the Partnership bank account such sums on account of his share of the profits as the Partners from time to time agree but if when the account is taken in respect of any Accounting Period it appears that any of the Partners has drawn from the said bank account any amount in excess of his share of the profits for that Accounting Period he shall (if so requested by any of the other Partners) immediately repay the amount of such excess into the said bank account

12. Time and attention

Each of the Partners shall devote [all of his time and attention to the Business] [such time and attention to the Business as is from time to time] [requisite] [from time to time agreed between them]

13. Partners' obligations

Each Partner shall at all times:-

13.1 punctually pay and discharge his separate debts and engagements and indemnify the other Partners and the Partnership assets against these and all proceedings costs claims and demands in respect of them and

13.2 be just and faithful to the other Partners in all matters relating to the Partnership and at all times give to the other Partners a true account of such transactions

14. Restrictions on Partners

No Partner shall without the consent of the others:-

14.1 engage or be concerned or interested either directly or indirectly in any other business or occupation

14.2 assign or charge the whole or any part of his interest in the Partnership

15. Retirement

A Partner may retire from the Partnership by giving to the other Partners not less than six months' written notice expiring at the end of any Accounting Period

16. Purchase Option

16.1 When a Determining Event occurs, the Remaining Partners shall have the option to purchase the share of the Outgoing Partner in the capital assets [and future profits] of the Partnership as at the date of such Determining Event in accordance with the terms contained in this clause

16.2 The Purchase Option may be exercised by notice in writing to the Outgoing Partner not later than three months after the Option Date

16.3 If the Purchase Option is exercised the share of the Outgoing Partner in the capital and assets [and future profits] of the Partnership shall vest in the Purchasing Partners (if more than one in the proportions in which they were entitled to share *inter se* in the net profits of the Partnership immediately prior to the Determining Event or such other proportions as they agree) and the purchase price of the said share ('the Purchase Price') shall be paid by the Purchasing Partners (if more than one in the like proportions)

[16.4 The Purchase Price shall be a sum equal to the Outgoing Partner's capital in the books of the Partnership [without any revaluation or addition as at the date of the Determining Event] or [determined as set out in clause 16.5]]

[16.5 Any freehold or leasehold property the herd and any milk quota or Entitlements which are Partnership assets shall be valued on the application of the Purchasing Partners or the Outgoing Partner on the date of the Determining Event. Such valuation shall be agreed between the Purchasing Partners and the Outgoing Partner or in default of agreement determined on the application of any party by a valuer appointed by agreement between the parties or in default of agreement within two months after the exercise of the option by a valuer appointed by the President of the Royal Institution of Chartered

Surveyors on the application of any party. Such valuer shall act as an expert and not as an arbitrator and his decision shall be final and binding on the parties. Any profit or loss on such revaluation shall be credited or debited to the Partners capital accounts in proportion to their respective shares in the [capital] [profits] of the Partnership immediately prior to the date of the Determining Event]

16.5 The Purchase Price shall be paid in accordance with the Payment Period specified in the Particulars

16.6 So much of the Purchase Price as is for the time being unpaid shall carry interest at the Interest Rate calculated on a daily basis in accordance with the Particulars and payable with each instalment of the Purchase Price

16.7 If any instalment of the Purchase Price or the interest on it is in arrear for more than twenty eight days the whole amount or the balance of it then outstanding shall become immediately due and payable

16.8 In the event of arrears interest shall be compounded at the Interest Rate on the instalment dates

16.9 Any payment on account of the moneys payable under these provisions shall in default of agreement be apportioned rateably as between the Purchase Price and interest outstanding at the date of payment

16.10 There shall also be paid to the Outgoing Partner by the Remaining Partners any undrawn balance of his share of the net profits of the Partnership for the relevant Accounting Period

16.11 There shall be paid by the Outgoing Partner to the Remaining Partners:-

16.11.1 any sum due from the Outgoing Partner in respect of losses of the Partnership for the relevant Accounting Period

16.11.2 any amount drawn by the Outgoing Partner on account of his share of the profits of the Partnership for the relevant Accounting Period in excess of his share

16.11.3 any sum by which his capital account is overdrawn

16.12 If the Determining Event does not fall on the last day of an Accounting Period the Outgoing Partner's share of profits or losses during the relevant Accounting Period shall be deemed to be a proportionate part calculated on a daily basis of the share of the net profits or losses of the Partnership for that Accounting Period to which the Outgoing Partner would have been entitled or which he would have borne if he had been a Partner throughout the relevant Accounting Period

16.13 Any sums payable under clauses 16.10 and 16.11 shall be payable as soon as ascertained and if not paid within fourteen days shall bear interest at the Interest Rate calculated on a daily basis

16.14 The Purchasing Partners shall indemnify the Outgoing Partner from and against all existing and future debts, liabilities and engagements of the Partnership

16.15 All necessary and proper instruments shall be executed for vesting the share of the Outgoing Partner in the Purchasing Partners

17. Dissolution

17.1 If by the end of the period limited for its exercise the Purchase Option referred to in clause 16 has not been duly exercised or if all the Remaining Partners agree in writing not to exercise the Purchase Option the Partnership shall then be dissolved and on such dissolution the affairs of the Partnership shall be wound up in accordance with the provisions of the Partnership Act 1890

17.2 If on the winding up of the Partnership any Partnership asset is to be offered for sale by public auction any Partner shall be entitled to bid for it

18. Partners' residences

Any Partner residing in any house on any holding occupied by the Partnership shall not be required to pay or reimburse any rent in respect of such residence

19. Voting

All matters relating to the management and conduct of the affairs of the Partnership shall be decided by a majority of the votes of the Partners each Partner being entitled to such number of votes as is directly proportionate to his share of the net profits of the Partnership for the time being

20. Property occupied by the Partners

In the absence of any contrary written agreement where any land is held by a Partner ('the Land holder') (whether as owner tenant or licensee) but farmed by the Partnership:-

[20.1 such land shall remain vested in the Land holder beneficially and the other Partners shall have no beneficial interest in it]

AND

[20.2 the other Partners shall occupy such land jointly with the Land holder as licensees only and shall vacate the same immediately on demand subject to the payment of compensation for any growing crops]

OR

[20.3 the [freehold property] [tenancy of the tenanted land] shall continue to be held on behalf of the Partnership and the Partnership shall indemnify the Landholder against the rent reserved by and the liabilities of the tenant under the said tenancy]

OR

[20.4 in respect of any period during which any land is held by the Landholder but farmed by the Partnership the Partnership shall indemnify the Landholder against the rent reserved by and the liabilities of the tenant under the tenancy of that land]

[21. Milk Quota

[*Similar provisions can be included to deal with the Entitlements. Reference will need to be made to the allocation of the Entitlements in the name(s) of whoever, the Entitlements being an asset of the Partnership, and the agreement between the Partners in connection with the division and consequent transfer of the Entitlements at the end of the Partnership, or, see clause 22 below.*]

21.1 The milk quota ('the Milk Quota') registered in the name of the Partners is an asset of the Partnership

21.2 If the Partners wish to dispose of all or any part of the Milk Quota (or on dissolution of the Partnership) each Partner will facilitate such disposal so far as practicable

21.3 If notwithstanding the terms of this Agreement any part of the Milk Quota becomes vested in one or more of the Partners without payment such Partner(s) shall pay to or reimburse the Partners the full value of it

21.4 If the Purchase Option is exercised the Milk Quota shall be valued and its value credited to the capital accounts of the Partners immediately prior to the Determining Event in their [capital] [profit-sharing] proportions at that time. The Outgoing Partner's share of it shall be added to the sum payable to him and shall be payable in accordance with clause 16.5 together with interest (if any) payable in accordance with clause 16.6

21.5 The Outgoing Partner's share of any value attributable to the Milk Quota in the books of the Partnership otherwise than in accordance with this clause shall be deducted in arriving at the sum payable to the Outgoing Partner in accordance with clause 16

21.6 If any Milk Quota is to be valued pursuant to this clause such value shall be agreed between the Outgoing Partner and the Purchasing Partners or in default of agreement determined by a surveyor appointed on the application of either party by the President of the Royal Institution of Chartered Surveyors. Such surveyor shall act as an expert and not as an arbitrator and his decision shall be final and binding on all persons interested]

OR

[21.1 The Milk Quota registered in the name of the Partners actually belongs to [*name*] as the Partners hereby admit and has been held by [him/them] on trust for [*name*]

21.2 The Partners shall take such steps as may reasonably be requested by [*name*] to comply with his wishes in respect of the Milk Quota including the disposal and transfer of it to one or more third party as [*name*] shall direct]

[22. The Single Payment Scheme

22.1 It is agreed by the Partners that:-

22.1 the Entitlements shall be held by the Partners to whom they were allocated and/or transferred (as the case may be) ('the SPS Partners') on trust for the Partners in shares equal to the Partners shares in the Partnership from time to time

22.2 if the Business wishes to dispose of all or any of the Entitlements the SPS Partners will do everything in their power to facilitate their transfer

22.3 the entirety of the proceeds of sale (or income from leasing as the case may be) of any of the Entitlements shall be payable to and belong to and be retained by the Partnership

22.4 the provisions of this clause shall so far as legally possible bind the respective successors in title of the SPS Partners

22.5 if after the date of this Agreement there are changes in the Regulations or in the interpretation of them by the Department for the Environment Food and Rural Affairs or the Rural Payments Agency or in any other way the intention of the parties regarding the Entitlements would be frustrated without amendment to this Agreement any of the Partners may give notice to the other or others of them, in the absence of agreement between them on the content of any subsequent Agreement to achieve their objectives as herein evidenced, to appoint an arbitrator to redraft the terms of this Agreement as far as they relate to the SPS in such a way as may fairly achieve the objectives of the Partners and the costs of such appointment shall be borne equally by the Partners]

23. Arbitration

If any dispute arises under this Agreement any of the Partners can give notice to the other or others of them in writing specifying the dispute and requesting that agreement be reached on the identity of a single arbitrator to be appointed to determine the dispute in accordance with the provisions of the Arbitration Act 1996 and if no arbitrator is appointed by agreement within two months of the notice any of the Partners may make a written request to the President for the time being of the Royal Institution of Chartered Surveyors for him to appoint an arbitrator to determine the matter in dispute

AS WITNESS the hands of the parties hereto the day and year first before written

SIGNED by
in the presence of:-

SIGNED by
in the presence of:-

SIGNED by
in the presence of:-

PRECEDENT 2 DECLARATION OF TRUST

THIS DECLARATION OF TRUST is made the [ ] day of [ ] 20[ ]

BY:-

[*names*] of [*addresses*] ('the Trustees')

WHEREAS:-

1. The Trustees own the freehold interest in the property described in the Schedule ('the Property') [*It could be that this Declaration will need to include milk quota and/or entitlements allocated under the Single Payment Scheme. If so, reference to the holding of these items should be included here and as far as the entitlements are concerned it is recommended that the short definitions used in the Partnership Agreement (1.14 and 1.15) be included.*]

2. The Trustees together with [*names*] (together called 'the Partners') carry on in partnership the business of farmers under the firm name of [*name*] ('the Partnership')

NOW THIS DEED WITNESSES that:-

The Trustees declare that from the date of this Declaration of Trust they will hold the Property [and the Milk Quota and/or Entitlements (*as defined*)] upon trust for the Partners as part of the property of the Partnership

IN WITNESS whereof the parties have set their hands to this Deed the day and year first before written

SCHEDULE
THE PROPERTY

ALL THAT the farmland situate at and known as [ ] in the County of [ ] as the same is delineated and edged red on the plan annexed to this Declaration

SIGNED as a Deed by
in the presence of:-

SIGNED as a Deed by
in the presence of:-

SIGNED as a Deed by
in the presence of:-

PRECEDENT 3 DEED OF DISSOLUTION

THIS DEED OF DISSOLUTION is made the [ ] day of [ ] 20[ ]

BETWEEN:-

(1) [*name*] of [*address*] ('the First Partner')

and

(2) [*name*] of [*address*] ('the Second Partner')

WHEREAS:-

1. The First Partner and the Second Partner ('the Partners') have for some years past carried on the business of farmers at [*name farm*] [*plus any other locations as appropriate*] in partnership under the name or style of [ ] ('the Partnership') on the terms of a Partnership Agreement made between them and dated the [*date*] ('the Partnership Agreement')

2. The Partners have agreed to dissolve the Partnership and to divide the Partnership assets between them in specie as provided in this Deed

NOW THIS DEED WITNESSES as follows:-

1. The Partnership [is hereby dissolved] [shall be dissolved on the [ ] day of [ ] next] [is hereby declared to have been dissolved on the [ ] day of [ ]]

2.1 The Partners shall with all reasonable speed:-

2.1.1 realise the book debts of the Partnership and the other Partnership assets not listed in the Schedule [except the goodwill which shall not be sold]

2.1.2 out of the Partnership cash and out of the net proceeds of the realisation of the book debts and other assets as aforesaid pay all the liabilities of the Partnership to persons other than the parties to this Deed

2.2 If the cash and net proceeds are not sufficient to meet the liabilities of the Partnership the Partners shall provide the additional money required in cash in equal shares [or as the case may be] as and when it is required. If the cash and net proceeds shall be more than sufficient to meet the liabilities of the Partnership the surplus shall belong to the Partners in [equal] [their Partnership] shares

3.1 The assets of the Partnership listed in the Schedule shall not be sold but shall be divided between the Partners hereto in specie as follows:-

3.1.1 the First Partner shall take in severalty the Partnership assets listed in Part I of the Schedule

3.1.2 the Second Partner shall take in severalty the Partnership assets listed in Part II of the Schedule

[3.2 For equality of division the First Partner/Second Partner shall pay the sum of £[ ] to the Second Partner/First Partner on or before the [*date*]]

4. Each of the Partners hereto shall be at liberty to carry on business anywhere in his own name using such trading name as he shall select which can be followed by the words 'formerly of [ ] & Co' but neither Partner shall otherwise trade under or use the name of the Partnership

5. Each of the Partners shall sign execute and do all such documents deeds and things as shall be necessary for carrying the provisions of this Deed into effect and completing the winding up of the Partnership affairs

6. Each of the Partners releases any claim he may have against the other in the Partnership Agreement but such releases shall not extend to claims arising under this Deed [or to any right of contribution or indemnity that may arise under the Partnership Agreement or otherwise in respect of any claim for negligence or breach of contract in the conduct of the Partnership business]

7. Due notice of the dissolution of the Partnership shall be given in the London Gazette and so far as may be reasonably practicable by circular letter to all persons firms and bodies with whom the Partnership has had dealings during the period of [ ] years preceding the date of this Deed. For this purpose each of the Partners shall sign all necessary documents and if either shall refuse or neglect to do so the other may sign the same on his behalf and in his name

8. All disputes and questions whatsoever which arise either during the Partnership or afterwards between the Partners relating to the Partnership Agreement or this Deed or the construction or application of either or any clause or matter contained or referred to in either of them or any account valuation or division of assets debts or liabilities to be made under this Deed or as to any act deed or omission of either Partner or as to any other matter in any way relating to the Partnership business or the affairs of the Partnership shall be referred to a single arbitrator in accordance with and subject to the provisions of the Arbitration Act 1996 [or any statutory modification or re-enactment of it for the time being in force]

IN WITNESS whereof the parties have set their hands to this Deed the day and year first before written

SCHEDULE

PART I

(Assets to be taken by the First Partner)

PART II

(Assets to be taken by the Second Partner)

SIGNED as a Deed by
in the presence of:-

SIGNED as a Deed by
in the presence of:-

SIGNED as a Deed by)
duly authorised on behalf of:-)
in the presence of:-)
Signature of Witness:
Name of Witness:
Address of Witness:

PRECEDENT 4 JOINT VENTURE PARTNERSHIP AGREEMENT

Date of Agreement	:		
Parties	:	(1)	....
		(2)	....
		(3)	....
Name of Firm	:		....
Business	:		....
Commencement Date	:		....
Minimum Termination Date	:		....
Place of Business	:		....
Bankers	:		....
Cheque Signing	:		....
The Profit and Loss Sharing Proportions	:		....
Account Date	:		....
Partnership Accountants	:		....

1. Definitions and Interpretations

The terms used in the Particulars shall unless the context otherwise requires have the same meaning throughout this Agreement

1.1 'The Partners' means the parties hereto

1.2 'The Business' means the business referred to in the Particulars and carried on by the Partners pursuant to the terms hereof

1.3 'The Account Date' means the Account Date specified in the Particulars or such other date as may be agreed by the Partners from time to time

1.4 'The Accounting Period' means the period of twelve months ending on the Account Date or such other period as the Partners agree

1.5 'An Outgoing Partner' means a partner who gives notice under clause 14 hereof or receives notice under clause 15 hereof

1.6 'determining event' means in relation to an Outgoing Partner the expiry of a notice under clause 14 or the receipt of a notice under clause 15

1.7 'the Remaining Partners' means the Partners other than the Outgoing Partner

1.8 'the option' means the option contained in clause 16 of this Agreement

1.9 'the Purchasing Partners' means such of the Remaining Partners as exercise the option

1.10 'the Option Date' means:-

> in the case of a Retiring Partner the date of the expiry of the notice given under clause 14 hereof
>
> in the case of an expelled Partner the date of receipt of the notice given under clause 15 hereof

1.11 'the relevant Accounting Period' means the Accounting Period during which the determining event occurs

1.12 References to the singular include the plural and vice versa references to one gender include the others and references to legislation include legislation by which it is amended or replaced

1.13 The clause headings are convenience only and do not form part of this Agreement

2. Commencement

The Partners shall carry on the Business and the Partnership shall commence on the Commencement date

3. Determination

The Partnership shall continue until determined in manner hereinafter provided

4. Place of Business

The Business shall be carried on at the Place of Business and/or at such other places as the Partners from time to time decide

5. **Bankers**

The bankers of the Partnership shall be the Bankers or such other bank as the Partners from time to time decide

6. **Cheque signing**

All cheques on the Partnership bank account shall be signed as set out in the Particulars

7. **Capital**

The capital for the time being of the Partnership shall belong to the Partners in the Profit and Loss sharing Proportions

8. **Profits and losses**

The profits and losses of the Partnership including capital profits and losses shall be divided among and borne by the Partners in the Profit and Loss Sharing Proportions

9. **Accounts**

As at the Account Date in each year an account shall be taken by the Partnership accountants of all assets and liabilities of the Partnership and of its profits and losses for the Accounting Period ended on that date

10. **Drawings**

Each of the Partners may draw out of the Partnership banking account such sums on account of his share of the profits as the Partners from time to time agree but if when the account is taken in respect of any Accounting Period it appears therefrom that any of the Partners has drawn from the said banking account any amount in excess of his share of the profits of that Accounting Period he shall (if so requested by any of the other Partners) forthwith repay the amount of such excess to the said banking account

11. **Time and attention**

Each of the Partners shall devote such time and attention to the Business as is from time to time requisite

12. **Partners' obligations**

Each Partner shall at all times:-

12.1 punctually pay and discharge his separate debts and engagements and indemnify the other Partners and the Partnership assets against the same and all proceedings costs claims and demands in respect thereof and

12.2 be just and faithful to the other Partners in all matters relating to the Partnership and at all times give to the other Partners a true account of such transactions

12.3 provide a representative at Partners' meetings approved by the Partners such approval not to be unreasonably withheld

13. Restrictions on Partners

No Partner shall without the consent of the others:-

13.1 engage or be concerned or interested either directly or indirectly in any business or occupation competing with the Business

13.2 assign or charge the whole or any part of his interest in the Partnership

13.3 give any security or promise for the payment of money on behalf of the partnership otherwise than in the ordinary course of business

13.4 enter into any bond or become bail or surety for any person or knowingly cause or suffer to be done anything whereby the partnership money may in any way be endangered

13.5 draw accept or endorse any bills of exchange or promissory notes on behalf of the partnership or in any way charge any asset of the partnership

13.6 give any guarantee on behalf of the partnership

13.7 compromise or compound or (except upon payment in full) release or discharge any debt due to the partnership)

14. Retirement

A Partner may retire from the Partnership by giving to the other Partners not less than six months' written notice expiring at the end of any Accounting Period but not before the Minimum Termination Date

15. Expulsion

If:-

15.1 any Partner shall commit any breach of the stipulations contained in clauses 12 and 13 hereof

15.2 in the case of [ ] any member of the partnership is adjudicated bankrupt or enters into an arrangements or composition for the benefit of his creditors or applies for an interim order under the Insolvency Act 1986

15.3 in the case of [ ] a resolution is passed or order made for the company to be wound up (other than for reconstruction or amalgamation) or an incumbrancer takes possession of or a receiver is appointed to all or any part of its undertaking the other Partners may by notice in writing forthwith expel that Partner from the Partnership

16. Purchase option

16.1 When a determining event occurs then the Remaining Partners shall have the option to purchase the share of the Outgoing Partner in the capital assets and future profits of the Partnership as at the date of such determining event in accordance with the terms in this clause

16.2 The option may be exercised by notice in writing to the Outgoing Partner not later than three months after the determining event

16.3 If the option is exercised the share of the Outgoing Partner in the capital and assets and future profits of the Partnership shall vest in the Purchasing Partners (if more than one in the proportions in which they were entitled to share inter se in the net profits of the Partnership immediately prior to the determining event or such other proportions as they agree) and the purchase price of the said share shall be paid by the Purchasing Partners (if more than one in the like proportions)

16.4.1 The purchase price to be paid by the Partners shall be a sum equal to the Outgoing Partner's capital in the books of the Partnership subject to the following clause 16.4.2 hereof

16.4.2 Every Partnership asset shall be valued on the application of the Purchasing Partners or the Outgoing Partner as at the date of the determining event Such valuation shall be agreed between the Purchasing Partners and the Outgoing partner or in default of agreement within one month after the exercise of the option determined on the application of any party by the expert in accordance with clause 18 hereof

16.5 The purchase price shall be paid interest free on the first anniversary of the date of the determining event

16.6 There shall also be paid to the Outgoing Partner by the Remaining Partners any undrawn balance of his share of the net profits of the Partnership for the relevant Accounting Period

16.7 There shall be paid by the Outgoing Partner to the Remaining Partners:-

16.7.1 any sum due from the Outgoing partner in respect of losses of the Partnership for the relevant Accounting Period

16.7.2 any amount drawn by the Outgoing Partner on account of his share of the profits of the Partnership for the relevant Accounting Period in excess of his share thereof

16.7.3 any sum by which his capital account is overdrawn

16.8 If the determining event does not fall on the last day of an Accounting Period the Outgoing Partner's share of profits or losses during the relevant Accounting Period shall be deemed to be a proportionate part calculated on a daily basis of the share of the net profits or losses of the Partnership for that Accounting Period to which the Outgoing Partner would have been entitled or which he would have borne if he had been a Partner throughout the relevant Accounting Period

16.9 Any sums payable under clauses 16.6 and 16.7 shall be payable as soon as ascertained

16.10 The Purchasing Partners shall indemnify the Outgoing Partner from and against all existing and future debts liabilities and engagements of the Partnership

16.11 All necessary and proper instruments shall be executed for vesting the share of the Outgoing Partner in the Purchasing Partners

17. Meetings

17.1 Representatives of all Partners shall meet to discuss and review the business a minimum of four times during each accounting period

17.2 All financial records are to be made available to the Partners and each of them at each such meeting and at more regular intervals if required by any Partner

17.3 Such meetings are to be chaired by a representative of each Partner in rotation

17.4 Each Partner or its representative is to have one vote with the chairperson having a casting vote on any issue

17.5 The approved representatives pursuant to clause 12.3 hereof at the date hereof are [ ]

18. Disputes

Any dispute arising between the parties shall be determined by an independent chartered accountant appointed by agreement between the parties or in default of agreement within two months of such dispute arising by a chartered accountant appointed on the application of any party by the President of the Institute of Chartered Accountants in England and Wales Such accountant shall act as an expert and not as an arbitrator and his decision shall be final and binding on the parties

19. New business

19.1 Any new farm contracting opportunities offered to any of the Partners shall be tendered for by the Partners

19.2 Any new farm business tenancy opportunity offered to any of the Partners may be taken up by that Partner in his own name and for his own benefit

20. Confidentiality

No Partner shall disclose the structure or running of the partnership to third parties without the consent of the other Partners

AS WITNESS the hands of the duly authorised representatives of the parties hereto

SIGNED by)
duly authorised on behalf of:-)
in the presence of:-)
Signature of Witness:
Name of Witness:
Address of Witness:

SIGNED by)
duly authorised on behalf of:-)
in the presence of:-)
Signature of Witness:
Name of Witness:
Address of Witness:

SIGNED by)
duly authorised on behalf of:-)
in the presence of:-)
Signature of Witness:
Name of Witness:
Address of Witness:

PRECEDENT 5 LLP AGREEMENT BETWEEN PARTNERSHIPS

Dated 20[ ]

[ ]

and

[ ]

and

[ ] LLP

AGREEMENT

THIS AGREEMENT is made on the [ ] day of [ ] 20[ ]

BETWEEN:-

(1) [ ]

(2) [ ]

and

(3) [ ] LLP (Partnership Number [ ]) whose Registered Office is at [ ]

WHEREAS:-

(A) The Members wish to set up a farming business to be operated by the LLP which they intend to manage in accordance with the terms set out below

(B) The Members have agreed to enter into this Agreement with the LLP to set out the basis on which the LLP is to be organised and the rights and obligations of the LLP

1. Definitions and interpretations

1.1 In this Agreement including the Recitals the following terms shall have the following meanings (unless the context requires otherwise):-

1.2 'the Account Date' means [ ] or such other date as may be determined by the Members from time to time in accordance with this Agreement

1.3 'the Accounting Year' means the period of twelve months ending on the Account Date or such other period as the Members agree

1.4 'the Act' means the Limited Liability Partnership Act 2000

1.5 'the Assets' means the farming machinery and other assets of the LLP

1.6 'the Bank' means [ ]

1.7 '[ ]' means the partnership of [ ]

1.8 'the Business' means the business of providing contract farming services as further described in clause 11 hereof carried on by the LLP pursuant to the terms hereof

1.9 'the Capital' means the capital of the LLP as shown in any balance sheet prepared in accordance with the provisions of the Agreement and belonging to Members and to include the Assets

1.10 'the Commencement Date' means [ ]

1.11 '[ ]' means the [ ]

1.12 'the Designated Members' means [ ] or such other persons as may from time to time be appointed to serve as Designated Members in accordance with this Agreement

1.13 'Determining Event' means in relation to an Outgoing Member the expiry of a notice under clause 15 or the receipt of a notice under clause 16

1.14 'the Farming Year' means a year to the [ ] but the first Farming Year shall be the period from the Commencement Date to [ ]

1.15 [ ] 'the LLP' means the limited liability partnership incorporated under the Name which the Members have registered at Companies House with number [ ]

1.16 'the LLP Accountants' means [ ] or such other firm of Accountants as may from time to time be appointed in accordance with the provisions of this Agreement

1.17 'the Members' means the parties hereto of the first to [ ] parts and such other or additional persons as may from time to time be appointed in accordance with this Agreement. Each Partnership shall be deemed to be one Member and shall appoint one member to attend meetings and shall have one vote

1.18 'the Member's Share' means a Member's share and interest of and in the Capital

1.19 'the Name' means [ ] LLP or such other name as shall from time to time be registered by the LLP at Companies House as its name

1.20 'the Option' means the option contained in clause 17 of this Agreement

1.21 'the Option Date' means:-

1.21.1 in the case of a retiring Member the date of the expiry of the notice given under clause 15 hereof

1.21.2 in the case of an expelled Member the date of receipt of the notice given under clause 16 hereof

1.22 'An Outgoing Member' means a Member who gives notice under clause 15 hereof or receives notice under clause 16 hereof

1.23 'Partnership' means [ ]

1.24 'the Places of Business' means the areas of farm land farmed by the Members selected by the Members from time to time to receive the services of the Business and any land farmed by the LLP

1.25 'the Profit and Loss Sharing Proportions' means [ ]

1.26 'the Purchasing Members' means such of the Remaining Members as exercise the Option

1.27 'the Registered Office' means [ ] or such other address as shall from time to time be registered by the LLP at Companies House as its registered office

1.28 'the Regulations' means the Limited Liability Partnership Regulations 2001 SI 1090 of 2001

1.29 'the Relevant Accounting Year' means the Accounting Year during which the Determining Event occurs

1.30 'the Remaining Members' means the Members other than the Outgoing Member

1.31 references to the singular include the plural and vice versa references to one gender include the others and references to legislation include legislation by which it is amended or replaced

1.32 the clause headings are for convenience only and do not form part of this Agreement

2. Commencement and duration

2.1 The provisions of this Agreement shall take effect on the Commencement Date

2.2 The LLP shall subsist until wound up in accordance with clause 23 and the provisions of the Act

2.3 In the event that any Member may be personally liable under any contract entered into by it prior to the incorporation of the LLP which was for the benefit of the LLP and with the express or implied consent of the other Members then the LLP shall on incorporation be deemed to ratify that contract and shall indemnify that Member from and against all claims liabilities and costs in connection with it

3. Name and Registered Office

3.1 The Members may from time to time determine upon a change in the Name and/or the Registered Office

3.2 Upon any change in the Name and/or the Registered Office it shall be the responsibility of the Designated Members to notify Companies House of any such change in accordance with the Act and to notify all Members

4. Places of Business

The Business shall be carried on at the Places of Business

5. Bankers

The bankers of the LLP shall be the Bank or such other bank as the Members shall from time to time agree

6. Cheque signing and transfers

6.1 All cheques on the LLP bank account shall be in the name of the LLP and signed by a minimum of [ ] persons each of whom must be a duly authorised representative of [ ]

6.2 All electronic transfers shall be authorised by a minimum of [ ] persons each of whom must be a duly authorised representative of a [ ]

6.3 Any payments in excess of £[ ] must be approved by duly authorised representatives of each of [ ]

7. Capital

7.1 Each of the Members as at the date of this Agreement shall acquire as at the Commencement Date and thereafter own a Member's Share in the Capital in the proportions from time to time shown in the accounts of the LLP

7.2 No Member shall be entitled to any interest on the amounts for the time being of his Member's Share

7.3 No capital shall be repaid without the consent of the LLP

8. Profits and losses

The profits and losses of the LLP including capital profits and losses shall be payable by the LLP to the Members or payable by the Members to the LLP (as the case may be) in the Profit and Loss Sharing Proportions

9. Accounts

9.1 It shall be the responsibility of the Members to ensure that proper books of account giving a true and fair view of the Business and the affairs of the LLP shall be properly posted and each Member shall duly and punctually make full and proper entries of all business transacted by him or at his direction on account of the LLP

9.2 Such books of account shall be:-

9.2.1 kept at the Registered Office or such other place as the Members may from time to time determine and

9.2.2 open to inspection by each Member and his agent and the LLP Accountants who shall be entitled to take copies thereof or extracts therefrom as set out in clause 18.2 hereof

9.3 If required by the Act the Designated Members shall (acting where appropriate in accordance with the requirements of the Act):-

9.3.1 be deemed to appoint the LLP Accountants as auditors of the LLP for the next ensuing Accounting Year

9.3.2 each year appoint auditors if required

9.3.3 have power to remove the accountants or auditors from office

9.3.4 have power to fix the remuneration of the accountants or auditors

9.4 The Members may from time to time determine to amend the Account Date

9.5 A profit and loss account shall be taken in every year on the Account Date and a balance sheet (taking no account of goodwill) shall be prepared and, if required by the Act the same shall be audited in accordance with all relevant Statements of Standard Accounting Practice and in such format and giving such information notes and disclosure of the interest of the Members of the LLP as may be required by the Act

9.6 The accounts to be prepared in accordance with clause 9.5 above shall be:-

9.6.1 approved by the Members in accordance with the Act which approval shall be given at a meeting and shall thereafter become binding on all the Members save that any Member may request the rectification of any manifest error discovered in any such accounts within three months of receipt of the same and

9.6.2 distributed to all Members as required by the Act

10. Drawings

Each of the Members may draw out of the LLP banking account such sums on account of its share of the profits as the Members from time to time agree but if when the account is taken in respect of any Accounting Year it appears therefrom that any of the Members has drawn from the said banking account any amount in excess of its share of the profits of that Accounting Year it shall (if so requested by any of the other Members) forthwith repay the amount of such excess to the said banking account

11. Business objects

11.1 The objects of the Business are to provide such farming services and carry on such farming activities as the Members shall agree

11.2 The LLP shall provide all the manpower and machinery additional to any supplied by the relevant Member required for the cultivation of the Places of Business and shall carry out such operations as shall reasonably be directed by such Member for the cultivation planting and harvesting of all arable crops to be grown on the Places of Business

11.3 The LLP shall keep such records and accounts in respect of work as are required in accordance with usual farming practice or as may reasonably be required by the Members

11.4 For the purposes of the Business, the Members undertake to ensure the provision in a timely fashion in accordance with good farming practice and at reasonable cost of:-

11.4.1 all necessary seeds fertilisers sprays and other variable inputs for the efficient performance by the LLP of the objects of the Business. Each of the Members authorises the LLP to act as its agent in the ordering and purchasing of such inputs required under this clause

11.4.2 all necessary farming labour in addition to any provided by the LLP necessary for the efficient performance by the LLP of the objects of the Business

11.5 For the purposes of the Business the Members shall procure insurance cover to the reasonable satisfaction of the LLP in respect of the following:-

11.5.1 buildings fixed machinery and plant located on the Places of Business

11.5.2 produce whilst growing and in store

11.5.3 seeds fertilisers and sprays and other sundry stores

11.5.4 all other property on the Place of Business made available by that Member owned by or in the custody or control of that Member and used in its business

11.5.5 all third party and occupiers liability properly insurable by the Members

11.6 For the purposes of the Business, the LLP shall at its own expense procure insurance cover to the reasonable satisfaction of the Members in respect of the following:-

11.6.1 the Assets and any machinery plant or other property in the LLP's custody against such loss or damage as the LLP shall consider appropriate and

11.6.2 the LLP's liability for death or personal injury or for loss or damage to property arising out of or in the course of fulfilling the objects of the Business

11.7 The LLP shall be entitled to charge the Members for its services provided on a monthly basis in arrears at such rate as shall be agreed between the Members. Such invoices shall be paid within [ ] days and in default shall bear interest from the date the invoice was delivered until payment at [ ]% per annum

12. Time and attention

Each of the Members shall devote such time and attention to the Business as is agreed between the Members

13. Members' obligations

Each Member shall at all times:-

13.1 punctually pay and discharge its separate debts and engagements to the LLP and indemnify the other Members and the LLP against the same and all proceedings costs claims and demands in respect thereof and

13.2 be just and faithful to the other Members in all matters relating to the LLP and the Business and at all times give to the other Members a true account of such transactions

13.3 provide a representative at Members' meetings approved by the Members such approval not to be unreasonably withheld

13.4 employ the LLP to provide farming services on the land as agreed by the Members

14. Restrictions on Members

No Member shall without the consent of the others:-

14.1 engage or be concerned or interested either directly or indirectly in any business or occupation competing with the Business

14.2 assign or charge the whole or any part of his interest in the LLP PROVIDED that where any person is admitted to a Partnership he shall subject to prior notification of the Members be admitted to the membership of the LLP and be entitled to an appropriate share therein

14.3 give any security or promise for the payment of money on behalf of the LLP otherwise than in the ordinary course of business

14.4 enter into any bond or become bail or surety for any person or knowingly cause or suffer to be done anything whereby the LLP money or Assets may in any way be endangered

14.5 draw accept or endorse any bills of exchange or promissory notes on behalf of the LLP or in any way charge any asset of the LLP

14.6 give any guarantee on behalf of the LLP

14.7 compromise or compound or (except upon payment in full) release or discharge any debt due to the LLP

14.8 be entitled to make an application to the Court under the Companies Act 2006, s 994

15. Retirement

15.1 A Designated Member may resign his designation upon giving not less than three months' notice to the LLP and to the other Members save that in the event that such resignation would reduce the number of Designated Members of the LLP to one then the notice shall not take effect until the Members shall have appointed a new Designated Member to fill the vacancy to be created by such notice

15.2 A Member may retire from the LLP by giving to the LLP and the other Members written notice, at any time to take effect as follows:-

15.2.1 in the event of the death of a partner in a Partnership on the expiration of eleven months thereafter

15.2.2 on the expiration of one year expiring on [ ] in any year PROVIDED that no such notice may be served to expire before [ ] if such retirement could adversely affect the viability of the LLP

15.3 On expiry of the notice under clause 15.2 the LLP shall be entitled and obliged to remove all its property (including machinery and equipment) from the Place of Business farmed by such Member serving the notice

16. Expulsion

If:-

16.1 any Member shall commit any breach of the stipulations contained in clauses 13 and 14 hereof

16.2 in the case of an individual partner in [ ] is adjudicated bankrupt or enters into an arrangement or composition for the benefit of his creditors or applies for an interim order under the Insolvency Act 1986

16.3 any Member should fail without just cause to pay any money owed by him to the LLP within [ ] days of being requested in writing by the LLP so to do

in any of such cases the LLP may by notice in writing to the Member forthwith expel that Member from membership of the LLP. For the avoidance of doubt a breach by any member of a Partnership party hereto shall authorise the service of notice hereunder on all the members of that Partnership

17. Purchase option

17.1 When a Determining Event occurs then the Remaining Members shall have the option to purchase the share of the Outgoing Member in the capital assets and future profits of the LLP as at the date of such Determining Event in accordance with the terms in this clause

17.2 The option may be exercised by notice in writing to the LLP and the Outgoing Member not later than three months after the Determining Event

17.3 If the option is exercised the share of the Outgoing Member in the capital assets of the LLP and future profits of the LLP shall vest in the Purchasing Members (if more than one in the proportions in which they owned the capital assets immediately prior to the Determining

Event and as to future profits in the proportions in which they were entitled to share among themselves in the net profits of the LLP immediately prior to the Determining Event or such other proportions as they agree) and the purchase price of the said share shall be paid by the Purchasing Members (if more than one in the like proportions)

17.3.1 the purchase price to be paid by the Purchasing Members shall be a sum equal to the Outgoing Member's capital in the books of the LLP as at the final day of the Farming Year in which the Determining Event occurs subject to the following clauses 17.3.2 and 17.3.3 hereof

17.3.2 every LLP asset shall be valued on the application of the Purchasing Members or the Outgoing Member as at the final day of the Farming Year in which the Determining Event occurs. Such valuation shall be agreed between the Purchasing Members and the Outgoing Member or in default of agreement within one month after the exercise of the Option determined on the application of any party by the expert in accordance with clause 19 hereof

17.4 The purchase price shall be paid in full interest free on the day falling two months following the agreement or determination of the value of the LLP assets pursuant to clause 17.3.2

17.5 There shall also be paid to the Outgoing Member by the LLP any monies owing to it by the LLP and any undrawn balance of its share of the net profits of the LLP for the Relevant Accounting Year

17.6 There shall be paid by the Outgoing Member to the LLP:-

17.6.1 any sum due from the Outgoing Member in respect of losses of the LLP for the Relevant Accounting Year

17.6.2 any amount drawn by the Outgoing Member on account of its share of the profits of the LLP for the Relevant Accounting Year in excess of its share thereof

17.6.3 any sum by which its capital account is overdrawn or owed by it to the LLP

17.7 If the Determining Event does not fall on the last day of an Accounting Year the Outgoing Member's share of profits or losses during the Relevant Accounting Year shall be deemed to be a proportionate part calculated on a daily basis of the share of the net profits or losses of the LLP for that Accounting Year to which the Outgoing Member would have been entitled or which it would have borne if it had been a Member throughout the Relevant Accounting Year

17.8 Any sums payable under clauses 17.6 and 17.7 shall be payable as soon as ascertained by the LLP Accountants and notified in writing by the LLP to the Outgoing Member

17.9 All necessary and proper instruments shall be executed for vesting the share of the Outgoing Member in the Purchasing Members

17.10 If following a Determining Event the Option is not exercised then the Members shall use their best endeavours to obtain the agreement of a third party or parties willing to be appointed as a Member and to purchase the Outgoing Member's capital in the books of the LLP and to execute all necessary and proper instruments for vesting the share of the Outgoing Member in such new Member

17.11 If following the expiry of a period of [ ] months from expiry of the Option, or such longer period as may unanimously be agreed between the Remaining Members and the Outgoing Member a third party has not been found who agrees to purchase the Outgoing Member's capital in the books of the LLP, then the Members shall resolve and carry out all steps required to effect the termination of this Agreement and the winding up of the LLP

18. Meetings

18.1 The Designated Members or their duly appointed representatives shall meet to discuss and review the Business a minimum of [ ] times during each Accounting Year to include agreement as to crop rotation and crops to be grown on the relevant farm land of each Member in each succeeding year. All Members shall be entitled to receive notice of and to attend and observe at such Meetings. There shall be no maximum number of Designated Members entitled to attend such meetings. The quorum for such meetings shall be such number as the Designated Members may agree in writing but in the absence of agreement shall be 2 persons each of whom must be a Designated Member or a duly authorised representative of each of the Member partnerships

18.2 All financial records are to be made available to the Members and each of them at each such meeting and at any other time upon the reasonable request of any Member

18.3 Such meetings are to be chaired by a Designated Member or their duly appointed representative in rotation

18.4 Except for the matters listed in clause 18.5 below, all decisions at meetings of the Designated Members will be decided by simple majority and each Designated Member or its representative is to have one vote. Where the matter being considered at such a meeting is within the matters listed in clause 18.5 below, no decision of the Designated Members shall be binding unless all the Designated Members or their duly appointed representatives shall have voted in favour of the decision

18.5 The subject matters referred to in clause 18.4 above are:-

18.5.1 amendments to this Agreement

18.5.2 appointment of a new Member or appointment of any Member as a Designated Member

18.5.3 any expenditure by the LLP not in a previously agreed budget

18.5.4 amendments to the Profit and Loss Sharing Proportions

18.5.5 approval of the accounts of the LLP

18.5.6 the borrowing or lending by the LLP or the giving of any guarantee or undertaking by the LLP in respect of any amount in excess of £10,000

18.5.7 renting or contract farming additional land

19. Disputes

Any dispute arising between the parties shall be determined by an independent professional or such other person as may be agreed by the Members and he shall act as an expert and not as an arbitrator and his decision shall be final and binding on the parties

20. New business

Any new farm business tenancy or farm contracting opportunity offered to any of the Members shall be notified to and discussed by the LLP

21. Confidentiality

No Member shall disclose the structure or running of the LLP or any affairs of the Business to third parties without the consent of the other Members

22. Indemnity and expenses

The LLP shall indemnify each Member from and against any claims, costs and demands arising out of payments made by it or liabilities incurred by it in the performance by it of its duties as a Member in the normal course of the operation of the Business

23. Winding up and termination

23.1 This Agreement will terminate immediately if the LLP shall be removed from the register of companies maintained by Companies House in accordance with the Act

23.2 This Agreement will terminate if all the Designated Members so resolve in writing on such date as they specify

23.3 In the event of the winding up of the LLP then any surplus of capital assets of the LLP over its liabilities remaining at the conclusion of the winding up after payment of all money due to the creditors of the LLP and all Members' expenses of the winding up shall be payable by the liquidator to the Members in the Profit and Loss Sharing Proportions

23.4 In the event that any Court makes a declaration or declarations under the Insolvency Act 1986 section 214A requiring any Member or

Members to make any contribution to the assets of the LLP then the other Member(s) shall indemnify the Member(s) in respect of whom the said declaration(s) shall have been made in such manner that the amount or aggregate amounts payable in accordance with the said declaration(s) shall be borne by the Members in the Profit and Loss Sharing Proportions

24. Changes in Members

24.1 The Members undertake to procure that on the appointment of a new Member all necessary and proper instruments shall be executed for the acquisition by such new Member of his share in the capital and assets of the LLP

24.2 The Designated Members undertake to notify Companies House of any changes in the Members of the LLP in accordance with the Act

25. Guarantees and indemnities

25.1 In the event that any Member shall have given any guarantee on behalf of the LLP and have obtained the necessary consent then if the person to whom it has been given shall call for any payment properly due under that guarantee the Member in question shall be entitled to be indemnified forthwith by the other members in such manner that the amounts or aggregate amounts payable in accordance with the said guarantee shall be borne by the Members in the Profit and Loss Sharing Proportions

25.2 Upon the dissolution, retirement or expulsion of that Member the other Members shall:-

25.2.1 use their best endeavours to procure that the person having the benefit of the guarantee shall release that Member from the guarantee

25.2.2 provide a substitute guarantor if required by that person as a condition of release

25.2.3 jointly and severally indemnify the Member in question from and against any liability under the guarantee arising after the date of dissolution, retirement or expulsion

AS WITNESS the hands of the duly authorised representatives of the parties hereto

SIGNED by)
in the presence of:-)
Signature of Witness:
Name of Witness:
Address of Witness:

SIGNED by)
in the presence of:-)
Signature of Witness:
Name of Witness:
Address of Witness:

SIGNED by)
in the presence of:-)
Signature of Witness:
Name of Witness:
Address of Witness:

SIGNED by)
in the presence of:-)
Signature of Witness:
Name of Witness:
Address of Witness:

SIGNED by)
in the presence of:-)
Signature of Witness:
Name of Witness:
Address of Witness:

SIGNED by)
in the presence of:-)
Signature of Witness:
Name of Witness:
Address of Witness:

SIGNED by)
on behalf of [ ] LLP)
in the presence of:-)
Signature of Witness:
Name of Witness:

Address of Witness:

PRECEDENT 6 AGREEMENT FOR JOINT OWNERSHIP OF SUGAR BEET HARVESTER

Dated 20[ ]

[ ]

-and-

[ ]

AGREEMENT
RELATING TO
SUGAR BEET HARVESTER

PARTICULARS

DATE OF AGREEMENT 20[ ]

PARTIES:-

(1) [ ]

(2) [ ]

SUGAR BEET HARVESTER	**Sugar Beet Harvester**	Serial Number	Registration Number

THE AGREED

PROPORTIONS

THIS AGREEMENT is made on the date of the Agreement between the Parties

NOW IT IS HEREBY AGREED as follows:-

1. Expressions defined in the Particulars have the same meanings in this Agreement
2. The Parties hereto purchased the Harvester in the agreed proportions
3. [ ] will be responsible for storing the Harvester in appropriate storage and for ensuring that it is insured in full replacement value in the joint names of the parties against normal insurable risks and appropriately serviced and maintained
4. The cost of insurance will be borne by the parties in the Agreed Proportions

5. The names of the parties will be painted on the Harvester

6. The parties will communicate as often as necessary to discuss the maintenance of the Harvester and its work schedule. The parties will work together in good faith to ensure that the Harvester is used as efficiently as possible during harvest for the benefit of all parties' farming businesses and in order to maximise the harvest of each party's crop in a timely basis and in good condition

7. Immediately before the Harvester is moved from one farm to another it shall be refuelled fully so that each party is responsible for fuel consumed on his farm

8. The parties hereto shall maintain records of:-

8.1 the acreage of crop harvested on the farm of each of the Parties

8.2 labour and machinery supplied by one of the Parties for work on the farm of another party

9.1 In the event of break-down the person on whose farm the Harvester is at the time shall be responsible for forthwith notifying [ ] of the breakdown

9.2 [ ] shall ensure that the Harvester is repaired and available for service as soon as possible and if commercially appropriate will hire in a replacement harvester

10. Any expenditure in respect of the Harvester shall in the first place be paid by the party incurring it and then reimbursed in the Agreed Proportions. Repairs and break-down expenses shall be met by the parties in the agreed proportions. If the Harvester is damaged while operating on the farm of one of the Parties by reason of negligence or improper use that Party shall be responsible for making good the damage at his own expense

11. Labour and machinery made available by one party to the others shall be charged at farmers cost rate set out in accordance with the Farm Management pocketbook edited by John Nix. Any balance due from one party to another will be invoiced monthly

12. The benefit of and burden of this Agreement shall not be assignable

13. Any income, net of agreed costs, from third party contracting will be split between the farming businesses to the Agreed Proportions

14. The Parties shall meet on a regular basis to keep each other informed of any relevant matters and to discuss and agree all matters relating to the efficient operation of the machine and any significant items of expenditure. A cropping schedule will be submitted by each member including varieties and expected markets in order to enable efficient planning of crop harvest. A provisional harvest plan will be drafted and agreed each year prior to 30 June

15. This Agreement may be terminated:-

15.1 forthwith by the Parties serving notice in writing on one of them if that party is guilty of any persistent breach or breaches of this Agreement

15.2 forthwith by notice in writing from the Parties to any of them in the event that that party or any member of a partnership is adjudicated bankrupt or enters into an arrangement or composition for the benefit of his creditors or applies for an interim order under the Insolvency Act 1986

15.3 by not less than six months' notice in writing expiring on 31 March in any year given by any party to the others provided that no such notice may be served to take effect before [*30 March 2012*]

16. If this Agreement is terminated in accordance with clause 15 then:-

16.1 the Parties serving notice under sub-clause 1 and 2 thereof or the Parties receiving notice under sub-clause 3 thereof shall have the option to be exercised by notice in writing within twenty eight days of service of the notice of acquiring the interest of the other party in the Harvester. The value of such interest shall be ascertained by a certificate given by [ ] as to the market value of the Harvester on the date of termination of this Agreement. If the option is exercised the share of the seller shall be purchased by the buyer forthwith on termination of this Agreement or if later on ascertainment of the purchase price of the share in the Harvester

16.2 if the Parties buy the share of another party in the Harvester all necessary documents shall forthwith be signed to transfer ownership

16.3 if such option is not exercised the Harvester shall be sold in the open market and the proceeds divided in the Agreed Proportions

17. Any dispute arising between the parties shall be determined by [ ] of [ ] who shall act as an expert and not as an arbitrator and whose decision shall be final and binding on the parties

AS WITNESS the hands of [ ], [ ], [ ]

and [ ] the duly authorised representatives of the parties hereto

SIGNED by)
for and on behalf of:-)
in the presence of:-)
Signature of Witness:
Name of Witness:
Address of Witness:

SIGNED by)
in the presence of:-)

Signature of Witness:
Name of Witness:
Address of Witness:

PRECEDENT 7 CONTRACTING AGREEMENT

Dated 20[ ]

[ ]

AND

[ ]

CONTRACTING AGREEMENT

RELATING TO

[ ]

CONTRACTING AGREEMENT

PARTICULARS

Date of Agreement	:	
Parties	:	
The Farmer	:	....
of	:	....
The Contractor	:	....
of	:	....
Farm	:	[ ] [as shown outlined in [red] for identification purposes only on the attached plan]
Total Hectares (Acres)	:	....
Commencement Date	:	1 October 20[ ]
Accounting Date	:	30 September
Farm Accountants	:	....
Contractor's Sum	:	For cultivated areas £[ ] per ha/ £[ ] per acre per Accounting Period (or pro rata if the Accounting Period is less than twelve months)

		For set-aside or fallow areas £[ ] per ha /£[ ] per acre per Accounting Period (or pro-rata if the Accounting Period is less than twelve months)
Bonus Percentage	:	[ ]
Base Figure	:	£[ ] per ha/ £[ ] per acre
Bankers	:	....
Farmer's Agent	:	....

DESCRIPTION OF THE FARM

[ha/acres] of land [and buildings] at [ ] (as outlined for identification purposes only on the attached plan) comprising the land identified in the Area Schedule and [list any buildings and fixed plant included]

AREA SCHEDULE

Description	**OS Ref/NG No**	**Gross Ha/Acres**	**Eligible SPS Area**
[*list*]			

1. Definitions and interpretations

1.1 The terms used in the Particulars shall unless the context otherwise requires have the same meaning throughout this Agreement

1.2 'the Farm' means the property described in the Particulars

1.3 'Accounting Period' means the twelve months ending with the Accounting Date in each year save that the first such period shall run from the Commencement Date to the 20[ ] Accounting Date and the last Accounting Date shall fall on the date upon which this Agreement is terminated in the manner provided below

1.4 'the Net Profits' means the profits of the Farming Business as determined in accordance with the provisions of the Schedule to this Agreement

1.5 'the Single Payment Scheme' means the Single Payment Scheme governed by Council Regulation (EC) No 73/2009 and all the relevant subsequent Commission and other regulations, administrative provisions, statutory guidance or codes of practice implemented by the EC and/or the UK Government ('the Regulations')

1.6 'the Farming Business' means the business of farming carried on by the Farmer on the Farm

1.7 'the Policy' means such works of cultivation and other works as may reasonably be required by the Farmer who will after consulting the Contractor establish a farming policy for the Farm

1.8 'Environmental Stewardship Schemes' means any national or European scheme for the conservation or enhancement of the environment under which the Farmer has assumed obligations or is subject to restrictions in respect of the Farm

1.9 'VAT' means Value Added Tax or any replacement or additional tax of a similar nature

1.10 The clause headings are for convenience only and do not form part of this Agreement

1.11 Words importing the singular number shall where appropriate include the plural and words importing the masculine gender shall where appropriate include the feminine and neuter genders

1.12 Where any party comprises more than one person the obligations and liabilities of that party under this Agreement shall be joint and several obligations and liabilities of those persons

2. Commencement

This Agreement shall take effect from the Commencement Date and shall continue until determined in the manner provided below

3. Engagement

The Farmer agrees to engage the Contractor and the Contractor agrees to be so engaged as Contractor for the duration and upon the terms of this Agreement to carry out the services hereinafter described

4. Contractor's responsibilities

4.1 The Contractor shall be responsible for carrying out the Policy

4.2 The Contractor shall provide all the manpower and machinery (unless otherwise agreed) required for the cultivation of the farm and shall carry out such operations as shall reasonably be directed by the Farmer or the Farmer's Agent for the cultivation planting harvesting [drying storage] and sale of all crops to be grown on the Farm using all reasonable means to maintain improve and extend the Farming business and to protect and further the reputation and interest of the Farmer and shall cultivate the Farm and carry out all other operations (as required by the Policy) so that the full and efficient use of the Farm is made for agriculture according to the nature of farming practised in the neighbourhood and so as to respect the statutory management requirements and the good agricultural and environmental conditions as defined under the Single Payment Scheme

[4.3 The Contractor shall ensure that in carrying out its activities under this Agreement the requirements of the Environmental Stewardship Schemes written details of which have been provided by the Farmer are not breached]

4.4 The Contractor shall comply with the requirements of any Crop Assurance Schemes for the time being applicable to the Farming Business

4.5 The Contractor shall as agent for the Farmer order and purchase such inputs as may from time to time be authorised by the Farmer or the Farmer's Agent. All records vouchers bills cheques and other documents required in connection with the Farming Business shall be in the name of the Farmer

4.6 The Contractor shall during the currency of this Agreement comply with all reasonable instructions in writing issued by the Farmer or the Farmer's Agent for the implementation of the Policy

4.7 The Contractor agrees to indemnify the Farmer and keep the Farmer indemnified on a continuing basis against all penalties reduction in subsidies costs claims proceedings or demands (including legal costs and expenses) arising from the act or default of the Contractor its employees or agents

5. Farmer's obligations

The Farmer shall at the expense of the Farming Business provide in a timely fashion in accordance with good farming practice and at reasonable cost all necessary seeds fertilisers sprays and other variable inputs for the efficient performance by the Contractor of its obligations hereunder

6. Insurance

6.1 The Farmer shall at the expense of the Farming Business procure insurance cover to the reasonable satisfaction of the Contractor in respect of the following:-

6.1.1 buildings fixed machinery and plant referred to in the Description of the Farm

6.1.2 produce whilst growing [and in store]

6.1.3 seeds fertilisers and sprays and other sundry stores

6.1.4 all other property on the Farm owned by or in the custody or control of the Farmer and used by the Farming Business

6.1.5 all third party and occupiers' liability properly insurable by the Farmer

6.2 The Contractor shall at its own expense procure insurance cover to the reasonable satisfaction of the Farmer in respect of the following:-

6.2.1 the Contractor's machinery plant and other property owned by it or in its custody against such loss or damage as the Contractor shall consider appropriate and

6.2.2 the Contractor's liability for death or personal injury or for loss or damage to property arising out of or in the course of the Contractor's implementation of this agreement

6.2.3 the Contractor's liability as an employer

7. Non exclusivity

Nothing in this Agreement shall prevent the Farmer from carrying out or from engaging other contractors to carry out any operations reasonably required on the Farm (not being operations in contemplation herein as normal obligations of the Contractor)

8. Remuneration

The Farmer shall in respect of each Accounting Period pay to the Contractor:-

8.1 a Variable Sum equal to the Bonus Percentage of the amount by which the Net Profits exceed in any Accounting Period the Base Figure such sum to be paid against a properly presented VAT invoice by a single instalment within twenty eight days after the making up of the annual accounts PROVIDED that payment of the Variable Sum may be withheld by the Farmer so far as reasonable either in whole or in part until such times as the crops for that Accounting Period shall have been actually sold and payment for them received and the Single Payment Entitlements paid

8.2 the Contractor's Sum payable by the Farmer quarterly in arrears within twenty eight days of the Contractor's properly presented VAT invoice

9. Banking

9.1 The Farmer shall maintain a separate banking account ('the No 2 Account') for the Farming Business in its name with the Bankers

9.2 The Farmer shall pay into the No 2 Account all income properly attributable to the Farming Business (as defined in clause 1 of the Schedule)

9.3 Any sums reasonably introduced by the Farmer to the No 2 Account for the furtherance of the Policy (other than those payments under clause 9.2) shall bear interest at 2% over the base lending rate of the Bankers such interest to be withdrawn by the Farmer on the Accounting Date

9.4 On the termination of each Accounting Period the Farmer shall withdraw from the No 2 Account the Base Figure

9.5 The Farmer shall withdraw from the No 2 Account the amount by which the Net Profits exceed the Base Figure for each Accounting Period within twenty eight days after the making up of the annual accounts and the sale of the crops and Single Payment Entitlements (if later) and shall pay any sums due to the Contractor under clause 8.1 where there is a delay such withdrawal shall be made rateably

10. Records and accounts

10.1 The Contractor shall keep such records in respect of the Farming Business as may be required by the Farmer or the Farm Accountants with sufficient details so that compensation aid subsidies and grants payable to the Farmer under any statute or statutory or local regulation or byelaw or otherwise may be readily determined so that all statutes regulations or byelaws are complied with and generally so that the rights of the Farmer can be accurately ascertained

10.2 Proper books of account giving a true and fair view of the Farming Business shall be kept properly posted by the Farmer and such books shall be available for inspection by each of the parties and their respective advisers at all reasonable times either at the Farm or at the Farmer's usual place of business or at the offices of any person nominated by the Farmer

10.3 A profit and loss account giving a fair and accurate record of the Farming Business shall be promptly prepared and certified by the Farm Accountants in respect of every Accounting Period during the currency of this Agreement and a copy of the same shall be supplied to the Contractor

11. Restrictions

The Contractor shall not without the prior written approval of the Farmer except when duly authorised by the Farmer or the Farmer's Agent in writing (such authority being if the Farmer so requires of a general nature) enter into any contract or agreement on behalf of the Farmer draw accept or endorse any bill on behalf of the Farmer or in any way pledge the credit of the Farmer or give any security or promise for the payment of money by the Farmer

12. Personal supervision

The Contractor hereby agrees with the Farmer that all work to be done or services to be performed by the Contractor under this Agreement shall be so done or performed by the Contractor [or under the personal supervision of [ ]]

13. Licence to enter

The Farmer hereby gives leave to the Contractor to enter on the Farm specifically and exclusively for the purpose of carrying out the Contractor's obligations hereunder the Farmer remaining at all times in full occupation of the Farm and nothing in this Agreement shall give the Contractor any right to enter use or occupy the Farm other than by the licence of the Farmer for the purposes hereof

14. Extra work

If the Contractor at the written request of the Farmer or the Farmer's Agent undertakes any operation which he is not obliged to carry out hereunder or if the Contractor with the written consent of the Farmer pays for any items properly payable by the Farmer hereunder the Farmer shall reimburse the Contractor for the costs thereof forthwith upon presentation by the Contractor of the appropriate invoice

15. Termination on short notice

This Agreement may be terminated:-

15.1 by the Farmer forthwith by giving notice to the Contractor in any of the following events:-

15.1.1 if there shall be any persistent breach or material breach or breaches by the Contractor of the terms of this Agreement

15.1.2 if the Contractor shall after proper warning fail to comply with any lawful directions given by the Farmer in writing in accordance with the provisions stated above subject to weather conditions permitting and such failure continues for a period of seven days after the Contractor has been required to comply with such direction by written notice from the Farmer

15.1.3 if the Contractor shall through illness, accident or any other cause be incapacitated or prevented from carrying on his duties under this Agreement for a continuous period of not less than three months or for not less than one hundred days in any continuous period of twelve months

15.1.4 if the Contractor shall commit any act of bankruptcy or become bankrupt or make any composition or voluntary arrangement with his creditors or have any distress levied or any process by way of execution ordered against his property which shall not have been discharged within fourteen days

15.1.5 if the Contractor becomes a patient within the meaning of the Mental Health Act 1983

15.1.6 if the Contractor dies

15.1.7 if the Contractor being a company enters into liquidation (whether compulsory or voluntary) other than for the purpose of amalgamation or reconstruction or if a receiver is appointed of the Contractor's undertaking

15.2 in the event of the death of the Farmer by his personal representatives giving to the Contractor not less than six months' notice in writing

15.3 by the Farmer giving to the Contractor not less than twelve months' notice in writing

15.4 forthwith by the Contractor giving to the Farmer notice in writing in any of the following events:-

15.4.1 if the Farmer being a Company enters into liquidation (whether compulsory or voluntary) other than for the purpose of amalgamation or reconstruction or if a receiver is appointed of the Farmer's undertaking

15.4.2 if the Farmer being an individual commits any act of bankruptcy becomes bankrupt or makes any compulsory or voluntary arrangement with his creditors or has any distress levied or any process by way of execution ordered against his property which shall not have been discharged within fourteen days

15.4.3 if there shall be any persistent or material breach or breaches by the Farmer of the terms hereof

15.5 by the Contractor giving the Farmer not less than three months' notice in writing should the Contractor contract any debilitating chronic illness

16. Termination on full notice

This Agreement may be determined by either party giving to the other not less than 12 months' notice in writing to expire on an Accounting Date in any year other than [20[ ] or 20[ ]]

17. Valuation

17.1 A valuation shall be taken at the Commencement Date on the following basis:-

17.1.1 stores shall be valued at actual cost

17.1.2 growing crops shall be valued at the cost of production based on the Contractor's Sum together with the cost of seeds fertilisers sprays and other variable inputs and a fair allowance for enhancement where appropriate to be determined by the Farmer's Agent after consultation with the Contractor

17.2 On the termination of each Accounting Period a valuation of growing and harvested crops shall be taken on the following basis:-

17.2.1 growing crops shall be valued at the cost of production based on the sums paid or payable to the Contractor pursuant to clause 8.2 together with the cost of seeds fertilisers sprays and other variable inputs and a fair allowance for enhancement where appropriate to be determined by the Farmer's Agent after consultation with the Contractor

17.2.2 harvested crops shall be valued if already contracted to be sold at contract prices less any anticipated storage and other costs and relevant levies and otherwise at open market value less any anticipated storage and other costs and relevant levies

17.3 On the termination of this Agreement a valuation shall be taken on the following basis:-

17.3.1 stores shall be valued at actual cost

17.3.2 the growing crops for harvest in the following year shall be valued at the cost of production based on the sums paid or payable to the Contractor pursuant to clause 8.2 together with the cost of seeds fertiliser sprays and other variable inputs and a fair allowance for enhancement where appropriate to be determined by the Farmer's Agent after consultation with the Contractor

17.3.3 harvested crops shall be valued if already contracted to be sold at contract prices less any anticipated storage and other costs and relevant levies and otherwise at open market value less any anticipated storage and other costs and relevant. When such crops have been harvested and/or sold then the net proceeds of sale thereof less the cost of cultivations harvesting storage dressing and sale shall be brought into account in the preparation of the final accounts and the Variable Sum shall be adjusted when the final results are known

17.4 The valuation shall be prepared by the Farm Accountants (unless otherwise agreed) provided that in the event of disagreement between the Farmer and the Contractor as to such valuation clause 22 shall apply

18. Compensation on early termination

If this Agreement shall be terminated other than pursuant to clause 15.1 on a date which is not an annual Accounting Date then the Contractor shall be entitled (in addition to a proportionate part of Contractor's sum) to a proportionate part of the Variable Sum referred to in clause 8.1 which would have been payable to the Contractor at the next making up of the annual accounts had this Agreement then been in existence Such last mentioned proportionate part to be such sum as shall be fair and equitable having regard to all the circumstances of the case and to be payable in like manner as the Variable Sum

19. Certificates

All certificates as to the accounts of the Farm or the Net Profits or of any other matter which are by this Agreement to be given by the Farm Accountants shall in the absence of manifest error be final and binding upon both parties

20. Relationship

It is hereby agreed and declared by the parties to this Agreement that nothing in this agreement shall constitute a partnership between the Contractor and the Farmer or create a tenancy of the Farm or a licence for the exclusive occupation of the same and for the avoidance of doubt the Contractor shall make no application in its own name to the Single Payment Scheme in respect of the Farm

21. Assignment

The Contractor shall not be entitled to assign or otherwise deal with the benefit of this Agreement

22. Disputes

22.1 Subject to clause 22.2, any dispute arising out of this Agreement is to be determined by an expert and where any issue is required to be dealt with by or submitted for the determination of an expert:-

22.1.1 the expert is to be an independent person of not less than ten years' post qualification experience at the date of appointment relevant to the dispute appointed by the parties jointly or if they cannot or do not agree on the appointment within twenty working days of one giving to the other notice of its nomination appointed by the President for the time being of the Royal Institution of Chartered Surveyors ('the President')

22.1.2 a person so appointed is to act as an expert

22.1.3 the expert so appointed is required to afford the parties the opportunity to make representations to him and permit each party to make submissions on the representations of the other

22.1.4 the expert as agreed or nominated above shall give notice of his appointment to the parties not later than seven days after his appointment

22.1.5 the expert shall assume that the parties wish to determine such dispute or difference as quickly as possible and shall:-

22.1.5.1 require the submission to him and the other party of any representations within ten working days of notice of his appointment

22.1.5.2 require the submission to him and the other party of any counter representation within ten working days of the submission of the original representation

22.1.5.3 (subject to payment of his fees which shall initially be paid on demand by the parties in equal shares) issue his decisions within twenty five working days of notice of his appointment

22.1.5.4 decide how the costs of such determination are to be borne

22.1.6 If the expert dies or refuses to act or becomes incapable of acting the resident may on the application of either party discharge the independent person previously appointed and appoint another in his place

22.2 No party may appoint an expert under this clause until it has attempted to settle any dispute by mediation and that mediation has terminated

23. Notices

Any notice or direction given in pursuance of the provisions of this Agreement shall be deemed to have been properly given to or served on the other party if it is delivered to it or left at the address shown in the Particulars or such other address as has been notified in writing No notice given by fax or by other electronic means will be valid unless a copy of the notice is also sent by post or delivered to the address of the recipient within seven days

24. VAT

Where any sums payable pursuant to this Agreement are properly chargeable with VAT they shall be paid upon production of a properly presented VAT invoice

25. Substitute contractors

If the Contractor does not perform the obligations on its part herein contained expeditiously then the Farmer may after notifying the Contractor in writing procure such person or persons as he may select to carry out such work and the reasonable cost of such work shall be charged against the remuneration due to the Contractor in accordance with this Agreement and any excess thereof over such remuneration then owing shall be forthwith recoverable from the Contractor

26. Meetings

Prior to the commencement of each Accounting Period the Farmer and the Contractor shall meet and the Farmer shall notify the Contractor of the farming programme for that period. A budget shall be prepared at the expense of the Farming Business and at least 3 meetings during each Accounting Period and any additional meetings as the Farmer or Contractor may reasonably request shall be held to review the progress of the Farming Business

AS WITNESS the hands of the parties hereto the day and year first before written

THE SCHEDULE

THE NET PROFITS OF THE FARMING BUSINESS

1. The Net Profits of the Farming Business means the income thereof arising from sales (excluding sales of fixed assets) and for the avoidance of doubt that income includes any relevant Single Payment Scheme payments payable to the Farmer [but excludes any payments under Environmental Stewardship Schemes] in respect of the Accounting Period and the proceeds from any insurances maintained pursuant to clause 6 by the Farmer certified by the Farm Accountants utilising normal accounting practices after providing for all expenses and outgoings subject to the provisions of this Schedule utilising the valuation at the expiration of each Accounting Period or the termination of this Agreement and the valuation at the Commencement Date referred to in clause 17

2. The expenses and outgoings of the Farming Business (without prejudice to the generality of the expression) include:-

2.1 all rates and taxes including owners and occupiers drainage rates and other outgoings payable by the Farmer in respect of the Farm including levies other than income or corporation tax

2.2 insurances maintained by the Farmer pursuant to clause 6 in respect of the Farm and the Farming Business

2.3 the cost of seeds fertilisers sprays and other items provided by the Farmer at the expense of the Farming Business

2.4 the expense of keeping the books of the Farming Business computing interest and preparing the valuations referred to in clause 17 and preparing the accounts of the Farming Business

2.5 any bank charges and interest payments incurred in respect of the No 2 Account

2.6 fees and charges incurred in connection with any marketing arrangements and crop sale agreements

2.7 the cost to the Farmer of labour and machinery provided by the Contractor and the cost of other work undertaken by other contractors or agents to carry out any operations (including agronomy) reasonably required on the Farm and which are not an improvement

SIGNED by or on behalf of)

in the presence of:-)

Signature of Witness:

Name of Witness:

Address of Witness:

SIGNED by or on behalf of)
in the presence of:-)
Signature of Witness:
Name of Witness:
Address of Witness:

SIGNED by or on behalf of)
in the presence of:-)
Signature of Witness:
Name of Witness:
Address of Witness:

SIGNED by or on behalf of)
in the presence of:-)
Signature of Witness:
Name of Witness:
Address of Witness:

PRECEDENT 8 LIVESTOCK SHARE FARMING AGREEMENT

Dated 20[ ]

[ ]

-AND-

[ ]

SHARE FARMING AGREEMENT
RELATING TO
[ ]

THIS AGREEMENT is made the [ ] day of [ ] 20[ ]

BETWEEN:-

[ ] ('the Landowner')

and

[ ] of [ ] ('the Share Farmer')

WHEREAS:-

(1) The Landowner is the owner and occupier of [ ] farm

(2) The Landowner and the Share Farmer have agreed to co-operate in the farming of certain parts of that farm for the benefit of their separate businesses

NOW IT IS AGREED as follows:-

1. Share Farming Agreement

The land described in Schedule 1 hereto and more particularly delineated and shown edged red on the plan annexed hereto (hereinafter called 'the land') shall be farmed by the Landowner and the Share Farmer for the separate benefit of each of them on the terms and conditions following

2. Woodlands game etc

2.1 No woodlands timber minerals game wildfowl wild birds or fish are included in this Agreement

2.2 This Agreement is subject to the exercise by or under the authority of the owner of rights in respect of the matters set out in sub-clause 2.1 of this clause and of sporting rights in on or over the land

3. Commencement, duration and termination

3.1 This Agreement shall commence on the [*date*] and shall continue until the 29th day of September 20[ ] [25 March 20[ ]] and unless terminated on that date by not less than twelve months' prior notice in writing given by either party to the other expiring on that date shall continue until terminated on any subsequent 29 September [25 March] by not less than twelve months' prior notice in writing as aforesaid

3.2 Notwithstanding sub-clause 3.1 of this clause either party may terminate this Agreement by giving the other twenty eight days' prior notice in writing in any of the following circumstances:-

3.2.1 Grave and persistent breach of the terms of this Agreement by the other provided that he has previously been given written notice specifying the breach and giving him a reasonable period to remedy the same if remediable

3.2.2 the other being prevented from carrying out his obligations under this Agreement for a continuous period of not less than six months during the subsistence of this Agreement by illness accident or any other cause

3.2.3 the other becoming bankrupt or making any composition or arrangement with his creditors

3.3 Notwithstanding sub-clause 3.1 of this clause this Agreement may be terminated in the event of the death of the Landowner by his executors or administrators giving the Share Farmer not less than six months' notice in writing expiring on the 29 September [25 March] in any year but in the event of the Share Farmer suffering loss by reason of the termination under this sub-clause being earlier than it could have been under sub-clause 3.1 of this clause he shall be entitled to compensation for the loss, such compensation being limited to a maximum of £[ ]

3.4 Notwithstanding sub-clause 3.1 of this clause in the event of the Share Farmer dying during the subsistence of this Agreement this Agreement shall terminate on the quarter day next occurring after a period of four months has expired from the date of his death

3.5 Notwithstanding sub-clause 3.1 of this clause this Agreement may be terminated in the event of the Landowner entering into a contract to sell or otherwise dispose of the ownership of the whole or a substantial part of the land by giving the Share Farmer not less than six months' notice in writing expiring on the 29 September [25 March] in any year but in the event of the Share Farmer suffering loss by reason of termination under this sub-clause being earlier than it could have been under sub-clause 3.1 of this clause he shall be entitled to compensation for the loss, such compensation being limited to a maximum of £[ ] provided always that if such sale or disposition is completed before expiry of the notice referred to in this sub-clause the Landowner while himself remaining liable to perform all his obligations under this Agreement throughout its subsistence shall nevertheless procure that the transferee shall enter into an agreement in writing with the Share Farmer that the transferee will be personally liable thenceforth for all such obligations

3.6 Notwithstanding sub-clause 3.1 this Agreement can be terminated by the Landowner at any time on giving eleven months' notice but in the event of the Share Farmer suffering loss by reason of termination under this sub-clause being earlier than it could have been under sub-clause 3.1 of this clause he shall be entitled to compensation for the loss, such compensation being limited to a maximum of £[ ]

4. Review in the event of change in cost structure of farming

Either the Landowner or the Share Farmer may if a significant change in the cost structure of farming occurs call for a review and revision of the terms of the Agreement as at 29 September [25 March] (the review date) in any year upon giving the other party not less than nine months' previous written notice and upon service of such notice the parties shall endeavour to agree by 1 September [1 March] of that year what amendment if any should take effect at the review date and in default of agreement the provisions of the Agreement shall remain unchanged

5. Occupation of the land

The land and every part of it shall remain in the possession and occupation of the Landowner throughout the subsistence of this Agreement The Share Farmer shall have the right to occupy and use the land jointly with the Landowner for the purposes and during the currency of this Agreement only but neither the Share Farmer nor his servants or agents shall have exclusive possession of the land or any part of it as against the Landowner

6. Indemnity

Neither party shall hold himself out as partner of or agent for the other party and each party agrees to indemnify and to keep indemnified the other against all claims demands proceedings costs and liabilities in respect of the activities or business of the indemnifier whether or not in relation to this Agreement

7. No tenancy partnership agency or employment

Nothing in this Agreement nor anything done in pursuance of this Agreement shall create or be deemed to create a tenancy, partnership, relationship of principal and agent or contract of employment between the parties

8. Farming policy

The parties shall carry on the share farming enterprise under this Agreement in accordance with the farming policy and cropping programme set out in Schedule 2 hereto

9. Share of inputs

9.1 The input costs including the provision of land buildings machinery and fixed equipment shall be provided by the parties in accordance with Schedule 3 hereto and any cost charge tax levy or other expense incurred by the Landowner or the Share Farmer shall be that of his respective business

9.2 Livestock provided at the commencement of this Agreement additional stock and replacements purchased and the number of calves reared annually shall be in accordance with Schedule 2 hereto and no livestock shall otherwise be brought onto the land except with the express consent of both parties

9.3 Each and every head of livestock on the land shall until sold under this Agreement belong as to [ ]% to the Landowner and as to [ ]% to the Share Farmer

10. Duties of Share Farmer

The Share Farmer agrees as follows:-

10.1 to devote adequate time to the share farming enterprise for its efficient running To manage the day to day farming of the land and animals efficiently and to a high standard of husbandry, and to keep the land clean and in good heart and to carry out the agreed farming policy and cropping programme referred to in clause 8 hereof including the livestock husbandry and all cultivations harvesting of crops and (subject to the agreed policy referred to in clause 14 hereof) marketing and sale of milk, livestock, crops and other produce and to provide such labour plant machinery implements and transport necessary therefor

10.2 to provide or pay for (as the case may be) the inputs required of him in accordance with Schedule 3 hereof and to purchase on behalf of and at the expense of the Landowner any items required from him in accordance with the said Schedule 3 in so far as the Landowner shall not have provided or purchased such items by the time they are required at commencement of the Agreement to purchase his share of the Landowner's livestock in accordance with Schedule 6. To purchase (i) livestock in accordance with Schedule 2 hereto additional to the initial herd of diary cows provided by the Landowner and (ii) any replacements from an agreed or otherwise reputable supplier at the most favourable price obtainable ensuring that the animals are healthy and of good quality suitable for the farming policy required by this Agreement

10.3 to carry out routine maintenance and repairs to the land buildings and fixed equipment (other than such as fall to be carried out by the Landowner under clause 11 hereof) including all fixtures and fittings together with all internal decorations and internal repairs to the houses and buildings, regular trimming and laying of hedges and cleaning of ditches. To carry out repairs and regular maintenance (the Landowner providing the materials thereof) to fences hedges gates culverts land drains and roads and to repair and fully maintain the plant and machinery owned by the Landowner set out in Part (a) of Schedule 4 hereto

10.4 to effect and maintain adequate policies of insurance to cover public and occupier's liability and to cover the plant and machinery and all property owned by him (whether solely or jointly) or in his possession and used or kept for the purposes of this Agreement (other than any required by this Agreement to be insured by the Landowner. To insure in the joint names of the Landowner (as to [ ]% ownership) all livestock kept or obtained under this Agreement To refund to the Landowner [ ]% of the cost of insurance of severed crops covered under clause 11.3 hereof

10.5 not without the prior consent of the Landowner to take down any trees or to take up any hedges or fences and to comply with the conservation policies set out in Schedule 5 hereto

10.6 to ensure to the best of his ability that the milk quota and all other quotas and subsidy allocations for the land and for production shall be maintained at the highest possible level and to avoid any act or omission that would prejudice the acquisition or maintenance of any quota at that level

11. Duties of Landowner

The Landowner agrees as follows:-

11.1 to provide or pay for (as the case may be) the inputs required from him in accordance with Schedule 3 hereto

11.2 to carry out routine external repairs to and painting of the buildings and within a reasonable time if reasonably necessary any major maintenance and repairs to the buildings and to provide materials for repairs to fences hedges gates culverts land drains and roads

11.3 to effect and maintain adequate policies of insurance to cover public and occupier's liability and to cover the buildings and the growing and severed crops on the land and to pay to the Share Farmer [ ]% of any insurance monies paid to the Landowner in respect of the growing or severed crops

11.4 to pay [ ]% of the cost of any livestock additional to the herd provided by him which is purchased to comply with the stocking policy set out in Schedule 2 hereto

11.5 to pay to the Share Farmer [ ]% of any premiums paid by the Share Farmer to insure livestock under this Agreement

11.6 to avoid any act or omission that would prejudice the maintenance of any milk or other quota for the land or any production quota at the highest possible level

11.7 to arrange for the performance of the share farming enterprise under this Agreement to be costed to provide a quarterly written schedule of performance and annual summary together with an annual inventory of live and dead stock The cost of this service to be borne as to ([ ]%) by the Landowner and ([ ]%) by the Share Farmer

12. Ownership of growing crops and machinery

Growing crops on the land shall belong to the Landowner until severed but cannot be sold or otherwise disposed of by the Landowner without the written consent of the Share Farmer and the plant machinery and equipment specified in Schedule 4 hereto shall belong as shown in that Schedule

13. Quotas

It is agreed by the Landowner and the Share Farmer as follows:-

13.1 throughout the term of this Agreement and any extension thereof a Quota Register for the Share Farming Agreement shall be maintained by the parties. The register shall record the initial quota provided by each party for the use in his separate business and any extra quota acquired by either party and all compulsory loss or suspension of quota

13.2 at the commencement of this Agreement [ ] litres of milk quota shall be provided by the Landowner and [ ] by the Share Farmer and shall be used throughout the term of the Share Farming Agreement on the land specified in the First Schedule hereto and on no other land

13.3 on the termination of the Agreement the Landowner shall be entitled to retain his [ ] litres and any additional quota (less permanent cuts) recorded as his in the Quota Register referred to in clause 13.1 above and to use such quota on any land in which he has an interest. The Share Farmer hereby undertakes to make no claim to that quota nor to any compensation in respect thereof

13.4 likewise on the termination of this Agreement the Share Farmer shall be entitled to retain his [ ] litres and any additional quota (less any permanent cuts) recorded as his in the said Quota Register and to use such quota on any land in which he has an interest. The Landowner hereby undertakes to make no claim to that quota nor to any compensation in respect thereof

OR

13.5 the Landowner shall make arrangements with the Rural Payments Agency to lease [ ] litres of quota to the Share Farmer on an annual basis throughout the term of this Agreement. Such quota shall be used by the Share Farmer in his own business but the quota shall be used only on the land specified in the First Schedule

13.6 if the regulations permitting such leasing should be changed or withdrawn then the Share Farmer may call for a review and revision of the terms of the Agreement by giving the Landowner not less than twelve months prior notice in writing expiring on the 25 March notwithstanding that the full fixed term agreed has not expired

13.7 both parties undertake to sign any documents required by the Rural Payments Agency or similar regulatory body to enable an entry to be made on the Rural Payments Agency Register (or that of a similar body) recording that the Landowner and the Share Farmer are joint suppliers under a Share Farming Agreement and to delete such an entry on the termination of the Agreement

13.8 should the Rural Payments Agency or similar regulatory body succeed in an action taken against the Landowner alone for breach of the undertakings warranties or indemnities given in the contract with the Rural Payments Agency or similar body, then the Share Farmer will pay [ ]% of any fine or payment unless the Landowner acted contrary to the express instructions of the Share Farmer

13.8.1 should the Rural Payments Agency or similar regulatory body succeed in an action taken against the Share Farmer alone for breach of the undertakings warranties or indemnities giving in the contract with the Rural Payments Agency or similar body, then the Landowner will pay [ ]% of any fine or payment unless the Share Farmer has acted contrary to the express instructions of the Landowner

14. Share of gross output

The produce of the farm after severance shall belong as to [ ]% to the Landowner and as to [ ]% to the Share Farmer and shall be marketed and sold by the Share Farmer in accordance with a policy to be agreed between the parties at the best price reasonably obtainable and the gross receipts for the produce shall be paid to the parties in those percentages

15. Grants and subsidies

Any production grants or subsidies paid to either party in respect of livestock shall be shared between the parties [ ]% being paid to the Landowner and [ ]% to the Share Farmer

16. Bank accounts

The parties shall maintain their own separate bank accounts and shall have no joint bank account in respect of this Agreement other than a joint account on which neither party shall be at liberty to draw opened at [*name and address of bank*] into which shall be paid gross receipts for sale of produce (other than such receipts which may from time to time be paid by purchasers directly to the parties in the shares referred to in clause 14 hereof); no other monies shall be paid into that joint account and the parties shall procure that all monies which are paid into that joint account shall be transferred forthwith by [*name of bank*] to the separate bank accounts of the parties in the shares referred to in clause 14 hereof

17. Records and accounts

The Share Farmer shall keep proper and true records of all the farming operations carried out under this Agreement and shall if requested make them available for inspection by the Landowner or his agent at any reasonable time and each party shall keep proper and separate books of account in respect of his expenditure and receipts under this Agreement and each shall be separately registered for Value Added Tax

18. Payment of accounts

18.1 Accounts in respect of inputs to be provided or paid for by each of the parties in accordance with Schedule 3 hereto shall be paid by the parties within one month of receipt of those accounts

18.2 The parties shall endeavour to secure that separate invoices are submitted to each of them by any suppliers in respect of inputs required by each part by Schedule 3 hereto

18.3 Where a supplier submits a single invoice in respect of inputs required by Schedule 3 to be provided by each of the parties separately then to comply with the requirements of HM Customs and Excise the party to whom the invoice is addressed must pay the bill. He shall then re-invoice the other party that party's share and Value Added Tax

19. Disclosure of insurance

A copy of any policy of insurance required to be effected by a party under this Agreement shall be furnished to the other party if requested

20. Meetings

The parties shall meet at least once in every three months to review the farming operations under this Agreement and after each harvest they shall review the farming policy for the following farming year. Notes shall be kept of decisions made at meetings

21. Dwellinghouses

21.1 The Share Farmer shall be entitled during the subsistence of this Agreement to occupy rent free the dwellinghouse known as [ ] for the purposes of carrying out his agricultural duties under this Agreement

21.2 The Share Farmer shall pay any rates levied on that dwellinghouse during his occupation to the authority levying them and shall pay all other outgoings and shall keep the dwellinghouse clean and in full decorative repair internally and shall keep the garden clean and tidy

21.3 the Landowner will provide the cottages known as [ ] for occupation of the Share Farmer's employees provided that the employees:-

21.3.1 pay the Landowner a rent of £251 (or greater agreed sum)

21.3.2 acknowledge receipt of the intention to grant Assured Shorthold Tenancies on Form 9 (or any prescribed form replacing Form 9)

21.3.3 sign the tenancy agreements relating to their properties in the form annexed hereto before entering into possession

22. Surplus land

Any land becoming surplus to the requirements of the contract due to Government/EC regulations or directives shall either remain within the contract or be released from the contract for the sole use of the Landowner on his giving not less than six months' notice to the Share Farmer

23. Arrangements at commencement

On the commencement of this Agreement the Share Farmer shall pay to the Landowner:-

23.1 in respect of the growing crops on the land the percentages of the costs as allocated to the Share Farmer in Schedule 3 hereto for any seeds sown fertilisers lime and sprays already expended in respect of those crops since severance or grazing of the last previous crop and likewise for costs so expended for labour machinery plant electricity and water

23.2 for the crops stocks and stores in accordance with Schedule 6(a) hereto

23.3 for any field machinery agreed to be taken over by the Share Farmer in accordance with Schedule 6(b) hereto

23.4 for any livestock agreed to be taken over by the Share Farmer in accordance with Schedule 6(c) hereto

24. Arrangements on termination

On the termination of this Agreement:-

24.1.1 any growing crop shall belong to the Landowner but the Landowner shall repay to the Share Farmer the costs borne by the Share Farmer in respect of the growing crop for seeds fertilisers lime sprays labour machinery plant electricity and water applied or used after the severance or grazing (as the case may be) of the last previous crop

24.1.2 any crop severed under this Agreement and not yet disposed of and any seeds fertilisers lime and sprays acquired under this Agreement and still held in stock shall belong to the parties in the percentages set out in clause 14 hereof

24.2 either of the parties may be given the first option to buy the interest in the livestock of the other

24.2.1 Mr [ ] shall have the option to purchase from Mr [ ] the interest of Mr [ ] in the livestock belonging to the parties under this Agreement provided that:-

> Mr [ ] gives notice in writing before the termination date to Mr [ ] of his intention to exercise the option and in exercising the option he purchases each and every head of livestock unless otherwise agreed

24.2.2 if due notice to exercise the option referred to in sub-clause 24.2.1 of this clause is not given by Mr [ ] the Mr [ ] shall have the option to purchase from Mr [ ] the interest of Mr [ ] in that livestock provided that:-

Mr [ ] gives notice in writing not later than two weeks after the termination date to Mr [ ] of his intention to exercise the option and in exercising the option he purchases each and every head of livestock unless otherwise agreed

24.2.3 any livestock not acquired by one or other party under sub-clause 24. 2.1 or 24.2.2 of this clause shall be sold by the Share Farmer on the expiration of two weeks from the termination date and the proceeds of sale shall be divided between the parties in the percentages referred to in clause 9.3 hereof

24.2.4 any unsold hay silage straw feeding stuffs or produce produced from the land under this Agreement and any purchased feeding stuffs and other stores on the land acquired for the purposes of this Agreement and belonging to the Share Farmer solely or jointly shall become the property of the Landowner on the termination date but the Landowner shall make payment to the Share Farmer in respect thereof as set out in Schedule 6(a) hereto

25. Rights and duties personal

Subject to sub-clause 3.5 hereof the rights and obligations of the parties under this Agreement shall be personal to them (or to their personal representatives) and shall not be assigned transferred mortgaged charged or otherwise disposed of

26. Disputes

In the event of any dispute arising as to any valuation required under this Agreement or as to any compensation payable under sub-clauses 3.3, 3.5 and 3.6 hereof either party may refer such dispute to be determined by arbitration under the Arbitration Act 1996 or any statutory modification or re-enactment thereof for the time being in force by a single arbitrator to be agreed upon by the parties or in default of agreement appointed by the President for the time being of the Royal Institution of Chartered Surveyors

27. Costs

Each party shall bear his own costs of preparation and execution of this Agreement

28. No variation unless in writing

This Agreement contains the whole of the agreement between the parties and no variation shall be made unless it is made in writing and signed by or on behalf of both parties

AS WITNESS the hands of the parties hereto the day and year first before written

SCHEDULE 1
LAND AND BUILDINGS INCLUDED IN THE AGREEMENT AND PLAN

SCHEDULE 2
FARMING POLICY AND CROPPING PROGRAMME

SCHEDULE 3
INPUTS REQUIRED FROM EACH PARTY

Input	**Landowner**	**Share Farmer**
Land, Buildings, Farmhouse plus certain fixed machinery		
Maintenance of the above, rates and insurance		
Insurance of Growing Crops		
Materials for fencing, road and gate repairs		
Costings		
Labour		
Machinery and Plant		
Electricity and Water		
Telephone – miscellaneous		
Livestock Sundries		
Ownership of livestock and insurance		
Seeds, fertilisers and sprays		
Lime		
Insurance of crops in store		
Purchased feed stuffs		
A.I. – Vet. & Med.		

SCHEDULE 4
OWNERSHIP OF PLANT AND EQUIPMENT
Part (a) The property of the Landowner
Part (b) the property of the Share Farmer

SCHEDULE 5
CONSERVATION POLICIES

SCHEDULE 6
VALUATION AT COMMENCEMENT AND TERMINATION OF THE

AGREEMENT
(a) Crops Stocks and Stores
(b) Field and other Machinery
(c) Livestock

SIGNED by
in the presence of:-

SIGNED by
in the presence of:-

PRECEDENT 9 ARABLE SHARE FARMING AGREEMENT

Dated 20[ ]

[ ] (1)

- and –

[ ] (2)

SHARE FARMING AGREEMENT

- relating to -

[ ]

THIS AGREEMENT is made the [ ] day of [ ] 20[ ]

BETWEEN:-

[ ] (Company Number [ ]) of [ ] ('the Landowner')

and

[ ] of [ ] ('the Share Farmer')

WHEREAS:-

(1) The Landowner is the owner and occupier of the freehold property known as [ ]

(2) The Landowner and the Share Farmer have agreed to co-operate in the farming of certain parts of that farm for the benefit of their separate businesses

NOW IT IS AGREED as follows:-

1. Definitions

1.1 'the Council Regulation' means Council Regulation (EC) No 73/2009 and any subsequent revisions of that Regulation

1.2 'Entitlements' means such entitlements to the single income payment and any other entitlement to financial support of whatever kind deriving in any way from the Single Payment Regulations

1.3 'the Land' means the land transferred to the Landowner by a Transfer dated [*date*] and made between (1) [ ] and (2) the Share Farmer and (3) the Landowner

1.4 'the Single Payment Regulations' means the Council Regulation and all subsidiary European or UK regulations that are applicable to England

2. Share Farming Agreement

The Land shall be farmed by the Landowner and the Share Farmer for the separate benefit of each of them on the terms and conditions following

3. Woodlands game etc

3.1 No woodlands timber minerals game wildfowl wild birds or fish are included in this Agreement

3.2 This Agreement is subject to the exercise by or under the authority of the owner of rights in respect of the matters set out in sub-clause 3.1 of this clause and of sporting rights in on or over the Land

4. Commencement and duration

This Agreement shall commence on the [*date*] and shall continue until [five years from allocation of entitlements]

5. Review in the event of change in cost structure of farming

Either the Landowner or the Share Farmer may if a significant change in the cost structure of farming or the Single Payment Regulations occur call for a review and revision of the terms of the Agreement as at 29 September (the review date) in any year upon giving the other party not less than nine months previous written notice and upon service of such notice the parties shall endeavour to agree by 1 September of that year what amendment if any should take effect at the review date and in default of agreement the provisions of the Agreement shall remain unchanged

6. Occupation of the Land

The Land and every part of it shall remain in the possession and occupation of the Landowner throughout the subsistence of this Agreement The Share Farmer shall have the right to occupy and use the land jointly with the Landowner for the purposes and during the currency of this Agreement only and shall have the land at its disposal for the purpose of the Single Payment Regulations but neither the

Share Farmer nor his servants or agents shall have exclusive possession of the land or any part of it as against the Landowner

7. Indemnity

Neither party shall hold himself out as partner of or agent for the other party and each party agrees to indemnify and to keep indemnified the other against all claims demands proceedings costs and liabilities in respect of the activities or business of the indemnifier whether or not in relation to this Agreement

8. No tenancy partnership agency or employment

Nothing in this Agreement nor anything done in pursuance of this Agreement shall create or be deemed to create a tenancy, partnership, relationship of principal and agent or contract of employment between the parties

9. Farming policy

The parties shall carry on the share farming enterprise under this Agreement in accordance with the farming policy and cropping programme as reasonably determined by the Landowner

10. Share of inputs

The input costs including the provision of land buildings machinery and fixed equipment shall be provided by the parties in accordance with Schedule 1 hereto and any cost charge tax levy or other expense incurred by the Landowner or the Share Farmer shall be that of his respective business

11. Duties of Share Farmer

The Share Farmer agrees as follows:-

11.1 to devote adequate time to the share farming enterprise for its efficient running

11.2 to provide or pay for (as the case may be) the inputs required of him in accordance with Schedule 1 hereof

11.3 to complete and submit claims for payment under the Entitlements and to permit the Landowner to review the content of all such documentation

12. Duties of Landowner

The Landowner agrees as follows:-

12.1 to provide or pay for (as the case may be) the inputs required from him in accordance with Schedule 1 hereto

12.2 to maintain the Land in good agricultural heart and condition as required by the Single Payment Regulations

12.3 to comply with the provisions of the Single Payment Regulations and not to do or omit to do anything that will or may jeopardise the Entitlements or any payments due to the Share Farmer thereunder

12.4 to effect and maintain adequate policies of insurance to cover public and occupier's liability and the growing and severed crops on the Land

12.5 to arrange for the performance of the share farming enterprise under this Agreement to be costed and to provide if reasonably requested by the Share Farmer a quarterly written schedule of performance and annual summary together with an annual inventory of live and dead stock. The cost of this service is to be borne by the Share Farmer

13. Ownership of growing crops and machinery

Growing crops on the Land shall belong to the Landowner until severed but cannot be sold or otherwise disposed of by the Landowner without the written consent of the Share Farmer and the plant machinery and equipment shall belong to the Landowner

14. Share of gross output

The produce of the farm after severance shall belong as to [ ]% to the Landowner and as to [ ]% to the Share Farmer and shall be marketed and sold by the Landowner in accordance with a policy to be agreed between the parties at the best price reasonably obtainable and the gross receipts for the produce shall be paid to the parties in those percentages

15. Grants and subsidies

[Any monies paid to either party in respect of the Entitlements shall be paid to the Landowner save for any payments referable to compensation for the sugar beet contract which shall be paid to the Share Farmer] or as agreed

16. Bank accounts

The parties shall maintain their own separate bank accounts and shall have no joint bank account in respect of this Agreement

17. Records and accounts

The Share Farmer shall keep proper and true records of all the farming operations carried out under this Agreement and shall if requested make them available for inspection by the Landowner or his agent at any reasonable time and each party shall keep proper and separate books of account in respect of his expenditure and receipts under this Agreement and each shall be separately registered for Value Added Tax

18. Payment of accounts

18.1 Accounts in respect of inputs to be provided or paid for by each of the parties in accordance with Schedule 1 hereto shall be paid by the parties within one month of receipt of those accounts

18.2 The parties shall endeavour to secure that separate invoices are submitted to each of them by any suppliers in respect of inputs required by each part[y] by Schedule 1 hereto

18.3 Where a supplier submits a single invoice in respect of inputs required by Schedule 1 to be provided by each of the parties separately then to comply with the requirements of HM Revenue and Customs the party to whom the invoice is addressed must pay the bill. He shall then re-invoice the other party that party's share and Value Added Tax

19. Disclosure of insurance

A copy of any policy of insurance required to be effected by a party under this Agreement shall be furnished to the other party if requested

20. Meetings

The parties shall meet at least once in every three months to review the farming operations under this Agreement and after each harvest they shall review the farming policy for the following farming year. Notes shall be kept of decisions made at meetings

21. Surplus land

Any land becoming surplus to the requirements of the contract due to Government/EC regulations or directives shall either remain within the contract or be released from the contract for the sole use of the Landowner on the Share Farmer giving not less than six months' notice to the Landowner

22. Arrangements on termination

On the termination of this Agreement:-

22.1 any growing crop shall belong to the Landowner but the Landowner shall repay to the Share Farmer the costs borne by the Share Farmer in respect of the growing crop for seeds fertilisers and sprays applied or used after the severance or grazing (as the case may be) of the last previous crop

22.2 any crop severed under this Agreement and not yet disposed of and any seeds fertilisers and sprays acquired under this Agreement and still held in stock shall belong to the parties in the percentages set out in clause 14 hereof

23. Rights and duties personal

The rights and obligations of the parties under this Agreement shall be personal to them (or to their personal representatives) and shall not be assigned transferred mortgaged charged or otherwise disposed of

24. Disputes

In the event of any dispute arising under this Agreement (including any valuation required or compensation payable) either party may refer such dispute to be determined by arbitration under the Arbitration Act 1996 or any statutory modification or re-enactment thereof for the time being in force by a single arbitrator to be agreed upon by the parties or in default of agreement appointed by the President for the time being of the Royal Institution of Chartered Surveyors

25. No variation unless in writing

This Agreement contains the whole of the agreement between the parties and no variation shall be made unless it is made in writing and signed by or on behalf of both parties

AS WITNESS the hands of the parties hereto the day and year first before written

SCHEDULE 1
INPUTS REQUIRED FROM EACH PARTY

Input	**Landowner %**	**Share Farmer %**
Land, Buildings, Farmhouse plus certain fixed machinery		
Maintenance of the above, rates and insurance		
Insurance of Growing Crops		
Materials for fencing, road and gate repairs		
Costings		
Labour		
Machinery and Plant		
Electricity and Water		
Telephone – miscellaneous		

Input	**Landowner %**	**Share Farmer %**
Livestock Sundries		
Ownership of livestock and insurance		
Seeds, fertilisers and sprays		
Lime		
Insurance of crops in store		
Purchased feed stuffs		
A.I. – Vet. & Med.		

SIGNED by
in the presence of:-

SIGNED by
in the presence of:-

INDEX

References are to page numbers.

Agistment
agreement for livestock to be taken in for feeding 385
meaning 376
Agreement for sale and purchase (long form) 69
Agricultural activity
meaning 3
Agricultural Holdings Act 1986
tenancies under 106
Alternative land uses 399
Arable share farming agreement 531
Arbitration
demand for arbitration of rent under AHA 1986, section 12 165
tenancy agreement 105
Arbitrator
notice of proposal to appoint 308
Archives
deposit with County Records Office 401
agreement for 414
Assignment of tenancy 141
oral 144
Assured agricultural occupancy
notice proposing new rent or licence fees 370
notice seeking possession 363
Assured periodic tenancy
landlord's notice proposing new rent 366
Assured shorthold tenancy 342, 344
landlord's notice proposing 354
tenant's notice proposing change from assured tenancy to 353
Assured tenancy
notice seeking possession 363

Bailment 414

Case G Notice 169
Common land 16
register 16
Contract farming agreements 471
Contracting agreement 507
County Records Office
deposit of archives 401
agreement for 414

Declaration of trust
milk quota 439
partnership 481
Deed of dissolution
partnership 482
Deed of surrender 158, 161
Deed of variation 150

Family partnership agreement 472
Farm business tenancies 201
break clause 203
improvement
compensation 204
landlord's notice complying with notice condition 205
landlord's notice demanding statutory review of rent 308
letting with entitlements 277
letting with transfer of entitlements 285
letting without entitlements 292
long form 205
mandatory provisions 202
notice exercising option to terminate 307
notice of reference of dispute to third party 309
notices 202
rent 203
rent review procedure 203
short form 259
short-term 202
specifiying compensation for tenant at valuation 270
specifying authorised entitlements and/or set-aside 298
tenant's notice complying with notice conditions 205
tenant's notice demanding statutory review of rent 307
termination 203
transfer back 285
Farmer
meaning 2
Film contract 400, 405
Fixed term tenancy
over 2 years
notice to terminate 306

Fixture
permission to erect under AHA 1986, section 10 164
Forfeiture
notice of intention 168

Grazing agreeements
common law tenancy 376
Grazing agreement
horses
alternative form 449
Grazing agreement (licence)
horses 445
Grazing agreement (tenancy)
non-business 390
Grazing agreements 375
capital taxation 375
inheritance tax 375
licence 376, 381

Highways Act 1980, s 31(6)
statement and declaration under 400, 412
Holding
meaning 3
Horse enterprise
planning permission 443
Horses 443
grazing agreement (alternative form) 449
grazing agreement (licence) 445

Improvement
tenant's
landlord's consent 166

Joint ownership
sugar beet harvester agreement 504
Joint venture partnership agreement 484
Joint ventures 469

Lease of farm buildings 313
Licence
grazing agreement 381
Licence for storage 312, 336
Limited liability partnerships 470
agreement 490
incorporation document 470
Livery agreement 452
landowner dealing with feeding and exercising of horses 459
Livery agreement (shorter form) 466
Livestock
tack/agistment agreement to be taken in for feeding 385
Livestock share farming agreement 519

Metal detecting licence 399, 402
Milk quota 431
apportionment of rules 432
Milk quota—*continued*
declaration of trust 439
deed regulating ownership between landlord, tenant and producer 437
fall in value 431
moving from one holding to another 432
supplemental deed apportioning between tenanted and owned property 433

Notice
pursuant to AHA 1986, section 6 163
quit
personal representatives, to 169
under LTA 1987, s 48 163

Occupation of farm buildings 311
capital gains tax 312
inheritance tax 312
rating 312
security of tenure 311
warehouse letting 311
Oral tenancy
assignment 144
Owner
meaning 400

Partnership
declaration of trust 481
deed of dissolution 482
family partnership agreement 472
joint venture agreement 484
meaning 469
use of 470
Periodic tenancy
landlord's notice to terminate 306
tenant's notice to terminate 306
Permissive path agreement 400, 407
Profit à prendre
deed 377
meaning 376
sporting rights 417
Protected shortholds 341
Public right of way 400

Re-entry
notice of intention 168
Rent
assured agricultural occupancy
notice proposing new rent 370
assured periodic tenancy
landlord's notice proposing new rent 366
increase
memorandum after landlord's improvement 166
notice to pay 170

Rent—*continued*
residential tenancy
landlord's letter proposing increase 356
Rent Assessment Committee
application referring notice proposing different terms in statutory periodic tenancy 359
Rent review 104
memorandum 165
Repair, maintenance or replacement
notice to tenant 171
Residential tenancies 341
landlord's letter proposing rent increase 356
obtaining possession 343
rent control 342

Sale and acquisition of land 15
aid schemes 16
enquiries before contract 18
agricultural 18
supplemental to commercial property enquiries 54
general enquiries
purchase of bare land with vacant possession 38
optional enquiries 16
Single Payment Scheme 15
Single Payment Scheme contract clauses 93
sale of land with entitlements 93
sale of land without entitlements 96
transfers of entitlements 16
use by members of public 16
Sale and acqusition of land
agreement for sale and purchase (long form) 69
Share farming 471
arable share farming agreement 531
concept, aim of 471
covenant not to share possession 471
independent businesses 471
livestock share farming agreement 519
Shorthold tenancies 341
Single Payment Scheme 1
calculation 3
contract clauses 93
sale of land with entitlements 93
sale of land without entitlements 96
cross compliance 4
deed of variation 8
modulation 4
sale of entitlements (without land) 5
Sporting rights 417
long-term agreement 419
profit à prendre 417
Stables
planning permission 443
Statutory periodic tenancy
notice proposing different terms 357
application referring to rent assessment committee 359
Succession tenancies 101
assignment 102
existing tenant's will 102
inheritance tax 102
short form by reference 140
succession by agreement 102
terms of new tenancy 101
Succession tenancy agreement 106
Sugar beet harvester
agreement for joint ownership 504
Surrender agreement 155

Tack/agistment agreement for livestock to be taken in for feeding 385
Tenancies
statutory amendments 100
Tenancy
application for direction giving entitlement 176
notice to landlord 198
notification to landlord 186
reply 187
retirement 190
Tenancy agreement
agreement to surrender where part of holding sub-let 159
arbitration 105
breach
notice to tenant to remedy 171, 175
compensation
notice of intention to claim 167
notice of intention to claim on termination 168
deed of surrender 158, 161
deed of variation 150
end of term compensation 104
notice under LTA 1987, section 48 163
notices 103
oral 103
prescribed forms 103
recovery of possession 104
rent review 104
request to enter into 163
suggested designation for inclusion in will 175
surrender agreement 155
written 103
Tenantright 84
Terms of tenancy
agreement recording 146

Village green 17
application for registration 17

Will
suggested designation for inclusion in 175